109 Springer Series in Solid-State Sciences
Edited by Peter Fulde

Springer Series in Solid-State Sciences

Editors: M. Cardona P. Fulde K. von Klitzing H.-J. Queisser

Managing Editor: H. K.V. Lotsch

Volumes 1–89 are listed at the end of the book

90 **Earlier and Recent Aspects of Superconductivity** Editors: J. G. Bednorz and K. A. Müller
91 **Electronic Properties of Conjugated Polymers III** Basic Models and Applications Editors: H. Kuzmany, M. Mehring, and S. Roth
92 **Physics and Engineering Applications of Magnetism** Editors: Y. Ishikawa and N. Miura
93 **Quasicrystals** Editors: T. Fujiwara and T. Ogawa
94 **Electronic Conduction in Oxides** By N. Tsuda, K. Nasu, A. Yanase, and K. Siratori
95 **Electronic Materials** A New Era in Materials Science Editors: J. R. Chelikowsky and A. Franciosi
96 **Electron Liquids** By A. Isihara
97 **Localization and Confinement of Electrons in Semiconductors** Editors: F. Kuchar, H. Heinrich, and G. Bauer
98 **Magnetism and the Electronic Structure of Crystals** By V. A. Gubanov, A. I. Liechtenstein, and A.V. Postnikov
99 **Electronic Properties of High-T_c Superconductors and Related Compounds** Editors: H. Kuzmany, M. Mehring, and J. Fink
100 **Electron Correlations in Molecules and Solids** By P. Fulde
101 **High Magnetic Fields in Semiconductor Physics III** Quantum Hall Effect, Transport and Optics By G. Landwehr
102 **Conjugated Conducting Polymers** Editor: H. Kiess
103 **Molecular Dynamics Simulations** Editor: F. Yonezawa
104 **Products of Random Matrices in Statistical Physics** By A. Crisanti, G. Paladin, and A. Vulpiani
105 **Self-Trapped Excitons** By K. S. Song and R. T. Williams
106 **Physics of High-Temperature Superconductors** Editors: S. Maekawa and M. Sato
107 **Electronic Properties of Polymers** Orientation and Dimensionality of Conjugated Systems Editors: H. Kuzmany, M. Mehring, and S. Roth
108 **Site Symmetry in Crystals** Theory and Applications By R. A. Evarestov and V.V. Smirnov
109 **Transport Phenomena in Mesoscopic Systems** Editors: H. Fukuyama and T. Ando

H. Fukuyama · T. Ando (Eds.)

Transport Phenomena in Mesoscopic Systems

Proceedings of the 14th Taniguchi Symposium, Shima, Japan, November 10–14, 1991

With 129 Figures

Springer-Verlag
Berlin Heidelberg New York
London Paris Tokyo
Hong Kong Barcelona
Budapest

Professor Hidetoshi Fukuyama
Department of Physics, Faculty of Science, University of Tokyo
7-3-1 Hongo, Bunkyo-ku, Tokyo 113, Japan

Professor Tsuneya Ando
Institute for Solid State Physics, University of Tokyo
7-22-1 Roppongi, Minato-ku, Tokyo 106, Japan

ISBN-13:978-3-642-84820-9 e-ISBN-13:978-3-642-84818-6
DOI: 10.1007/978-3-642-84818-6

Library of Congress Cataloging-in-Publication Data. Transport phenomena in mesoscopic systems : proceedings of the 14th Taniguchi symposium, Shima, Japan, November 10-14, 1991 / H. Fukuyama, T. Ando. p. cm. – (Springer series in solid-state sciences ; 109) Includes bibliographical references and index. ISBN 0-387-55794-6 1. Mesoscopic phenomena (Physics)–Congresses. 2. Transport theory–Congresses. I. Fukuyama, H. (Hidetoshi), 1942. II. Ando, Tsuneya, 1945- . III. Series. QC176.8.M46T73 1992 530.4'1–dc20 92-28119

Typesetting: Camera-ready by authors/editors
54/3140 - 5 4 3 2 1 0 - Printed on acid-free paper

Preface

This volume contains the proceedings of the Fourteenth Taniguchi Symposium on the Theory of Condensed Matter, which was held from November 10 to 14, 1991, at the Shima Kanko Hotel, Shima, Japan. The topic of the symposium was *Physics of Mesoscopic Systems.*

Mesoscopic systems have been developed hand in hand with the recent progress in nanotechnology and are the melting pot of basic science and technology. In nanostructures, the quantum effect of the electron wave manifests itself because of the limited dimensionality of the structure. The most typical features of these structures are the discreteness of the energy spectrum and the interference effect of electron waves, which have led to various fascinating phenomena. The purpose of this symposium was to discuss the latest developments in mesoscopic systems, especially transport phenomena, from the viewpoint of basic physics.

This volume starts with an introduction to the field of mesoscopic systems together with the paper by Prof. R. Kubo, who was the first to note the existence of particular features of discrete energy levels in small metallic particles. In Part II the electronic states of quantum dots and the conductance through them are discussed. Tunneling via small structures and junctions is studied in Part III. Part IV discusses some features of conductance fluctuations, while the various properties of electron waves in quantum wires are treated in Part V. The features of quantum Hall currents in quantum wires are explored in Part VI. In Parts II–VI, the electronic current with dissipation, i.e. transport currents, is of main interest, while in Part VII the nature of the persistent currents induced by an external magnetic field is studied in the presence of boundaries. Finally, in Part VIII, another type of current flow in small structures, the Josephson current, is discussed.

The symposium was financially supported by the Taniguchi Foundation, which sponsors 15 series of similar symposia, including the present one on the theory of condensed matter. The symposia were conceived by Mr. Toyosaburo Taniguchi, whose donation established the Taniguchi Foundation. The speech he delivered at the reception in Kyoto after the symposium is given in the Foreword, from which the philosophy behind his thoughtful activity is evident. On behalf of all the participants we would like to express our sincere thanks to Mr. Taniguchi and the Taniguchi Foundation. We especially wish good health to Mr. Taniguchi, who has an ever inspiring and positive attitude to basic science.

Tokyo,
May 1992

Hidetoshi Fukuyama
Tsuneya Ando

Participants in the Taniguchi International Symposium

November 10–14, 1991 (at Kanko Hotel, Shima, Japan)

Seated (left to right)
E. Hanamura, J.E. Mooij, Mrs. Imry, Mrs. Mooij, Y. Nagaoka, R. Kubo, Y. Imry, H. Fukuyama, A.D. Stone, P.A. Lee, T. Ando

Second row (left ro right)
D.E. Khmelnitzkii, T. Chakraborty, H. Ebisawa, D. Yoshioka, S. Katayama, H. Lotsch, A. Kawabata, A. Okiji, K. Kawamura, H.U. Baranger

Third row (left to right)
A. Furusaki, S. Iwabuchi, Y. Ono, Y. Isawa, M. Ueda, H. Akera, H. Yoshioka, H. Tamura, C.W.J. Beenakker

Foreword

Speech by Toyosaburo Taniguchi

Dr. Kubo, Distinguished Guests, and Friends:

The Taniguchi Foundation wishes to welcome the participants of the Fourteenth International Symposium on the Theory of Condensed Matter, who have come from within this country and from different parts of the world. I feel privileged, as a Japanese industrialist, to be given the opportunity of meeting with the world-famous scholars attending this symposium.

The concept of the symposium is unique in that participants, both Japanese and from abroad, are limited in number so as to form small discussion groups, a result of my desire to see scientists living together, although for a short period, as a close-knit community. It is easy to talk about, but difficult to realize, the ideal of international friendship and understanding in a world which is steadily growing smaller.

We are deeply indebted to Dr. Kubo and his associates for their efforts which made this particular symposium possible. Any praise, therefore, should be directed to all those who have lent their support to this program.

So far, I have met in person a total of 103 participants in this Division from 22 foreign countries and 212 participants from Japan. Whenever I meet with participants of our symposia, both young and old, I am always impressed by the unselfishness and sincerity of the dedication they show in the pursuit of their aims. I hope and trust that, even after they have reached the heights of academic fame during the coming decades, they will continue to join forces and help to forge closer bonds of friendship and cooperation that will make major contributions, not only to academia, but also towards world peace and the welfare of mankind.

In conclusion, we hope that all the participants will return home with warm memories of this symposium and the pleasant times that we have shared.

Thank you.

Contents

Part IV Conductance Fluctuation

Part V Quantum Wires

Part VI Quantum Hall Effects

Part VII Persistent Current and Orbital Magnetism

Part I

Introduction

Quantum Transport in Mesoscopic Systems: An Introduction

T. Ando[1] *and H. Fukuyama*[2]

[1]Institute for Solid State Physics, University of Tokyo,
7-22-1 Roppongi, Minato-ku, Tokyo 106, Japan
[2]Department of Physics, Faculty of Science, University of Tokyo,
7-3-1 Hongo, Bunkyo-ku, Tokyo 113, Japan

A brief review is given of quantum transport in mesoscopic systems, mainly from a theoretical point of view.

1. INTRODUCTION

Owing to the recent developments in microfabrication technology, artificial structures as small as a few hundreds angstrom can be fabricated such as metallic wires, quantum wires, dots, and point contacts. These mesoscopic structures have revealed interesting quantum-mechanical nature of transport at low temperatures. Typical examples observed in metallic wires are conductance fluctuations, the Aharonov-Bohm (AB) oscillation, and the Coulomb-blockade effect in micro-tunnel junctions. On the other hand, quantum wires at semiconductor heterostructures have shown fascinating new phenomena associated with the ballistic transport in which electrons can move without being scattered by impurities or random potential fluctuations. Typical examples include the conductance quantization across a quantum point contact and the anomaly in the weak-field Hall effect in crossed quantum wires. The purpose of this paper is to give a short introduction on quantum transport in such mesoscopic systems.

Various length scales characterizing the system are discussed in Sec. 2. In Sec. 3 conductance fluctuations and the AB oscillation in metallic diffusive systems are discussed together with intriguing interplay of the phase of Cooper-pairs in a superconductor and those of electrons in normal metals. The ballistic transport in quantum wires and point contacts is discussed in Sec. 4. In Sec. 5 a brief introduction is given on Landauer's formula which constitutes the major tool for understanding mesoscopic quantum transport. Quantum wires in a strong localization regime and topics related to tunneling are discussed briefly in Secs. 6 and 7, respectively. Section 8 deals with quantum-dots and anti-dots arrays.

2. RELEVANT LENGTH SCALES

2.1 Mean Free Path

The classical motion of electrons in external fields is described by the Boltzmann transport equation [1], which leads to the important length scale called the mean free path Λ. The mean free path is the distance covered by an electron within the relaxation time τ inversely proportional to the probability of scattering from a state with a particular momentum $\mathbf{p}$ to other state with a different momentum. In metallic systems at sufficiently low temperatures, electrons with an energy close to the Fermi level contribute to the transport and hence $\Lambda = v_F \tau$, where v_F is the Fermi velocity. In terms of τ the conductivity is written as

$$\sigma = \frac{ne^2\tau}{m}, \tag{2.1}$$

where n is the concentration, m is the effective mass, and $-e$ is the charge of an electron.

It should be noticed that the relaxation time determining the conductivity can sometimes be quite different from the relaxation time τ. In fact, a scattering which does not result in an appreciable change in the direction of momentum is not effective in the reduction of the current. Such a low angle scattering becomes much stronger than large angle scattering when the potential of scatterers is slowly varying in space. In such a case, we have to distinguish the *transport relaxation time* τ_{tr} from τ and replace τ in Eq. (2.1) by τ_{tr}. Usually we have $\tau_{\mathrm{tr}} > \tau$, but for short-range scatterers $\tau_{\mathrm{tr}} = \tau$.

2.2 Fermi Wavelength

Another relevant length is the Fermi wavelength $\lambda_F = 2\pi/k_F$ where k_F is the Fermi wave vector. Typically $\lambda_F \sim 1$ Å in typical metals such as Cu and Ag, whereas $\lambda_F \sim 400$ Å in a two-dimensional system at GaAs/AlGaAs heterostructures with the carrier density $n \sim 3\times10^{11}$ cm^2. In metallic systems the mean free path Λ is much longer than the wavelength λ_F. In dirty systems where $\Lambda \sim \lambda_F$, multiple scattering from impurities starts to play an important role in the electron transport and sometimes leads to the Anderson localization [2-7]. The Anderson localization depends strongly on the dimension and the symmetry of the system. It is known that states are all localized exponentially in one dimension however small randomness may be [8]. The two-dimensional system is marginal. In fact, there exists the weakly localized regime where the wave function decays logarithmically as a precursor of the complete localization. This is the regime where detailed comparison of experiments and perturbation theories has been carried out. In the following we shall assume $\Lambda \gg \lambda_F$ if not stated otherwise.

2.3 System Size

Let us consider a system with $L_x \times L_y \times L_z$ as shown in Fig. 1 and assume $L_x < L_y < L_z$ without loss of generality. If this system is free of randomness, the electronic motion is ballistic and the one-particle energy spectrum consists of discrete levels whose spacings are characterized by $\hbar^2/2mL_x^2 \equiv \varepsilon_x$, $\hbar^2/2mL_y^2 \equiv \varepsilon_y$, and $\hbar^2/2mL_z^2 \equiv \varepsilon_z$, corresponding to the quantization of momentum in each direction. Depending on the system size relative to the Fermi wavelength, one can think of various cases:

(1) $\lambda_F \ll L_x$: three-dimensional (bulk)

(2) $\lambda_F \sim L_x \ll L_y < L_z$: two dimensional (film)

(3) $L_x < L_y \sim \lambda_F < L_z$: quasi-one-dimensional

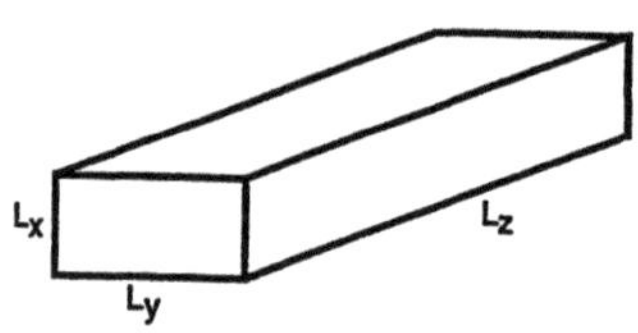

Fig. 1 Schematic illustration of a system with size L_x, L_y, and L_z.

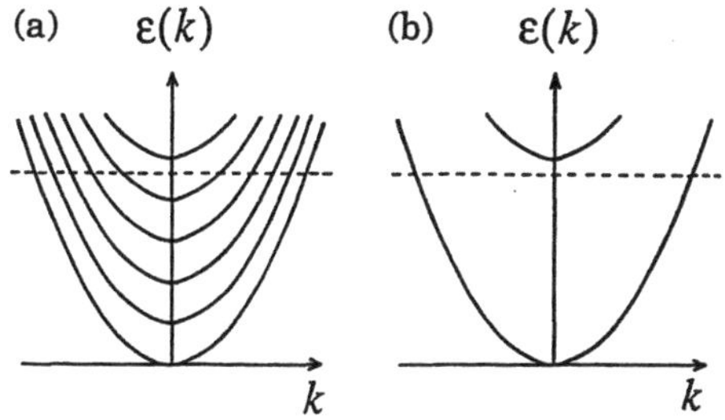

Fig. 2 Schematic illustration of the energy levels as a function of wave vector k in a quasi-one-dimensional wire and in a one-dimensional wire. The horizontal dashed lines represent the Fermi level E_F.

(4) $L_x < L_y < \lambda_F < L_z$: one-dimensional (quantum wire)

(5) $L_x < L_y < L_z < \lambda_F$: zero-dimensional (quantum dot).

For example, in Fig. 2 is shown the crossover of the spectrum of the kinetic energy, $\varepsilon(k)$ from the case (3) to (4) in the case that $\varepsilon_x \gg E_F$ with E_F being the Fermi energy. If the system has a complicated structure as is usually the case in actual experiments, the energy spectrum does not have a systematic structure as discussed above and can be quite complicated (quantum chaos). The discreteness of the one-particle energy spectrum is reflected in various physical quantities once the temperature becomes comparable to it.

2.4 Thermal Diffusion Length

In the presence of randomness, a finite mean free path Λ introduces a different physics. In this case, although the electron motion is ballistic in the length scale shorter that Λ, it can be considered as diffusive in the scale longer than Λ, which is characterized by the diffusion constant $D = v_F^2\tau/d$, d being the dimensionality of one-particle state, *i.e.* $d=3$ if $\lambda_F < L_x$ and $d=2$ if $L_x < \lambda_F < L_y$.

In the diffusive regime the energy of a Bloch state with a particular wave vector $\mathbf{k}$ is characterized by a finite broadening $\hbar/\tau$ corresponding to the fact that each eigenstate has an appreciable spatial overlap with the eigenstates having energies in the range of the order $\hbar/\tau$ around the unperturbed kinetic energy $\varepsilon(\mathbf{k})$. It is important to note that each wave function belonging to these eigenvalues has a well-defined spatial variation, *i.e.*, the phase of the wave function has a definite correlation as schematically shown in Fig. 3a. The phase coherence of a particular eigenstate is not destroyed by random scatterers.

The new length scale in the diffusive regime is $L_T = \sqrt{D\hbar/k_BT}$ called the thermal diffusion length, where k_B is the Boltzmann constant. This represents the characteristic length over which the electron diffuses in the time scale corresponding to the thermal broadening of the Fermi distribution function $\hbar/k_BT$. It can be considered as a length over which the effects of the phase coherence persists when averaged over states which lie near the Fermi energy within the thermal broadening and thus contribute to the transport phenomena.

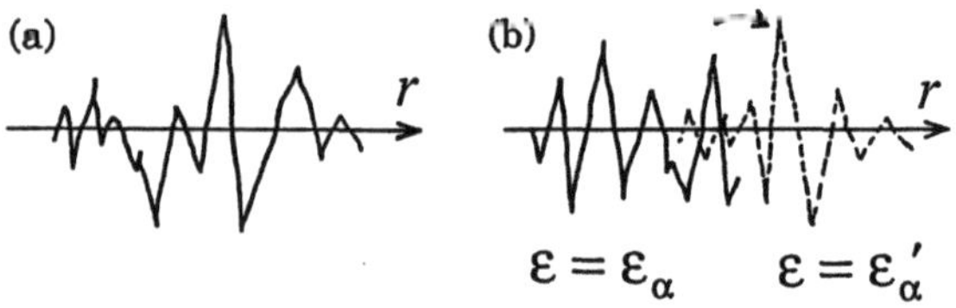

Fig. 3 Schematic illustration of a wavefunction in diffusive regime (a). In the presence of inelastic scattering an electron in state α is scattered into state β with energy E_β different from E_α.

2.5 Phase Coherence Length

So far various characteristic length scales are introduced which are relevant to the mesoscopic systems within the one-particle picture, *i.e.* in the absence of dynamical interactions such as the mutual Coulomb and electron-phonon interaction. These dynamical interactions cause inelastic scattering of an electron with a particular energy to a state with different energy as schematically shown in Fig. 3b. The rate of such an inelastic scattering is given by $\hbar/\tau_\phi$ in terms of the inelastic scattering time τ_ϕ. This is one of the key concepts introduced through the course of the study of the Anderson localization and is known to be strongly dependent on temperature as $\tau_\phi \propto T^{-p}$ (p=1∼2) [3]

The characteristic length defined by $L_\phi = \sqrt{D\tau_\phi}$, called a phase coherence length, is the scale over which an electron stay in a particular eigenstate and its phase coherence is maintained. Since $\tau_\phi \gg \tau$ in general, $L_\phi \gg \Lambda$. It is a remarkable fact that as the temperature is lowered L_ϕ increases and can be comparable to the system size even if the (elastic) mean free path Λ is much shorter than the system size and of a microscopic scale.

3. DIFFUSIVE REGIME

3.1 Aharonov-Bohm Effect

According to the classical electromagnetism the electromagnetic field is described by the electric **E** and magnetic field **H** obeying Maxwell's equation. The scalar potential $\phi(\mathbf{r})$ and the vector potential $\mathbf{A}(\mathbf{r})$ are introduced because of convenience. The equation of motion for an electron with charge $-e$ is determined by the force $\mathbf{F} = -e(\mathbf{E}+\mathbf{v}\times\mathbf{H}/c)$ and does not contain the scalar and vector potential explicitly. There are infinite numbers of ways in choosing these potentials giving the same **E** and **H** and physically observable quantities are independent of these different gauges (the gauge invariance). Quantum mechanically, however, the potentials appear in the Schrödinger equation explicitly and therefore their effect can appear directly in physical quantities [9].

Let us consider a vector potential, for example. The Schrödinger equation is given by

$$\frac{1}{2m}\left[\left(\mathbf{p}+\frac{e}{c}\mathbf{A}(\mathbf{r})\right)^2 + V(\mathbf{r})\right]\psi(\mathbf{r}) = E\psi(\mathbf{r}). \tag{3.1}$$

We assume that an electron moves along a contour in a region where a magnetic field is absent. Let $\psi_0(\mathbf{r})$ be a solution of the above equation for $\mathbf{A}=0$ for some particular energy. Then, the wave function

$$\psi(\mathbf{r}) = \exp[-i\theta(\mathbf{r})]\psi_0(\mathbf{r}), \quad \theta(\mathbf{r}) = \frac{2\pi}{\Phi_0}\int^{\mathbf{r}} A(\mathbf{s})\cdot d\mathbf{s}, \tag{3.2}$$

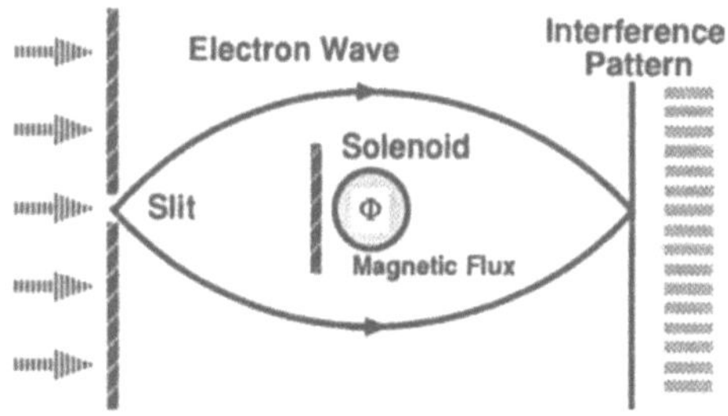

Fig. 4 Schematic illustration of the AB effect corresponding to the vector potential. The interference pattern at the screen on the right hand side varies as a function of the magnetic flux Φ with period Φ_0, although the magnetic field is totally confined inside the solenoid.

satisfies the Schrödinger equation where $\Phi_0 = ch/e$ is the magnetic flux quantum and $\mathbf{s}$ is the vector along the contour. Thus, the vector potential appears as a phase of the wave function.

By applying Stokes' theorem for a closed loop C and corresponding surface S, we have

$$\oint_C \mathbf{A}(\mathbf{s})\cdot ds = \int_S [\vec{\nabla}\times\mathbf{A}(\mathbf{r})]\cdot d\mathbf{S} = \Phi, \tag{3.3}$$

where Φ is the magnetic flux passing through the total area. This means that $\theta(\mathbf{r})$ is independent of the pass and a function of $\mathbf{r}$ only in singly connected systems. In such a case the phase of the wave function associated with the vector potential does not give rise to any observable effects. In fact, the intensity of the electron beam associated with this particular eigenstate is given by $|\psi(\mathbf{r})|^2 = |\psi_0(\mathbf{r})|^2$ independent of the vector potential.

Aharonov and Bohm [9] proposed interference experiments schematically shown in Fig. 4. There is no magnetic field outside the solenoid because the magnetic flux is confined inside the solenoid. Because of the magnetic flux Φ inside the solenoid, however, a vector potential exists with strength $A(r) = \Phi/2\pi r$ and direction perpendicular to the radial vector from the center of the solenoid to the position $\mathbf{r}$. In the absence of the flux the wave function $\mathbf{r}_0$ at the interference plane is given by $\psi_0(\mathbf{r}_0)$ independent of the electron path. In the presence of a magnetic flux, the phase of the wave function θ_j depends on path j and we have $\psi_j(\mathbf{r}_0) = \exp(-i\theta_j)\psi_0(\mathbf{r}_0)$. Therefore, the beam intensity becomes

$$|\psi_1(\mathbf{r}_0)+\psi_2(\mathbf{r}_0)|^2 = 2[1+\cos(\theta_1-\theta_2)]|\psi_0(\mathbf{r}_0)|^2, \tag{3.4}$$

where $\theta_1 - \theta_2 = 2\pi(\Phi/\Phi_0)$. This means that the intensity oscillates as a function of flux Φ with period Φ_0. This Aharonov-Bohm oscillation was experimentally confirmed immediately after the theoretical prediction [10] and also demonstrated in electron holograph [11]. The basic physics here is same as the Fraunhofer pattern in the Josephson junction [12], though the flux quantum in the latter is $\Phi_0/2$ because of the Cooper-pair condensate.

An AB oscillation can be observed in the conductance across a ring structure schematically shown in Fig. 5. The first actually observed magnetoresistance exhibited irregular oscillation as a function of the magnetic field [13,14]. This oscillation, later called the magneto-finger print, varies from sample to sample but is not a time dependent noise but is reproducible.

The magneto-finger-print is now known to be a manifestation of the universal conductance fluctuations. Altshuler [15] and Lee and Stone [16] showed that the conduc-

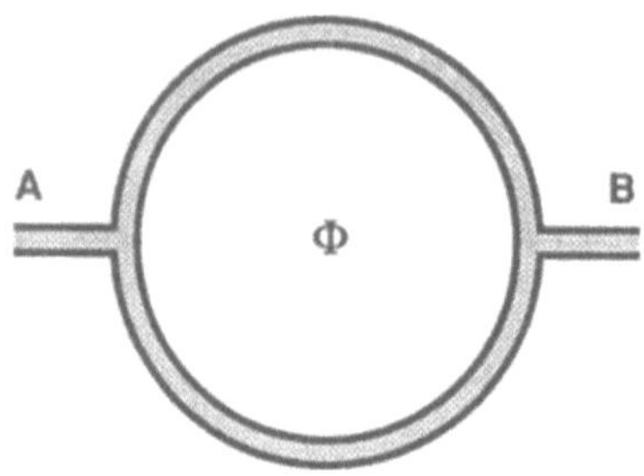

Fig. 5 A metallic ring for observing the AB effect in the conductance. The voltage drop between A and B is measured in the presence of current flowing from A to B. In actual experiments some magnetic flux penetrates the region where electrons exist.

tance fluctuations are independent of system sizes and strength of scattering in contrast to the common belief that fluctuations become smaller with increasing system size. An electron is scattered from many impurities and affected by complex interferences. The impurity configuration and the interference patterns vary from system to system. This difference in interferences leads to large fluctuations of the conductance. In the presence of a weak magnetic field the electron wave function acquires an extra phase factor due to the vector potential. This phase plays in interferences a role similar to that due to a change in the impurity configuration and the conductance fluctuates as a function of the magnetic field. The amount of the fluctuations due to the change of magnetic field is of the same order as that when the sample is varied and the typical magnetic field corresponding to a period of oscillations is determined by the condition that the total flux passing through the system is of the order of the flux quantum Φ_0. The fluctuations depend on the symmetry of the system [17].

Unfortunately, the first experiments were made for metallic rings with small diameter because of the limitation required by the condition that the size should be smaller than the phase coherence length [13]. Consequently the magnetic flux passing though the central hole and that passing through the conducting region were almost the same, which made unable the observation of the AB oscillation. The AB oscillation of the conductance was later observed in metallic wires with diameter much larger than the wire width [18].

At nonzero temperatures the amplitude of the AB oscillation is reduced by two different mechanisms. Because an electron keeps its phase memory over the distance L_ϕ, the amplitude is reduced by a factor $\exp[-\pi R/L_\phi]$ where R is the ring diameter. Another temperature effect arises due to the increase in the number of states in the vicinity of the Fermi energy E_F in proportion to k_BT. Because $k_BT \ll E_F$, each state gives almost the same amplitude but a different phase $(\theta_1 - \theta_2)$ because of slightly different path. There is a strong correlation between the wave functions of the states with energy close to each other and this correlation decreases with increasing energy difference. Let E_c represent a typical energy where the correlation becomes negligible $(E_c \sim \hbar D/R^2)$. Then, the amplitude of the AB oscillation is given by a sum of terms consisting of cosine functions with almost equal amplitude and random phases. The total number of the states contributing to the total intensity is given by k_BT/E_c when $k_BT \gg E_c$. Therefore the AB oscillation is reduced by the factor $N(T)^{-1/2}\exp(-\pi R/L_\phi)$ where $N(T) \propto \sqrt{k_BT/E_c}$ is the number of independent states contributing to the conductance. This explains the strong temperature dependence observed experimentally [18]. This averaging of the AB oscillation has been proved by experiments directly [19].

In metallic thin films deposited on a cylinder surface with diameter $\sim$ 1 μm schematically shown in Fig. 6 a different kind of oscillation appears as a function of

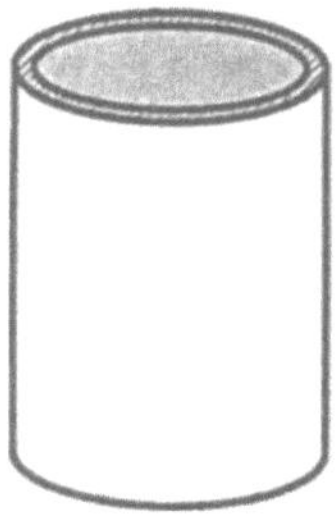

Fig. 6 A metallic film on a cylinder surface. In the presence of a magnetic field parallel to the cylinder direction the conductivity exhibits the AAS oscillation [$\propto \cos(4\pi\Phi/\Phi_0)$], where Φ is the flux passing through the cylinder and Φ_0 is the flux quantum.

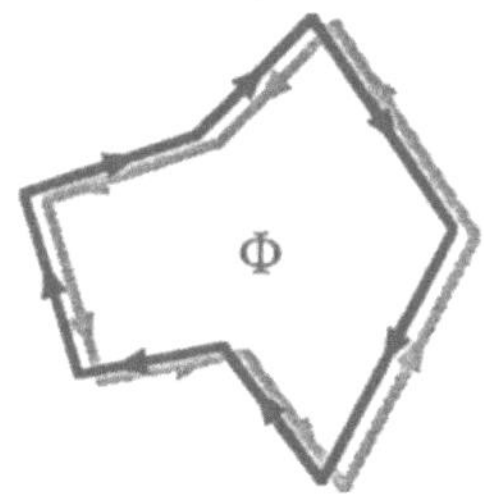

Fig. 7 A pass and corresponding reverse pass. These two paths lead to an interference term proportional to $\cos(4\pi\,\Phi/\Phi_0)$. Consequently, a quantum correction to the conductivity exhibits the AAS oscillation as a function of Φ with period $\Phi_0/2$.

Φ with period $\Phi_0/2$. This effect was first predicted theoretically by Altshuler, Aronov, and Spivak [20] and later observed in Mg and Cd films experimentally by Sharvin and Sharvin [21]. The oscillation, which is called AAS oscillation, arises in a quantum correction of the conductivity in the weak localization regime and is due to the ensemble average. Since many states are contributing to the transport in this case because of the macroscopic size of the cylinder, the AB oscillation disappears after the ensemble average. On the other hand for each path passing through a certain point, there exists a path in the reverse direction as shown in Fig. 7 [5]. These two paths lead to an interference term proportional to $\cos(4\pi\Phi/\Phi_0)$, where the energy-dependent phase factor vanishes because two waves correspond to the states with the same energy and the difference in the magnetic flux of two paths is now twice as large as that in Fig. 5. At nonzero temperature, this oscillation amplitude should be multiplied by $\exp(-2\pi R/L_\phi)$.

3.2 Persistent Current

The Aharonov-Bohm effect manifests itself not only in such a transport phenomenon as conductance but also in a persistent current, which is a current flow without dissipation [22]. Typically one can think of a ring geometry similar to that in Sec. 3.1 but without any leads and the persistent current can be probed as a magnetization. Actually this is the particular case, where the orbital magnetism is expected to have periodic oscillations with respect to the magnetic flux Φ.

The current induced by the flux is given at $T=0$ by

$$J = -c\frac{\partial E}{\partial \Phi} = -c\sum_n{}' \frac{\partial \varepsilon_n}{\partial \Phi}, \tag{3.5}$$

where E is the ground state energy, ε_n is the energy of each eigenstate, and the summation with respect to n is over the occupied states. In an ideal one-dimensional ring in the sense $L_x \sim L_y < \lambda_F \ll L$, ε_n is given by

$$\varepsilon_n = \frac{\hbar^2}{2m}\left(\frac{2\pi}{L}\right)^2\left(n+\frac{\Phi}{\Phi_0}\right)^2, \tag{3.6}$$

since the wave function $\psi(r)$ satisfies the boundary condition, $\psi(r+L)=\exp(-i2\pi\Phi/\Phi_0)$ $\psi_0(r+L)=\psi_0(r)$, with $\psi_0(r)$ being the simple plane wave, as shown in Eq. (3.2). Then for the case $E_F \gg (\hbar^2/2m)(2\pi/L)^2$, the typical value of J is given by ev_F/L [23]. Experiments have been carried out by Levy *et al.* [24] and Chandrasekhar *et al.* [25].

Especially in the latter experiment the magnetization due to a single ring of Au has been measured. The diameter is 2.4 μm, the loop linewidth is 90 nm, and the thickness is 60 nm, whereas the mean free path was estimated to be 70 nm. In this case there exist roughly several hundreds transverse channels which are affected by the elastic scattering. There are several interesting problems associated with the actual magnitude of a persistent current in such a ring geometry: the roles of many channels in the presence of a randomness, mutual Coulomb interaction, and the fact that the system is a canonical ensemble [22]. This subject is discussed in Part VI.

The orbital magnetism, typically known as Landau diamagnetism, is due to the persistent current induced by the magnetic field. Since the way current flows is very susceptible to the spatial variation of the potential, one can expect appreciable fluctuations of orbital magnetism, which is actually the case theoretically [26-28]. This problem is discussed in this volume.

3.3 Proximity Effects

Besides the transport and persistent currents, there exists another unique type of current, the Josephson current. In contrast to the former two which are due to one-particle effects, the Josephson current is carried by Cooper pairs in the superconducting state, in which the whole macroscopic system is described by a wave function and the phase coherence of Cooper pair amplitude is maintained over the macroscopic scale (macroscopic quantum phenomena).

The Josephson current is generally investigated in the geometry of a junction [12]. A supercondutor-normal metal-superconductor (SNS) structure is of the particular interest in the present context [29] because the N-region can be of a mesoscopic size. In an SNS structure the Cooper pairs generated at the boundaries of SN interfaces penetrate into the N-region and then the amplitude of the Cooper pair wave function is nonzero as schematically shown in Fig. 8. The characteristic length ξ over which the pair amplitude penetrates into the N-region is called the coherence length. If $\xi > L$, the N-region will behave like a superconductor whereas two superconductors are decoupled once $\xi < L$.

The fluctuations of the superconducting critical current are of natural interest [30]. It is also of interest how the phase coherence of wave functions of one-particle and that

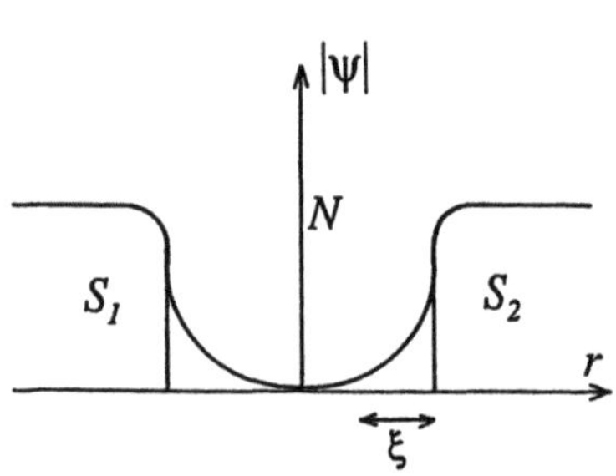

Fig. 8 The Cooper-pair amplitude in an SNS junction. In the region of a normal metal (N) the pair amplitude does not vanish but decays exponentially away from the interface.

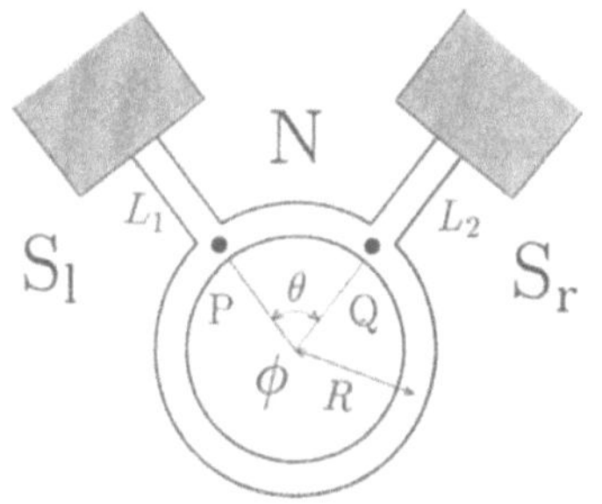

Fig. 9 Schematic illustration of an NSN structure with a mesoscopic AB ring (radius R) consisting of a normal metal. The magnetic flux passing the ring is denoted by ϕ and the difference in the Cooper-pair phase of superconductor S_l and S_r is given by θ.

of the pair condensate come into physical quantities. A mesoscopic structure with a AB ring for the N-region as shown in Fig. 9 will be most suitable for the study of such a problem [31,32]. There exists a competition of two different kinds of the interference effect: "2e"-oscillation and "e"-oscillation. Another interesting case is the one where the N-region consists of a narrow wire with few channels for which the critical current has been predicted to have a stepwise variation as the number of occupied channel is varied [33,34]. These and related problems are examined in Part VII of this volume.

4. BALLISTIC REGIME

In the diffusive regime considered in the previous section, scattering from impurities distributed randomly determines the transport properties. In the ballistic regime, on the other hand, the impurity scattering is not important and scattering from the system boundary plays a dominant role. In quantum wires fabricated at GaAs/AlGaAs heterostructures, the confining potential constituting the boundary is usually introduced through an electrostatic potential created either by split gates or the Schottkey barrier at the surface. Some examples of structures used for creating quantum wires are shown in Fig. 10.

Usually, the origin of the confining potential (split gate or surface) is located quite far away from the 2D system and the external potential is expected to vary smoothly in spite of the presence of roughness caused by the limitation imposed by fabrication processes. Because of the large screening due to 2D electrons, the confining potential is determined in a manner self-consistent with the electron density distribution. The resulting confining potential is expected to vary smoothly with small variation hav-

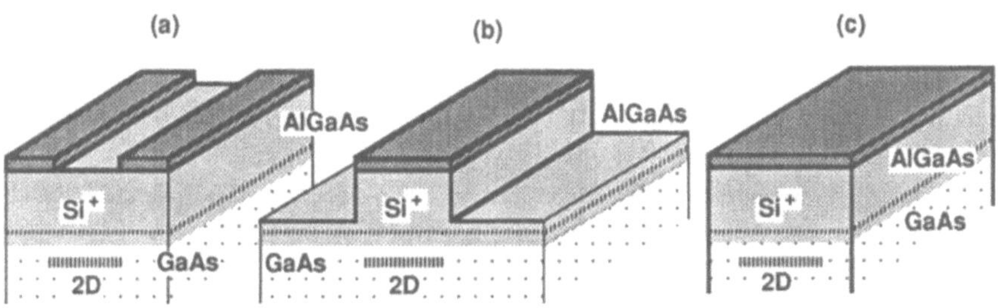

Fig. 10 Schematic illustration of confinement of 2D electrons at GaAs/AlGaAs heterostructures fabricated by (a) split-gate, (b) shallow-mesa etch, and (c) deep-mesa etch techniques.

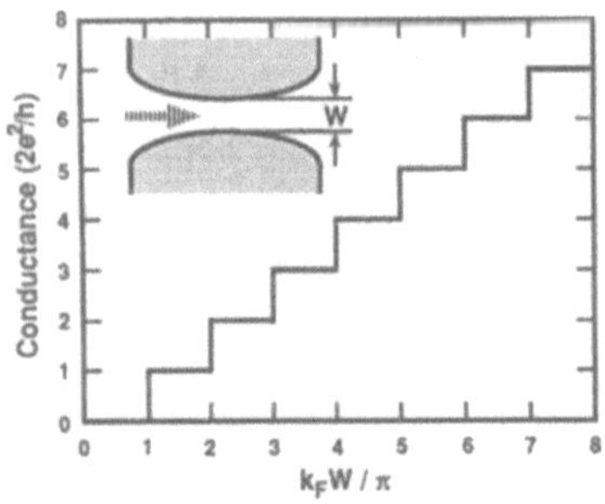

Fig. 11 Conductance across a narrow constriction versus electron wavelength in an ideal case. The inset illustrates an ideal quantum point contact in which the boundary varies much more slowly than the electron wavelength.

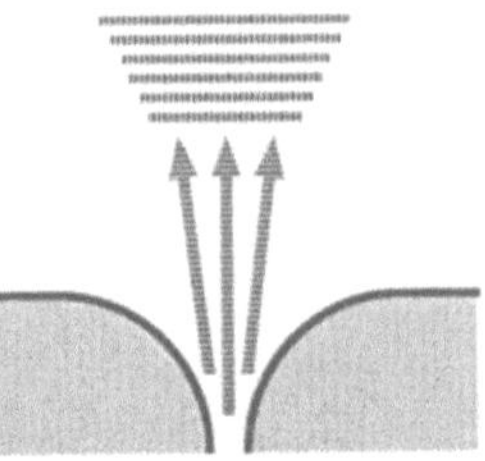

Fig. 12 Collimation of electron beam injected by a quantum point contact. Because of the large rounding of the corners of the boundary, the electron waves are collimated in the forward direction.

ing the length scale of the order of the screening length or the electron wavelength. Therefore, effects of rough boundaries are usually less important than one might expect except in some special systems where boundary roughness is introduced intentionally. A strong boundary roughness gives rise to a peculiar positive magnetoresistance observed experimentally [35] and analyzed theoretically [36] in quantum wires fabricated at GaAs/AlGaAs heterostructures. This topics is treated in Part IV in this volume.

One of the most typical phenomena observed in the ballistic regime is the conductance quantization across a quantum point contact [37-39]. By the metallic gate fabricated on top of the two dimensional system at a GaAs/AlGaAs heterostructure and by a negative bias voltage we can create a narrow path connecting two spatially separated 2D electron systems. The conductance across this quantum point contact was observed to be quantized into integer multiples of $e^2/\pi\hbar$ as shown in Fig. 11.

The electron wave emitted from a quantum point contact is collimated in such a way that the electron propagates mainly in one direction instead of being emitted as a point source from which electrons propagate as a spherical wave. This collimation is a result of a considerable rounding of the corner of the confining potential as illustrated in Fig. 12 and has been observed in the magnetic-field dependence of the conductance across series of point contacts [40] and also in magnetic focusing experiments [41]. The nature of this point contact can be used as electron emitters and collectors in ballistic quantum structures (electron optics) [42]. Interference of electron waves passing through regions with different electron concentrations has also been observed and used for the determination of the strength of electron-electron scattering of 2D electrons both as a function of energy and temperature [43]. Nonlinear conduction has also been a subject of recent study but is not included in this volume.

Usually, the transport coefficients in quantum wires are measured using a crossed-wires geometry. Various interesting phenomena arising from ballistic transmission across junctions of quantum wires have been observed. One most typical example is the anomalies in the low-field Hall effect such as the quenching and last plateau [44-46]. In weak magnetic fields, the Hall resistance has been observed to be much smaller than classical $\rho_H = H/nec$ (quenching) but increases rapidly and stays almost independent of the field (last plateau) beyond a certain magnetic field as shown in Fig. 13.

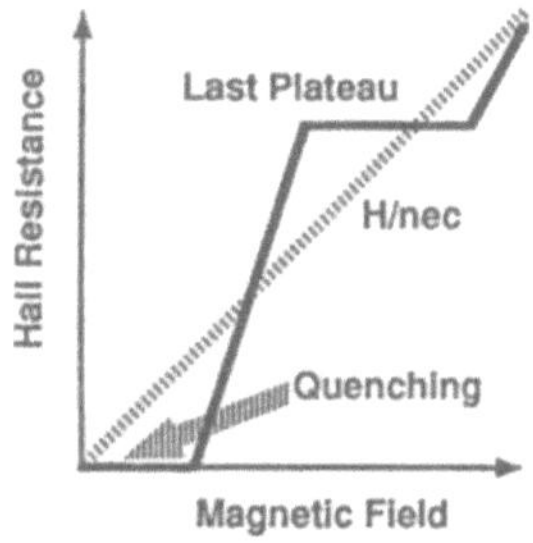

Fig. 13 Schematic illustration of the anomaly of the Hall effect in weak magnetic fields. The dotted line represents the Hall resistance for classical 2D electrons and the solid line represents that in narrow wires showing quenching and last plateau.

These anomalies arises from the presence of large rounding of the corner of the junction. In addition to the anomalies in the weak-field Hall effect, various other interesting effects were observed in this crossed-wires geometry such as the negative bend resistance and the transfer resistance [47,48]. For more details about the ballistic transport, the reader should refer to a comprehensive review by Beenakker and van Houten [34].

5. THEORY OF QUANTUM TRANSPORT

5.1 Landauer's Formula

Landauer derived the relation between the conductance of a one dimensional wire and the transmission and reflection probabilities at the Fermi level [49]. Let us consider the system schematically illustrated in Fig. 14. A wire is connected at both ends to an ideal wire which is infinitely long and eventually connected to an electron reservoir. The ideal reservoir satisfies the following two conditions: (1) All incident electrons are absorbed by the reservoir irrespective of their energy and phase. (2) It constantly provides electrons with energy below chemical potential μ. The energy and phase of these electrons are independent of those of absorbed electrons. Let μ_1 and μ_2 represent the chemical potential of the reservoir 1 and 2, respectively, and T and R be the transmission and reflection probability, respectively, of the wire. The conservation of the number of particles requires $T+R=1$. Further, let v be the electron velocity in the ideal lead and $\partial n_+/\partial E = 1/\pi\hbar v$ be the density of states for electrons moving from left to right. The total current flowing across the system is given by

$$I = (-e)v\frac{\partial n_+}{\partial E}T(\mu_1-\mu_2). \tag{5.1}$$

Because the voltage difference between 1 and 2 is $(-e)V_{21} = \mu_1-\mu_2$, the *two terminal* conductance measured between 1 and 2 is given by

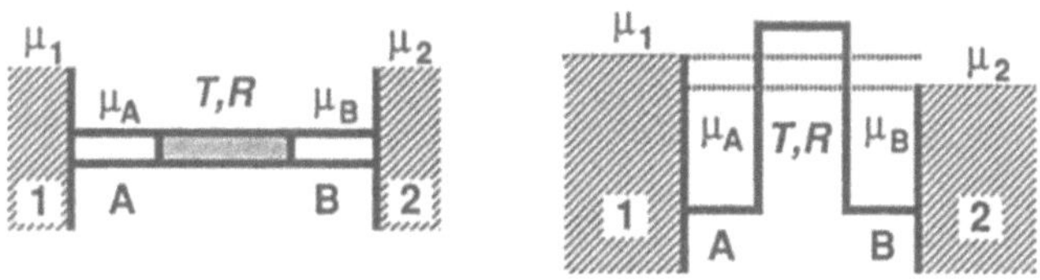

Fig. 14 A one-dimensional wire connected at both ends by an ideal lead to reservoirs and its schematic energy diagram.

$$G = \frac{I}{V_{21}} = \frac{e^2}{\pi\hbar}T. \tag{5.2}$$

This relation between the conductance and the transmission probability is frequently used as the basis of understanding the transport in mesoscopic systems.

The original version of Landauer's formula is slightly more complicated, because the above conductance is not that of the wire but that of the system containing the ideal leads. Usually, the conductance represents the ratio of the current to the voltage applied across the system. This voltage is the difference of the effective Fermi level of the ideal leads. When $T=0$, each ideal lead is in equilibrium with the connected reservoir and $(-e)V=\mu_1-\mu_2$. When $T=1$, on the other hand, we have $(-e)V=0$. It has been shown that for general T and R

$$(-e)V = R(\mu_1-\mu_2). \tag{5.3}$$

Therefore, the final expression of the conductance is

$$G = \frac{e^2}{\pi\hbar}\frac{T}{1-T}. \tag{5.4}$$

In analyzing the transport in mesoscopic systems, the formula (5.2) and its multi-channel and multi-terminal versions are more appropriate rather than the original formula (5.4).

5.2 Universal Conductance Fluctuations

The universality of the conductance fluctuations has been shown first by perturbational calculation based on the Kubo formula as has been discussed in Sec. 3.1. This can be understood as follows in terms of Landauer's formula as has been shown by Imry [50].

Let us consider a hyper cube with size L in dimension d. The conductance is written as

$$G = \frac{e^2}{\pi\hbar}\sum_{i,j}^{N}|t_{ij}|^2, \tag{5.5}$$

where $N\sim(k_FL)^{d-1}$ is the total number of the incoming and outgoing channels. It is a reasonable assumption that $|t_{ij}|$ itself does not vary so much from channel to channel in the diffusive regime. The most important point is that $|t_{ij}|$'s for different channels are correlated to each other because the system is described by a single Schrödinger equation. Let N_c be the effective number of channels independent of each other and t_0 be the typical value of $|t_{ij}|$. Then, the conductance and its fluctuation are given by

$$G \sim \frac{e^2}{\pi\hbar}N_c^2t_0^2, \quad \Delta G \sim \frac{e^2}{\pi\hbar}N_ct_0^2. \tag{5.6}$$

Using the usual expression of the conductivity

$$\sigma = \frac{ne^2\tau}{m} \sim \frac{e^2}{\pi\hbar}k_F^{d-1}\Lambda, \tag{5.7}$$

with Λ being the mean free path, we have

$$G = \sigma L^{d-2} = \frac{e^2}{\pi\hbar}(k_FL)^{d-1}\frac{\Lambda}{L}, \tag{5.8}$$

and

$$\Delta G \sim \frac{G}{N_c} \sim \frac{e^2}{\pi\hbar}\frac{N}{N_c}\frac{\Lambda}{L}. \tag{5.9}$$

Because $L \gg \Lambda$, the fluctuation would be much smaller than the average if the effective channel number N_c is of the same order as the total channel number N.

Let $\exp(-2\alpha L)$ denote an eigenvalue of the $N \times N$ matrix $(|t_{ij}|^2)$. Then, α distributes in the region $0 < \alpha < \Lambda^{-1}$. Because of repulsion between eigenvalues (corresponding to the strong correlation among $|t_{ij}|$'s of different channels) they distribute almost uniformly. Since the eigenvalues satisfying the condition $\alpha L \lesssim 1$ contribute to the conduction, the effective channel number is given by $N_c \sim (\Lambda/L)N$. The substitution of this into the above equation immediately leads to the universality of the fluctuations, $\Delta G \sim e^2/\pi\hbar$.

5.3 Conductance Quantization

It is quite straightforward to show the quantization of the conductance across a quantum point contact in the adiabatic limit where the confining potential varies slowly within the Fermi wavelength. In this case scattering between different channels can be neglected completely and the confining potential is represented by an effective one-dimensional potential for each channel. For a confinement in the y direction into a well with width $W(x)$ slowly varying in the x direction, for example, the effective potential for the jth channel ($j = 1, 2, \ldots$) is given by

$$V_j(x) = \frac{\hbar^2}{2m}\left[\frac{\pi j}{W(x)}\right]^2. \tag{5.10}$$

When the maximum value of the effective potential is below the Fermi level ($V_j^{\max} < E_F$), an electron passes through the point contact with probability unity. When the maximum is above the Fermi level ($V_j^{\max} > E_F$), an electron is reflected back with probability unity. Therefore, the multi-channel version of Landauer's formula immediately leads to the conductance quantization into an integer multiples of $e^2/\pi\hbar$, where the integer is the number of 1D channels below the Fermi level at the point contact.

When the potential varies abruptly in comparison with the Fermi wavelength, reflection becomes appreciable and leads to an oscillation of the conductance due to interference effects. There have been reported many calculations of the conductance, some of which seem to suggest that the quantization does not necessarily require the adiabatic limit [51]. Transport quantities other than the conductance have also been studied [52-56].

5.4 Weak-Field Hall Effect

In systems with many terminals Landauer's formula can be generalized into

$$I_i = \frac{e^2}{\pi\hbar}\Big[(1-R_{ii})\mu_i - \sum_{j\neq i} T_{ij}\mu_j\Big], \tag{5.11}$$

where I_i is the current flowing into the system from terminal i with chemical potential μ_i, T_{ji} is the transmission probability from terminal i to j, and R_{ii} is the reflection

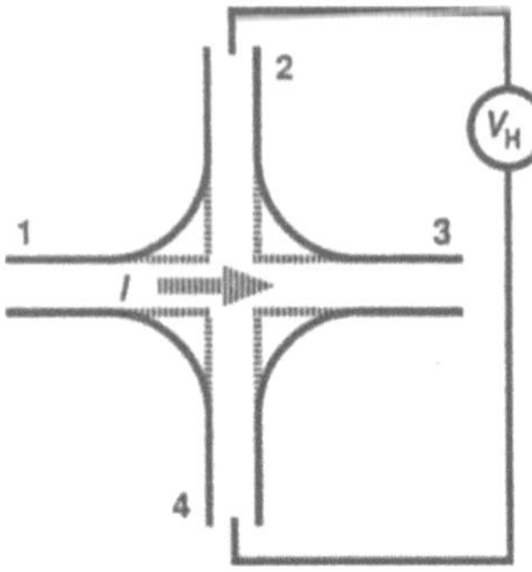

Fig. 15 A junction of two quantum wires used for Hall-effect measurements. The current I flows from terminal 1 to 3 and the voltage V_H is measured between terminal 2 and 4.

probability from i to i. The condition for the flux conservation leads to

$$R_{ii} + \sum_{j \neq i} T_{ij} = 1, \tag{5.12}$$

for every i. The Onsager's reciprocity theorem is expressed as

$$R_{ii}(H) = R_{ii}(-H), \quad T_{ij}(H) = T_{ji}(-H), \tag{5.13}$$

where H is the magnetic field.

Consider the crossed-wires geometry shown in Fig. 15. In order to obtain the Hall resistance in weak magnetic fields, we put

$$I_1 = -I_3 = I, \ I_2 = I_4 = 0, \ V_{24} = V_H, \tag{5.14}$$

and

$$\begin{aligned} &T_{31} = T_{13} = T, \ T_{24} = T_{42} = T', \\ &T_{21} = T_{32} = \cdots = T''(1+\beta H), \ T_{12} = T_{23} = \cdots = T''(1-\beta H). \end{aligned} \tag{5.15}$$

Then, the Hall resistance is calculated as

$$R_H = \frac{V_H}{I} = \frac{h}{2e^2} \frac{\beta H T''}{(T+T'')(T'+T'')} \tag{5.16}$$

Note that this is applicable to multi-channel cases if we replace T's by the sum over incoming and outgoing channels.

There can be two different possibilities leading to the quenching of the Hall effect, (1) $\beta H \ll 1$ and (2) $T'' \ll 1$. Many authors calculated the transmission and reflection probabilities at the junction [57-59]. The results show that $T \sim T' \sim T''$ or $R_H \sim (h/2e^2)\beta H \sim H/nec$ for junctions with no rounding of the corners of the confining potential. Thus, the quenching does not take place except in the presence of strong boundary roughness scattering [59]. In quantum wires with strong boundary roughness, the current is mainly carried by the lowest 1D subband and electrons in this subband have a large probability of going in the forward direction at the junction ($T'' \ll 1$). This is essentially the same as the nozzle effect proposed by Baranger and Stone [60], who assumed that the confining potential near the junction becomes wider very gradually.

Beenakker and van Houten proposed a billiard model in which the electron motion is replaced by a classical billiard ball [61]. Using this model, they could successfully explain both quenching and last plateau assuming a considerable rounding of the corners at the junction. More recent numerical calculations based on quantum mechanics seem to show that $T \sim T' \sim T''$ at realistic junctions with rounded corners and therefore the main origin of the quenching is $\beta H \ll 1$ [62].

Most electrons are scattered from the junction boundaries many times before being transmitted into voltage probes (electrons perform a chaotic motion classically). Effects of a magnetic field on electrons following such complicated trajectories are quite different from those in the absence of rounding and lead to the quenching of the weak-field Hall effect (sometimes even a change in the sign). When the magnetic field becomes stronger and the cyclotron radius becomes comparable to the wire width, most electrons are transmitted into the left voltage probe. This leads to a saturation of the Hall effect or the last plateau.

6. STRONG LOCALIZATION REGIME

Historically, a quantum wire was first fabricated using a Si space-charge layer [63]. Figure 16 shows a schematic illustration of a MOSFET in which electrons are confined within a narrow region sandwiched between p^+ regions at the Si/SiO_2 interface. The wire width is determined by both gate voltage and bias voltage applied on p^+ region. For a fixed bias voltage, the width becomes wider with the increase of the gate voltage proportional to the 2D electron density. Under appropriate conditions the wire width can be as small as 200–1000 Å. Fowler *et al.* observed that the conductance of such a wire exhibits a strong and irregular oscillation as a function the gate voltage [63]. Narrow wires with channel width of 200–1000 Å were also fabricated on Si MOSFET's using the electron beam lithography [64]. They also exhibited irregular oscillation of the conductance.

Because of the presence of strong randomness at the Si/SiO_2 interface and the reduced dimensionality, states are believed to be localized exponentially in such wires. In such a strongly localized regime the conduction takes place due to the so-called variable range hopping at low temperatures. Fowler *et al.* [63] fitted the temperature dependence of the minimum of the fluctuating conductance to the formula:

$$\sigma(T) = \sigma_0 \exp\left[-\left(\frac{T}{T_0}\right)^n\right], \tag{6.1}$$

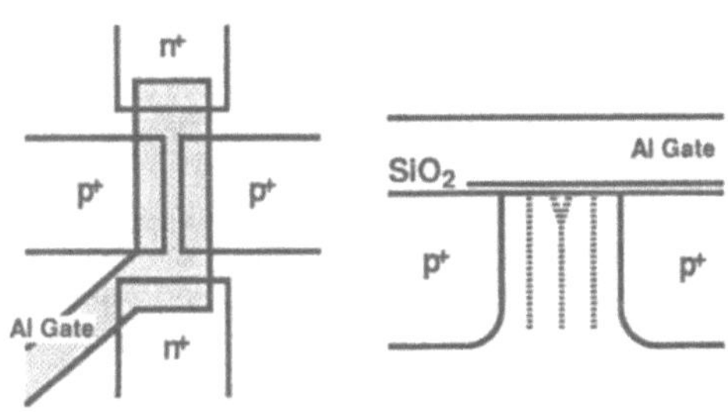

Fig. 16 Schematic illustration of a quantum wire fabricated at a Si/SiO_2 interface. 2D electrons are confined within a narrow region sandwiched between p^+ regions and the wire width is determined by both gate voltage applied between the metallic gate and the source electrode and the bias voltage applied on p^+ region. The left figure shows the top view and the right the side view.

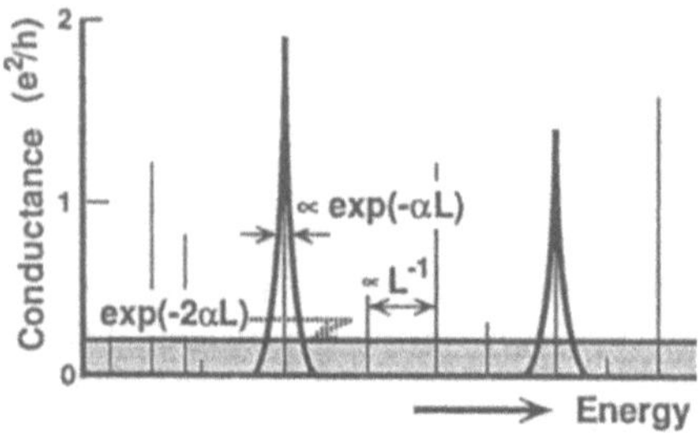

Fig. 17 Schematic illustration of the conductance in a 1D wire in the strong localization regime.

where the exponent n is related to the dimensionality of the system through $n = 1/(d+1)$. For small gate voltages the dimensionality is close to unity and for larger voltages the dimensionality approaches two, as is expected.

One possible mechanism leading to such irregular oscillations is a resonant tunneling. Let α be the inverse localization length. Then the transmission probability is usually of the order of $T \sim \exp(-2\alpha L)$, which gives the exponentially small conductance $G(L) \sim (e^2/\pi\hbar)\exp(-2\alpha L)$. When the energy coincides with a localized eigenstate, the transmission increases drastically. The width of this resonance enhancement is of the order of $\exp(-\alpha L)$, which is much smaller than the level separation proportional to $1/L$. Consequently, the conductance consists of sharp spikes and exhibits irregular oscillations as a function of energy or electron concentration as schematically shown in Fig. 17. However, the actual experimental results are understood as fluctuations due to a change in optimum paths of variable range hopping as a function of the electron concentration [65].

7. TUNNELING

7.1 Resonant Tunneling

Tunneling through a barrier is one of most typical quantum phenomena that are forbidden in classical mechanics. A resonant tunneling occurs in double-barrier diodes consisting, for example, of GaAs/AlGaAs heterostructures when the energy of an incident electron is the same as that of bound states in the well region. This resonance effect leads to a negative differential resistance shown in Fig. 18, which has been observed experimentally [66]. In the conventional picture of the resonant tunneling, the electron wave remains coherent between the emitter and collector electrodes. However, essentially the same negative differential resistance can be explained by the so-called sequential tunneling, in which an electron tunnels into a virtual bound state in the well region sandwiched by the barriers, loses its phase memory completely, and then tunnels out of the bound state into the collector electrode [67]. This problem of the coherent versus sequential tunneling has been a subject of considerable interest and intensive investigations. One of the papers in Part II in this volume is devoted to this problem.

In tunneling the Coulomb interaction between electrons can play an important role. One typical example is an electrical instability due to a charge-build-up in the well region observed experimentally in double-barrier structures [68]. There have been various efforts to control resonant tunneling by using gate structures. One typical approach has been to reduce the lateral size of the double barrier structure and realize 3D-0D-3D resonant tunneling [69-72]. With the decrease in the size of the tunnel structure, the

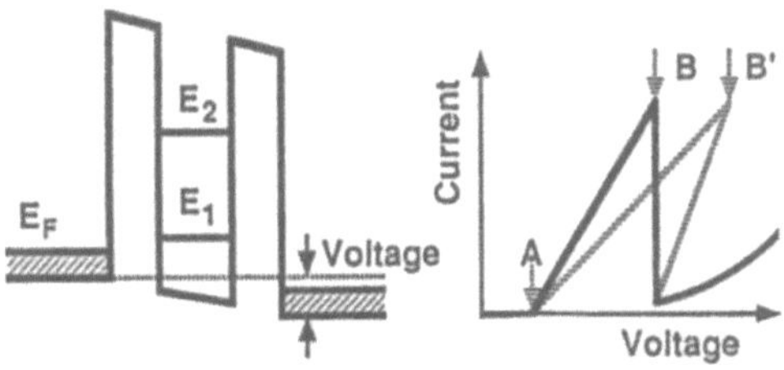

Fig. 18 A conventional GaAs/AlGaAs/GaAs/AlGaAs/GaAs double barrier structure showing negative differential resistance due to resonant tunneling and a typical current-voltage characteristics. The tunneling current starts to increase linearly, when the Fermi level of the emitter reaches E_1 (denoted as A). The resonance enhancement of the current disappears and only small non resonant current remains when the bottom energy of the emitter exceeds E_1 (denoted as B). The same resonant tunneling occurs also for the first excited level E_2. In actual tunneling diodes, an appreciable amount of electronic charges is accumulated in the well region under the resonant condition and the I-V characteristic is deformed as described by the dashed line (B → B'). This leads to an intrinsic instability.

Coulomb effect becomes more and more important and eventually leads to the single-electron tunneling discussed in the next section.

7.2 Single Electron Effect

In structures with dimensions below 100 nm, their capacitance C can be smaller than 10^{-15} F. As a consequence the charging energy associated with a single electron charge, $E_C = e^2/2C$, can be significant at temperatures considerably below 1K where the thermal energy is much less than E_C. The present nanotechnology has enabled fabrication of devices in which the tunneling of a single electron is manipulated.

Let us consider a tunnel junction consisting of two metallic electrodes with capacitance C. Because the electron tunneling is accompanied by a transfer of at least charge e, each tunneling requires energy ΔE given by

$$\Delta E = \frac{(Q-e)^2}{2C} - \frac{Q^2}{2C} = \frac{e(e-2Q)}{2C}, \tag{7.1}$$

where Q is the charge accumulated at the junction before tunneling. Thus, the tunneling is prohibited for $|Q| < e/2$ as shown in Fig. 19, leading to an I-V characteristic schematically shown in Fig. 20 [29]. With use of this Coulomb blockade in series of small metallic junctions various single electron transistors have been realized such as turnstiles and pumps [73,74]. Quasi-periodic oscillation of the conductance due to the Coulomb charging effect has been observed in series of narrow constrictions on GaAs /AlGaAs heterostructures [75-77]. This Coulomb blockade effect is one of the major topics discussed from both experimental and theoretical points of view in many papers contained in this volume.

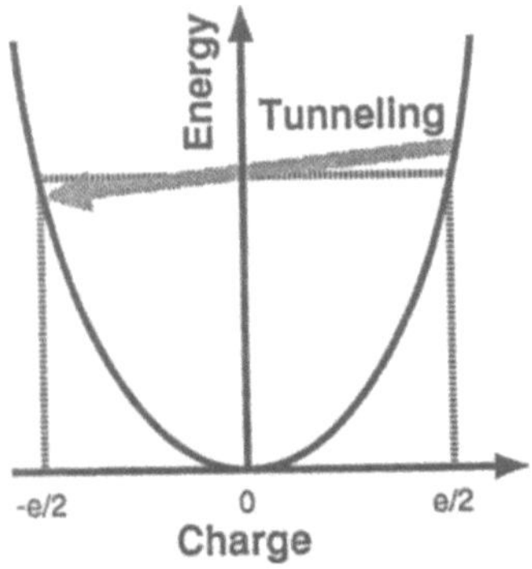

Fig. 19 Energy vs accumulated charge at a small tunnel junction. Tunneling is not allowed in the region $|Q| < e/2$.

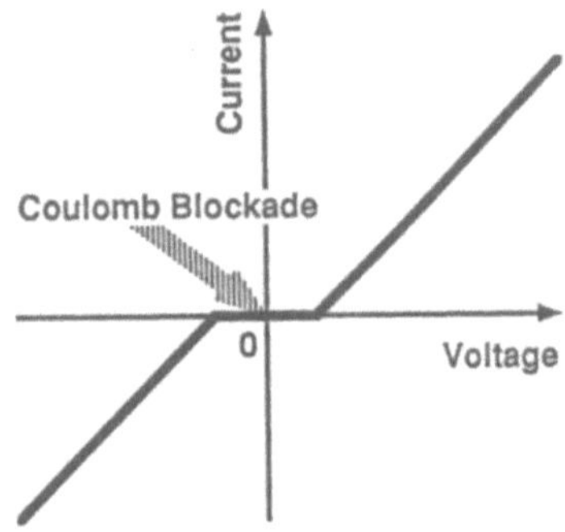

Fig. 20 Schematic illustration of current versus applied voltage in a small tunnel junction.

8. QUANTUM DOTS AND ANTI-DOTS

Arrays of quantum dots can be fabricated at GaAs/AlGaAs heterostructures by, for example, a holographic double exposure of laser light [78,79]. The period of the dots is of the order of 1 μm and the geometrical dimension can be as small as 300 Å. The number of electrons confined in each dot varies from 20 to 300. Quantum dots containing only a few electrons have been fabricated at an InSb surface [80]. Far-infrared (FIR) resonances are the major way of observing electronic states of such quantum dots. Self-consistent calculations of energy levels have been performed and shown that the confinement is well approximated by a parabolic harmonic potential [81].

The Schrödinger equation for an electron confined by a harmonic potential in a magnetic field has been solved [82]. The energy eigenvalues are given by

$$E_{nm} = (2n+|m|+1)\sqrt{(\hbar\Omega_0)^2+\left(\frac{\hbar\omega_c}{2}\right)^2} + \frac{1}{2}m\hbar\omega_c, \tag{8.1}$$

where Ω_0 is the frequency characterizing the confining potential, $n=0,1,\ldots$ is the radial quantum number, and $m = 0, \pm1, \pm2, \ldots$ is the azimuthal quantum number. Dipole allowed transitions have energies

$$E_{\pm} = \sqrt{(\hbar\Omega_0)^2+\left(\frac{\hbar\omega_c}{2}\right)^2} \pm \frac{1}{2}\hbar\omega_c. \tag{8.2}$$

These are shown in Fig. 21.

The energy level structure containing many electrons becomes quite complicated because of an interplay of the confining potential and the mutual Coulomb interaction which tends to the formation of an incompressible quantum-liquid-state leading to the fractional quantum Hall effect. This problem is discussed by some authors in Part I in this book. In spite of such complications, however, experimentally observed FIR spectra resemble that shown in Fig. 21. This can be understood in terms of the so-called generalized Kohn's theorem stating that optically active collective modes of interacting many electron systems in a parabolic confinement are those of a single particle with mass

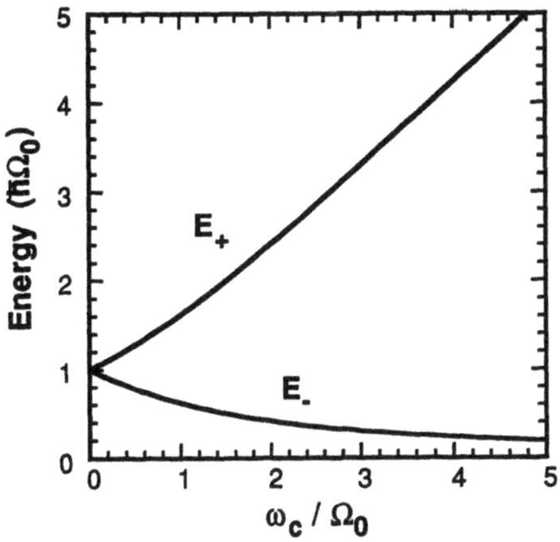

Fig. 21 Resonance energies versus the magnetic-field strength in a quantum dot confined by a parabolic confining potential. In the limit of high magnetic fields, E_+ approaches $\hbar\omega_c$ (cyclotron energy) while E_- which corresponds to an edge mode vanishes.

and charge given by the total number of electrons [83,84]. The spectra deviate from that shown in Fig. 21 only when the confinement becomes different from a harmonic potential, for example, due to couplings between neighboring dots.

Anti-dots arrays are the structure opposite to those of quantum dots. There are two regimes: In the weak modulation regime, 2D electrons are subject to periodically varying potential smaller than the Fermi energy. The energy spectra of such a system in magnetic fields are particularly interesting because of their self-similar nature (Hofstadter's butterfly) [85]. Until now only some oscillation considered as a precursor of such interesting level structure has been observed in magnetoresistance and a clear evidence has not been reported so far. When the amplitude of the periodic potential exceeds the Fermi energy, electrons are excluded from the region with high potential barrier and can move in the region of narrow valleys. A characteristic oscillation appears in the magnetoresistance in anti-dots lattices corresponding to the cyclotron orbit surrounding a single anti-dot, four anti-dots, etc. [86,87].

9. SUMMARY

This paper briefly reviewed the basic physics in mesoscopic systems, where the interference of electron waves and the quantization of energy levels are playing major roles as explained and as will be discussed in detail in all the articles to follow. The mesoscopic physics has rapidly advanced together with developments in device technology and materials science. The systems constitute an interdisciplinary research field where strong interactions among basic science, application, and technology are indispensable. We expect that this field will continue to develope further in future.

REFERENCES

1. See, for example, *Solid State Physics*, N.W. Ashcroft and N.D. Mermin (HRW International Editions, 1976), Chapt. 16.
2. P.W. Anderson, Phys. Rev. **109**, 1492 (1958).
3. E. Abrahams, P.W. Anderson, D.C. Licciardello, and T.V. Ramakrishnan, Phys. Rev. Lett. **42**, 673 (1979).

4. *Anderson Localization*, ed. Y. Nagaoka and H. Fukuyama (Springer, Heidelberg, 1982).
5. G. Bergmann, Phys. Rev. **107**, 1 (1984).
6. P.A. Lee and T.V. Ramakrishnan, Rev. Mod. Phys. **57**, 287 (1985).
7. *Anderson Localization*, ed. T. Ando and H. Fukuyama (Springer, Heidelberg, 1988).
8. V.L. Berezinskii, Sov. Phys. JETP **38**, 620 (1974).
9. Y. Aharonov and D. Bohm, Phys. Rev. **115**, 485 (1959).
10. R.G. Chambers, Phys. Rev. Lett. **5**, 3 (1960).
11. A. Tonomura, N. Osakabe, T. Matsuda, T. Kawasaki, J. Endo, S. Yano, and H. Yamada, Phys. Rev. Lett. **56**, 792 (1986).
12. M. Tinkham, in *Introduction to Superconductivity*, (McGrow-Hill, New York, 1975) Chapt. 6.2.
13. C.P. Umbach, S. Washburn, R.B. Laibowitz, and R.A. Webb, in *Proceedings of 17th International Conference on Low Temperature Physics* (North-Holland, Amsterdam, 1984) p. 717.
14. S. Washburn, *Mesoscopic Phenomena in Solids*, ed. B.L. Altshuler, P.A. Lee, and R.A. Webb (North-Holland, Amsterdam, 1991), p. 1.
15. B.L. Altshuler, JETP Lett. **41**, 648 (1985).
16. P.A. Lee and A.D. Stone, Phys. Rev. Lett. **55**, 1623 (1985).
17. P.A. Lee, A.D. Stone, and H. Fukuyama, Phys. Rev. B **35**, 1039 (1987).
18. R.A. Webb, S. Washburn, C.P. Umbach, and R.B. Laibowitz, Phys. Rev. Lett. **54**, 2696 (1985).
19. C.P. Umbach, C. Van Haesendonck, R.B. Laibowitz, S. Washburn, and R.A. Webb, Phys. Rev. Lett. **56**, 386 (1986).
20. B.L. Altshuler, A.G. Aronov, and B.Z. Spivak, JETP Lett. **33**, 101 (1981).
21. D.Yu. Sharvin and Yu.V. Sharvin, JETP Lett. **34**, 285 (1981).
22. Y. Imry, in *Quantum Coherence in Mesoscopic Systems*, ed. B. Kramer (Plenum Press, New York, 1991), p. 221.
23. H.-F. Cheung, Y. Gefen, E.K .Riedel, and W.H. Shih, Phys. Rev. B **37**, 6050 (1988).
24. L.P. Levy, G. Dolon, J. Dunsmiur, and H. Bouchiat, Phys. Rev. Lett. **64**, 2074 (1990).
25. V. Chandrasekhar, R.A. Webb, M.J. Brady, M.B. Ketchen, W.J. Gallagher, and A. Kleinsasser, Phys. Rev. Lett. **67**, 3578 (1991).
26. O.D. Cheishvili, JETP Lett. **48**, 225 (1988).
27. H. Fukuyama, J. Phys. Soc. Jpn. **58**, 47 (1989).
28. R.A. Serota and S. Oh, Phys. Rev. B **41**, 10523 (1990).
29. K.K. Likharev, Rev. Mod. Phys. **51**, 101 (1979).
30. B.L. Altshuler, and B.Z. Spivak, Sov. Phys. JETP **65**, 343 (1987).
31. M. Büttiker and T.M. Klapwijk, Phys. Rev. B **33**, 5114 (1986).
32. D. Yoshioka and H. Fukuyama, J. Phys. Soc. Jpn. **59**, 3065 (1990).
33. A. Furusaki, H. Takayanagi, and M. Tsukada, Phys. Rev. Lett. **67**, 132 (1991).
34. C.W.J. Beenakker and H. van Houten, Solid State Phys. **44**, 1 (1991).
35. T.J. Thornton, M.L. Roukes, A. Scherer, and B.P. Van de Gaag, Phys. Rev. Lett. **63**, 2128 (1989).
36. H. Akera and T. Ando, Phys. Rev. B **43**, 11676 (1991).
37. B.J. van Wees, H. van Houten, C.W.J. Beenakker, J.G. Williamson, L.P. Kouwenhoven, D. vander Marel, and C.T. Foxon, Phys. Rev. Lett. **60**, 848 (1988).
38. D.A. Wharam, T.J. Thornton, R. Newbury, M. Pepper, H. Ahmed, J.E.F. Frost, D.G. Hasko, D.C. Peacock, D.A. Ritchie, and G.A.C. Jones, J. Phys. C **21**, L209 (1988).

39. Y. Hirayama and T. Saku, Phys. Rev. B **39**, 5535 (1989); Appl. Phys. Lett. **54**, 2556 (1989); Y. Hirayama, T. Saku, and Y. Horikoshi, Jpn. J. Appl. Phys. **28**, L701 (1989).
40. L.W. Molenkamp, A.A.M. Staring, C.W.J. Beenakker, R. Eppenga, C.E. Timmering, J.G. Wiiliamson, C.J.P.M. Harmans, and C.T. Foxon, Phys. Rev. B **41**, 1274 (1990).
41. H. van Houten, C.W.J. Beenakker, J.G. Williamson, M.E.I. Broekaart, P.H.M. van Loosdrecht, B.J. van Wees, J.E. Mooij, C.T. Foxon, and J.J. Harris, Phys. Rev. B **39**, 8559 (1989).
42. J. Spector, H.L. Stormer, K.W. Baldwin, L.N. Pfeiffer, and K.W. West, Surf. Sci. **228**, 283 (1990); Appl. Phys. Lett. **56**, 967 (1990); Appl. Phys. Lett. **56**, 1290 (1990); Appl. Phys. Lett. **56**, 2433 (1990); Appl. Phys. Lett. **58**, 263 (1991).
43. A. Yacoby, U. Sivan, C.P. Umbach, and J.M. Hong, Phys. Rev. Lett. **66**, 1938 (1991).
44. G. Timp, A.M. Chang, P.M. Mankiewich, R.E. Behringer, J.E. Cunningham, T.Y. Chang, and R.E. Howard, Phys. Rev. Lett. **59**, 732 (1987).
45. M.L. Roukes, A. Scherer, S.J. Allen, Jr., H.G. Craighead, R.M. Ruthen, E.D. Beebe, and J.P. Harbison, Phys. Rev. Lett. **59**, 3011 (1987).
46. C.J.B. Ford, S. Washburn, M. Büttiker, C.M. Knoedler, and J.M. Hong, Phys. Rev. Lett. **62**, 2724 (1989).
47. G. Timp, H.U. Baranger, P. de Vegvar, J.E. Cunningham, R.E. Howard, R.E. Behringer, and P.M. Mankiewich, Phys. Rev. Lett. **60**, 2081 (1988).
48. Y. Takagaki, K. Gamo, S. Namba, S. Takaoka, K. Murase, S. Ishida, K. Ishibashi, and Y. Aoyagi, Solid State Commun. **68**, 1051 (1988); **69**, 811 (1989).
49. R. Landauer, IBM J. Res. Dev. **1**, 223 (1957); Philos. Mag. **21**, 863 (1970).
50. Y. Imry, Europhys. Lett. **1**, 249 (1986).
51. See, for example, T. Ando, Phys. Rev. B **44**, 8017 (1991) and references cited therein.
52. U. Sivan and Y. Imry, Phys. Rev. B **33**, 551 (1986).
53. P. Streda, J. Phys. Conden. Matter **1**, 1025 (1989).
54. P.N. Butcher, J. Phys. Condens. Matter **2**, 4869 (1990).
55. C. Proetto, Phys. Rev. B **44**, 9096 (1991); Solid State Commun. **80**, 909 (1991).
56. Y. Okuyama, T. Sakuma, and N. Tokuda, Surf. Sci. **263**, 258 (1992).
57. D.G. Ravenhall, H.W. Wyld, and R.L. Schult, Phys. Rev. Lett. **62**, 1780 (1989).
58. G. Kirczenow, Phys. Rev. Lett. **62**, 1920 (1989); **62**, 2993 (1989).
59. H. Akera and T. Ando, Phys. Rev. B **41**, 11967 (1990).
60. H.U. Baranger and A.D. Stone, Phys. Rev. Lett. **63**, 414 (1989).
61. C.W.J. Beenakker and H. van Houten, Phys. Rev. Lett. **63**, 1857 (1989).
62. H.U. Baranger, D.P. DiVincenzo, R.A. Jalabert, and A.D. Stone, Phys. Rev. B **44**, 10637 (1991).
63. A.B. Fowler, A. Hartstein, and R.A. Webb, Phys. Rev. Lett. **48**, 196 (1982).
64. W.J. Skocpol, L.D. Jackel, E.L. Hu, R.E. Howard, and L.A. Fetter, Phys. Rev. Lett. **49**, 951 (1982).
65. A.D. Stone and P.A. Lee, Phys. Rev. Lett. **54**, 1196 (1985).
66. L.L. Chang, L. Esaki, and R. Tsu, Appl. Phys. Lett. **24**, 593 (1974).
67. S. Luryi, Appl. Phys. Lett. **47**, 490 (1985).
68. A. Zaslavsky, V.J. Goldman, D.C. Tsui, and J.E. Cunningham, Appl. Phys. Lett. **53**, 1408 (1988).
69. M.A. Reed, J.N. Randall, R.J. Aggarwal, R.J. Matyi, T.M. Moore, and A.E. Westel, Phys. Rev. Lett. **60**, 535 (1988).
70. S. Tarucha, Y. Hirayama, T. Saku, and T. Kimura, Phys. Rev. B **41**, 5459 (1990).
71. P. Gueret, N. Blanc, R. Germann, and H. Rothuizen, Phys. Rev. Lett. **68**, 1896 (1992).

72. N.W. Dellow, P.H. Beton, P.C. Main, T.J. Foster, L. Eaves, A.F. Jezierski, W. Kool, M. Henini, S.P. Beaumont, and C.D.W. Wilkinson, Semicond. Sci. Technol. **7**, B442 (1992).
73. L.J. Geerligs, V.F. Anderegg, P.A.M. Holweg, J.E. Mooij, H. Pothier, D. Esteve, C. Uribina, and M.H. Devoret, Phys. Rev. Lett. **64**, 2691 (1990).
74. C. Urbina, H. Pothier, P. Lafarge, P. Orfila, D. Esteve, M.H. Deveoret, L.J. Geerligs, V.F. Anderegg, P.A. Holweg, and J.E. Mooij, IEEE Trans. Magn. **27**, 2578 (1991).
75. L.P. Kouwenhoven, A.T. Johnson, N.C. van der Vaart, C.J.P.M. Harmans, and C.T. Foxon, Phys. Rev. Lett. **67**, 1626 (1991).
76. U. Meirav, M.A. Kastner, and S.J. Wind, Phys. Rev. Lett. **65**, 771 (1990).
77. P.L. McEuen, E.B. Foxman, U. Meirav, M.A. Kaster, Y. Meir, N.S. Wingreen, and S.J. Wind, Phys. Rev. Lett. **66**, 1926 (1991).
78. T. Demel, D. Heitmann, P. Grambow, and K. Ploog, Phys. Rev. Lett. **64**, 788 (1990).
79. A. Lorke, J.P. Kotthaus, and K. Ploog, Phys. Rev. Lett. **64**, 2559 (1990).
80. Ch. Sikorski and U. Merkt, Phys. Rev. Lett. **62**, 2164 (1989).
81. A. Kumar, S.E. Laux, and F. Stern, Appl. Phys. Lett. **54**, 1270 (1989); Phys. Rev. B **42**, 5166 (1990).
82. V. Fock, Z. Phys. **47**, 446 (1928).
83. L. Brey, N. Johnson, and B. Halperin, Phys. Rev. B **40**, 10647 (1989).
84. P. Maksym and T. Chakraborty, Phys. Rev. Lett. **65**, 108 (1990).
85. D.R. Hofstadter, Phys. Rev. B **14**, 2239 (1976).
86. D. Weiss, M.L. Roukes, A. Menschig, P. Grambow, K. von Klitzing, and G. Weimann, Phys. Rev. Lett. **66**, 2790 (1991).
87. T. Yamashiro, J. Takahara, Y. Takagaki, K. Gamo, S. Namba, S. Takaoka, and K. Murase, Solid State Commun. **79**, 885 (1991).

Meso-scopic Physics of Small Metallic Particles

R. Kubo

Department of Physics, Faculty of Science and Technology,
Keio University, 3-14-1 Hiyoshi, Kohoku-ku, Yokohama 223, Japan

Abstract. The mesoscopic physics of small metallic particles is briefly reviewed from a historic point of view.

1. Introduction

Mesoscopic physics of metals and semiconductors in nanometer extension has called much attention in recent years. Thirty years ago, I published a paper [1] in which I considered some peculiar properties of small metallic particles with submicron size. This was a theoretical prediction of quantum effects to be seen in such small systems and was probably the very earliest work of mesoscopic physics.

In this talk, I am going to review on some important points raised in my old paper, which I believe to be relevant to the present and future developments of mesoscopic physics. Since I made the last week the same talk at Nara Symposium on Science and Technology of Mesoscopic Systems [2], this talk will be rather brief [3].

2. Quantized Electron States in a Small Metal Particle and Statistical Distribution of Quantum Levels

Ordinarily we are interested in physical properties of materials as Bloch electrons occupying one-electron quantum states which make continuous energy bands. In a small particle with a finite extension, the energy levels make discrete sets. Consider a particle of N atoms of sodium, the valence band states are discrete, and the average spacing of successive levels is roughly of the order of

$$\delta = E_F/N \,, \tag{1}$$

where E_F is the Fermi-energy measured from the bottom of the valence band. E_F is several eV and therefore δ can easily be much larger than the thermal energy kT at low temperatures or larger than an energy quantum of radiation with a centimeter wave number. This is not really a surface effect, but is due only to the presence of boundary.

When we make an observation of a physical property of an ensemble of particles, in most cases we shall not be able to control the particle size, shape and lattice defects with ideal exactness. We must consider the ensemble consisting of particles with statistical distribution of these parameters. If the energy levels in a sample particle are denoted by $(\varepsilon_0, \varepsilon_1, \varepsilon_2, \varepsilon_3, \varepsilon_4, \ldots, \varepsilon_j, \ldots)$, we regard this set of numbers as statistical. The statistics of level spacing

$$\Delta_j = \varepsilon_j - \varepsilon_{j-1}$$

is particularly important. In my first paper I assumed a purely random distribution, namely a Poisson distribution,

$$P(\varepsilon)d\varepsilon = C\exp(-\varepsilon/\delta)d\varepsilon \tag{2}$$

to find a spacing in the interval $(\varepsilon, \varepsilon + d\varepsilon)$. However, this assumption is not really true. When two levels accidentally coincide, they repel each other. Gor'kov and Eliashberg [4] pointed out that the theory of random matrix [5] can be applied here. Then Eq.(2) is replaced by

$$P(\varepsilon) = \text{Const} \cdot \left(\frac{\varepsilon}{\delta}\right)^n \tag{3}$$

for $\varepsilon \ll \delta$. According to Dyson [6], the number n is determined by the basic symmetry of the electronic Hamiltonian. Namely,

(1) *orthogonal ensemble*: if the Hamiltonian is real and invariant against orthogonal transformation, then $n = 1$.

(2) *unitary ensemble*: if the Hamiltonian is hermitian, $n = 2$.

(3) *symplectic ensemble*: if the spin-orbit interaction is strong, the matrix elements of Hamiltonian are quartanions, and $n = 4$.

Case (1) applies when electron states are only orbital, case (2) when a magnetic field is present. The case (1) is understandable, if we consider, for instance, a matrix of the form,

$$\begin{pmatrix} x & y \\ y & -x \end{pmatrix}$$

with uniformly random real numbers, x and y, which give $\varepsilon_{1,2} = \pm\sqrt{x^2 + y^2}$. Other cases can be considered similarly.

3. Charge Neutrality Condition

Really interesting mesoscopic effects come out if another important factor characteristic to small particles is considered. This is the Coulomb interaction of electrons.

When a metal particle with radius a is charged by Q, the work to be done is

$$W = Q^2/2a\,. \tag{4}$$

If $Q = e$ is the electron charge and a is the Bohr radius, W is equal to 13.7eV. When a is hundred times larger, W is still a large energy compared to the thermal energy at room temperatures. This means that, in a collection of small particles, each particle keep its electron number strictly constant. Charge fluctuation seldomly occurs. In fact, electron conduction of thin film consisting of granular platinum particles was known to be of activation type [7]. Also the Coulomb blockade phenomena is related to this effect.

For a collection of particles, the condition of charge neutrality imposes a stringent effect on the statistics. The well known Fermi-distribution function does not apply to electrons in a particle if the temperatures is low, namely,

$$kT \ll \delta, \tag{5}$$

because the Fermi distribution is valid only when number fluctuation is sufficiently large.

4. Anomalies at Low Temperatures

When we observe a collection of many particles at such low temperatures as the condition Eq.(5) is satisfied, we expect some peculiar anomalies of heat capacity and the Pauli susceptibility. We consider the case where the one electron states are orbital. Each of them below the Fermi level is occupied by electrons with two spin orientations. If the total number of electrons in a given particle is even, the top level is occupied by two spins. If the electron number is odd, the top level contains only one electron. In the latter case, the particle behaves as a free spin and the collection of particle shows the Curie law susceptibility. In contrast, in the former case, the spin susceptibility is quenched below the Pauli value. Because of the strict conservation of electron number in each particle, the energy fluctuation is greatly suppressed and so the heat capacity becomes smaller than the Sommerfeld value. It is recovered to the normal value if temperature is high in comparison to the average level spacing. Also a strong spin-orbit interaction will mix orbital and spin states and smears out the above mentioned anomalies [8].

These mesoscopic effects are interesting. The reduction of heat capacities is difficult to prove experimentally [9]. On the other hand, nuclear resonance observation of small particles has given clear indications of the anomalies of spin susceptibilities [11]. Denton *et al* [10] made detailed calculation of thermodynamic properties of small particles assuming for simplicity equally spaced energy levels instead of randomly distributed levels. The calculation shows the use of canonical ensemble with fixed number of electrons in each particle produces the above mentioned low temperature anomalies.

5. Quenching of Relaxation Processes

Korringa mechanism of nuclear spin relaxation and Elliott mechanism of conduction electron spin relaxation are caused by metal electrons when they exchange their orbital energies with the spin Zeeman energy when they are scattered by the contact interaction of metal electron and nuclear spin in the former case and by scattering of metal electrons by phonons or imperfections. In both cases, the discreteness of the electronic levels makes it difficult to get a proper matching of energy exchange. Therefore, these relaxation processes will be quenched in small particles. In reality, the particles will be in contact with other degree of freedom and relaxation is made possible through other channels. An important point about this is that the presence of boundaries or crystal defects is not an agent causing scattering of metal electrons but it acts as interactions determining each of one-electron states.

6. Electron Conduction

It is well known that small metallic particles show beautiful colors. In modern terminology, it is due to the plasma resonance absorption of light. The width of the absorption band is intuitively interpreted as the electrical resistance which is caused by random reflections of electrons at the surface boundaries. As was mentioned already in the previous section, this interpretation is not right from our point of view. It was shown by Kubo and Kawabata [12] that the broadening of resonance can be seen as the coupling of plasma mode and the individual motion of electrons. The plasma oscillation induces in a particle an oscillating electric field which induces transitions between one-electron levels. This was shown to be almost equivalent to the classical interpretation.

At infrared frequencies, metal particles will show absorptions of infra-red radiation with structures which reflect the random level distributions. Gor'kov and Eliashberg [4] once asserted that the dielectric response will be anomalously large in small particles. However, there is no experimental indication for this. It is canceled by the depolarization field effect which they did not take properly into account.

7. Conclusion

The above is a brief retroscopic review of mesoscopic physics of small particles. In recent years great progress has been made in experimental techniques. Experimentalists can prepare particles, which are called quantum dots, with remarkably good control. Now theorists can feel free to exert more imagination for possible mesoscopic effects which may cast light on the very basic aspects of nature.

References

[1] R. Kubo: J. phys. Soc. Jpn. **17** 975 (1962).

[2] R. Kubo: *Proceedings of Int. Symposium on Science and Technology of Mesoscopic Structures Nara, Nov.6-8, 1991* (Springer Verlag, Tokyo, 1991).

[3] For more detailed review articles, see: a) R. Kubo: J. de Physique. Colloq. **38** C 1977. b) R. Kubo: in *Polarisation, Matiere et Rayonnement*, volume jubilaire en honneur d' Alfred Kastler, Paris, Press Univ. France (1969)p.325. c) R. Kubo, A. Kawabata and S. Kobayashi: Ann. Rev. Mater. Sci. 1984, 14. d) W.P.Halperin: Rev. Mod. Phys. **58** 533 (1986).

[4] L. P. Gor'kov and G. M. Eliashberg: Zh.Eksp. Teor. Fiz **48** 1407 (1965).

[5] M. L. Mehta: *Random Matrices and Statistical Theory of Energy Levels* (Academic Press, New York and London, 1967).

[6] F. J. Dyson: J. Math. Phys. **3** 140 (1962).

[7] C. A. Neugebauer, M. B. Webb: J. Appl. Phys. **33** 74 (1962).

[8] J. Sone: J. Phys. Soc. Jpn. **43** 1457 (1977).

[9] G. R. Stewart: Phys. Rev. **B15** 15 (1977).

[10] R. Denton, B. Muhlschlegel and D. J. Scalapino: Phys. Rev. **B7** 3589 (1973).

[11] See [3] c).

[12] A. Kawabata and R. Kubo: J. Phys. Soc. Jpn. **21** 1765 (1968)

Part II

Quantum Dots

Small Structures: A Laboratory for Studying Strongly Interacting Systems

P.A. Lee

Department of Physics, Massachusetts Institute of Technology, Cambridge, MA 02139, USA

Abstract. We review some recent experimental and theoretical work on transport through a quantum dot. We emphasize the importance of Coulomb interaction between electrons in a confined geometry and point out that these small structures may serve as new probes of correlation effects in strongly interacting systems.

I. Introduction

The study of conductance through a quantum dot has revealed the importance of interaction between electrons in the confined geometry. The most dramatic example is the experiment on a narrow channel between two dimensional electron gas which displays periodic conductance peaks as a function of gate voltage. The basic physical origin of the phenomenon has been attributed to Coulomb blockade. In this model the energy of the N electron state on the quantum dot is written as

$$E(N) = \frac{e^2}{C}N^2 - \mu_G N \,, \tag{1}$$

where μ_G is the chemical potential proportional to the gate voltage. The first term is the charging energy where C is the capacitance of the dot. It expresses the Hartree energy associated with the long range Coulomb interaction of the electron. From equation (1) it is clear that as μ_G is varied, the energy difference between an N+1 and an N electron state E(N+1) - E(N) vanishes periodically with μ_G. When E(N+1) - E(N) vanishes, an electron can tunnel through the quantum dot without any activation energy cost, thus explaining the periodic conductance peaks. In this talk I review some recent theoretical developments which build further on this idea. They are:

1) The inclusion of energy level structures inside the quantum dot. This is necessary to explain some anomalous temperature dependence of the conductance peaks.
2) The extension of these considerations to include the presence of a strong magnetic field, so that the quantum dot is in the integer or the fractional quantum Hall regime.

3) To go beyond the Hartree theory. In the low temperature limit the most important correlation effect is the fact that the electron in the quantum dot acts like a localized impurity coupled to the electron gas in the leads. Kondo effect and mixed valence effects become important and lead to a modification of the line-shape.

We shall briefly discuss these topics in what follows.

II. Level Structure and the Anderson-Wolff model of the quantum dot

Let us examine the energy scales relevant to a quanutm dot. The typical capacitive charging energy is estimated to be $e^2/C \approx 0.5$ meV. For a 100-nm by 1-μm quantum dot in GaAs, the energy level spacing is roughly $\Delta\mathrm{E} \sim 0.05$ meV. Thus, even though $\mathrm{kT} \ll e^2/C$ in the experiment, kT may be either greater or smaller that ΔE. Thus, the level structures need to be taken into account in considering the temperature dependence of the conductance.[3]

The model we treat is the simplest one that takes account of the two dominant quantizations: charge and energy. It consists of two ideal leads coupled to a single site on which there is a Coulomb interaction of energy $\mathrm{U} \equiv e^2/C$. (The isolated 1-μm segment of the channel can be treated as a site, or "quantum dot," since the wave functions are coherent over lengths $\gg 1\mu$m.) The levels on the site are labeled by an index α, with energies ϵ_α. The Hamiltonian is

$$H = \sum_{\alpha;k\in L,R} \epsilon_{k\alpha} c^\dagger_{k\alpha} c_{k\alpha} + \sum_{\alpha} \epsilon_\alpha c^\dagger_\alpha c_\alpha + U n_\uparrow n_\downarrow + \sum_{\alpha;k\in L,R} (V_{k\alpha} c^\dagger_{k\alpha} c_\alpha + H.c.). \tag{2.1}$$

In (2.1), the states in the leads, labeled L and R, have energies ϵ_{ka} and are connected to the site by hopping matrix elements V_{ka}.

Equation (2.1) can be recognized as the Anderson-Wolff Hamiltonian for an impurity in a metal, if α labels spin up and down and ϵ_α is degenerate. In our problem we not only have the possibility of including many states on the "impurity site," but we also can tune the chemical potential of the electron gas in the left and rigth leads separately, an option not available to the conventional problem of an impurity in metals. The possibility leads to new physics which potentially provide new probes of the strongly interacting many-body problem.

A third important energy scale in this problem is the tunnelling rate of the electron state α to the left or the right

$$\Gamma_{L(R)_\alpha}(\omega) = 2\pi \sum_{k\in L(R)} |V_{ka}|^2 \delta(\omega - \epsilon_{ka}). \tag{2.2}$$

For the experiment we are clearly in the regime $U \gg \Delta E \gg \Gamma_L, \Gamma_R$. For simplicity we shall assume that $\Gamma_L = \Gamma_R$. In the experiment conducted so far, we also have $kT \gg \Gamma_L$. In this case more subtle correlations such as the Kondo effect can be ignored and a simple equation of motion solution to equation (2.1) suffices.

The conductance through the dot is given by a Landauer-type formula generalized to interacting systems,

$$\sigma = \frac{e^2}{h} \sum_{\alpha} \int_{-\infty}^{\infty} d\epsilon f'_{FD}(\epsilon) \frac{2\Gamma_{L\alpha}(\epsilon)\Gamma_{R\alpha}(\epsilon)}{\Gamma_{L\alpha}(\epsilon) + \Gamma_{R\alpha}(\epsilon)} Im[G_\alpha(\epsilon)]. \tag{2.3}$$

Equation (2.3) expresses the linear-response conductance as the sum of the elastic-transmission probabilities in each channel weighted by the derivative of the Fermi function. The elastic-transmission probabilities are constructed as a product of the elastic coupling to the leads and the interacting DOS on the site, $-Im[G_\alpha(\omega)]/\pi$. Here $G_\alpha(\omega)$ is the Fourier transform of the retarded Green's function,

$$G_\alpha(t) = -i\theta(t)\langle\{\mathbf{c}_\alpha(t), \mathbf{c}_\alpha^\dagger(0)\}\rangle, \tag{2.4}$$

where the curly brackets denote the anticommutator. We expect (2.3) to apply even in the presence of phonons or other inelastic-scattering mechanisms provided that the inelastic broadening is smaller than k_BT.

In reference (3) we worked out the equation of motion solution for G_α with the further approximation that the interacting density of states for a level n can be approximated by a sum of delta functions $\sum_m P_n(m)\delta(\epsilon_n + mU - \epsilon)$. The weight $P_n(m)$, obtained exactly from the equations of motion for multiple levels, turns out to be the probability that m levels, other than level n, will be occupied for the *isolated* site, and is therefore given by the corresponding Boltzmann weight. From equation (2.3) for the conductance, we find the multilevel conductance formula

$$\sigma = -\frac{\pi e^2}{2h} \sum_n \sum_m \Gamma_n P_n(m) f'_{FD}(\epsilon_n + mU). \tag{2.5}$$

The strengths of coupling to the leads, Γ_n, are still essential in (2.5) in setting the magnitude of the conductance even though their effect of broadening the DOS is negligible for $k_BT \gg \Gamma_n$.

In Fig. 1a, we have plotted experimental conductance data for a narrow-channel GaAs structure with lithographic barriers.[1] For comparison, in Fig. 1b, we have plotted the theoretical conductance (2.5) as a function of chemical potential for a system of 10 nondegenerate levels. The coupling streghts Γ_n of successive levels increase by a factor of

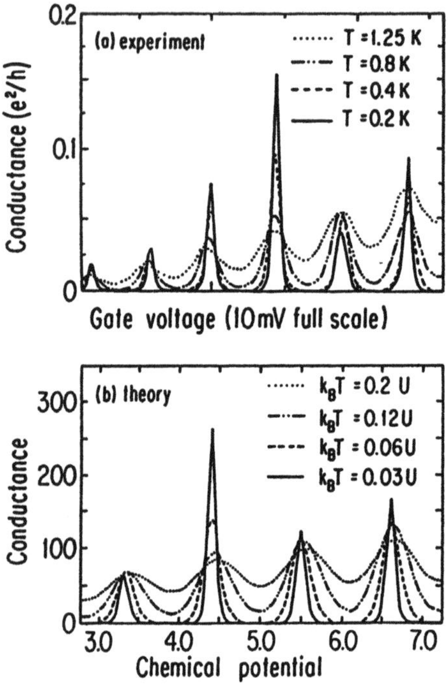

Fig. 1 (a) Experimental conductance of a narrow GaAs channel with two lithographically defined barriers plotted against gate voltage for four temperatures (Ref. 1). (b) Theoretical conductance, in units of $(\Gamma_1/U)(e^2/h)$, for ten levels, space by $\Delta\epsilon = 0.1U$, vs. chemical potential, in units of U, at four temperatures, calculated from (2.5). The elastic couplings of the levels increase geometrically, $\Gamma_n = 1.5^n\Gamma_0$ (to simulate disorder Γ_4 is increased by an additional factor of 4). While for $k_BT \ll \Delta\epsilon$ only are bare level contributes to each conductance peak, for $k_BT \simeq \Delta\epsilon$ many levels contribute, permitting the conductance to rise with temperature.

1.5 to reflect the increase in the tunneling matrix elements near the top of the tunnel barriers. The constant, bare-level spacing is taken to be $\Delta\epsilon = 0.1U$. (These parameters are obtained assuming parabolic barriers with the lithographic width of 100 nm and a height of 0.75 meV, estimated from the product of the level spacing, 0.05 meV, and the fifteen oscillations observed before the top of the barrier is reached. A similar rapid increase of couplings is also expected near the parabolic maxima of the screened-impurity-potential barriers in the Si structures.) At low temperatures ($k_BT \ll \Delta\epsilon$), each conductance resonance corresponds to transport through a single level. As μ increases and each level is filled, the peaks in the density of states of the remaining levels are pushed up by U, leading to conductance peaks spaced by $U + \Delta\epsilon = 1.1U$. For temperatures comparable to $\Delta\epsilon$, multiple levels participate in transport, and the stronger coupling of higher levels leads to an increase of conductance with temperature. The efects of random fluctuations in coupling strength, represented by a factor-of-4 enhancement of Γ_4, are smoothed out for $k_BT > \Delta\epsilon$ by the averaging over multiple levels. Similarly, random fluctuations in the bare-level spacing lead to fluctuations in the conductance-peak spacing. However, these fluctuations would no longer be noticeable for $k_BT > \Delta\epsilon$. The phenomenology of the experimental conductance is well reproduced by the theoretical results at a realistic ratio of the bare-level spacing to the Coulomb-interaction energy.

III. Effect of a strong magnetic field

The experiment described in the last section has been extended to include a strong magnetic field so that only the lowest few random levels are occupied.[4] This system has the advantage that the single particle states are well understood to be edge states and the dependence of E_α on magnetic field is well understood. The experimentalist observed small shifts on the resonant energy as well as dips in the magnitude of the conductance peaks as the magnetic fields is varied. These structures are satisfactorily explained as being due to the transfer of a single electron from the edge of a second Landau level to the edge of the first Landau level. Strictly speaking, what is being observed is the energy difference E(N+1) - E(N) which includes the Coulomb term plus a second term which depends on the magnetic field in a well defined way.

Motivated by the success of these experimetns, Kinaret et al.[5] went ahead to consider what would happen if the magnetic field is further increased so that only the lowest Landau level is occupied and the system enters the fractional quantum Hall regime. The calculation was performed numerically for a parabolic confinement potential for up to 8 particles. The new feature is that now the wavefunction on the dot is a strongly correlated many body state. The tunnelling probability Γ must be modified by the square of the single particle matrix element

$$T_N \equiv \langle N+1|c_\alpha^\dagger|N\rangle$$

for adding an electron in an angular momentum state α beteween ground state wavefunctions $|N\rangle$ and $|N+1\rangle$. This matrix element will modify the magnitude of the conductance peaks via the modifiation of $\Gamma_{L,R}$ in equation (1.2). We found two interesting new results. First for a filling factor $\nu = 1/3$, we find that

$$T_N \sim N^{-1/2}$$

it turns out that this result can be understood analytically based on the chiral fermion formulation of the fractional quantum Hall edge state.[6] Secondly, we find that for $\nu = 2/3$, the matrix element would alternate between being of order unity independent of N and being small as N increases. In our simulation, the original of this alternation comes from the structure of the $\nu = 2/3$ quantum dot. We have evidence that within our model of confining potential, the $\nu = 2/3$ state can be interpreted as a $\nu = 1/3$ hole drop within a $\nu = 1$ background. Then it is possible to add an electron to the outer edge of the $\nu = 1$ state which would involve an N independent overlap factor of order unity. A second possibility is to add a hole to the $\nu = 1/3$ drop and 2 electrons to the $\nu = 1$ edge. This would involve an overlap factor less than unity which decreases with N. The two possibilities must alternate in order to preserve the edge structure. More generally, for the $\nu = 2/3$ state the magnetic flux increases by 3 for each 2 added electrons. Thus it is not surprising that the basic periodicity upon adding electrons is two. This alternating overlap would show up experimentally as an alternation in the conductance peaks as a function of gate voltage. It will clearly be interesting to observe it experimentally.

IV. Effects of spin correlations

The discussion in section II was restricted to the region $\Gamma \ll kT$. In the opposite limit, the correlation between the spin of the state on the quantum dot and the spin in the leads become important in the analogous Anderson impurity problem.[7,8] This is the familiar Kondo effect, if the energy level E_α is much below the Fermi level. As E_α goes through the Fermi level, we enter the mixed valence regime, where the charge on the impurity may fluctuate. For our problem this corresponds precisely to the condition of conduction via resonant tunneling through the quantum dot. Compared with the impurity problem, the quantum dot system is unique in that by varying gate voltage, we can tune from the Kondo to the mixed valence regime continously. In fact, at $T = 0$ and in the Kondo regime, there exists a Kondo resonance at the Fermi level and the tunnelling probability is in principle unity per spin channel. Of course this Kondo temperature is exponentially small when E_α is far from the Fermi energy and in practice one is always above T_K. However as one enters the mixed valence regime, ie. $E_\alpha - \mu \approx \Gamma$, the coherence temperature becomes of order Γ and many body correlation effects become important. Thus, according to this picture, the line-shape of the conductance peak as a function of gate voltage actually provides a map of the cross-over from the high temperature limit of the Kondo problems to the low temperature limit of the mixed valence problem. The result is a non-Lorentzian line-shape which is asymmetrical, with temperature dependence on one side of the line. To observe this behavior one would like to produce structures where Γ is reasonably large. The predicted anomalous line-shape has so far not been observed.

The support by the Joint Services Electronic Program under contract number DAAL 03-89-C-0001 is gratefully acknowledged.

References

1. U. Meirav, M.A. Kastner and S.J. Wind, Phys. Rev. Lett. **65**, 771 (1990), and references therein.

2. H. van Houten and C.W.J. Beenaker, Phys. Rev. Lett. **63**, 1893 (1989).

3. Y. Meir, N.S. Wingreen and P.A. Lee, Phys. Rev. Lett. **66**, 3048 (1991).

4. P.L. McEuen, E.B. Foxman, U. Meirav, M.A. Kastner, Y. Meir, N.S. Wingreen, S.J. Wind, Phys. Rev. Lett. **66**, 1926 (1991).

5. J. Kinaret, Y. Meir, N. Wingreen, P.A. Lee and X.G. Wen, preprint.

6. X.G. Wen, Phys. Rev. B **41**, 12838 (1991).

7. T.K. Ng and P.A. Lee, Phys. Rev. Lett. **61**, 1768 (1988).

8. L.I. Glazman and M.E. Raikh, JETP Lett.**47**, 452 (1988).

Statistical Theory of Coulomb Blockade and Resonant Tunneling Oscillations in Quantum Dots

A.D. Stone[1], *R.A. Jalabert*[1;*], *and Y. Alhassid*[2]

[1]Applied Physics, Yale University,
P.O. Box 2157, New Haven, CT 06520, USA
[2]Center for Theoretical Physics, Sloane Physics Laboratory,
Yale University, New Haven, CT 06511, USA
*Present Address: Service de Physique de l'Etat Condense,
CEA Saclay, F-91191 Gif-sur-Yvette Cédex, France

Abstract. We develop a statistical theory of the amplitude of Coulomb blockade and resonant tunneling oscillations in semiconductor quantum dots based on the hypothesis that chaotic dynamics in the dot potential leads to behavior described by random matrix theory. We obtain a universal one-parameter distribution describing the fluctuations of the resonance amplitudes in the regime where a single thermally-broadened level dominates. As with other mesoscopic fluctuation phenomena, changing the fundamental symmetries of the dot hamiltonian(e.g. breaking time-reversal symmetry with a weak magnetic field) leads to changes in this distribution which should have experimentally-observable effects. The theory is tested numerically by calculating the resonance-width distribution of a very weakly disordered ensemble of dots and good agreement is found.

1. Introduction

Coulomb blockade effects in tunneling through metallic junctions have been studied for some time both experimentally and theoretically [1]. In these systems the single-particle level spacing is much less than kT and the tunneling process is always of the sequential or incoherent type. Recently it has become possible to observe single-charge tunneling through semiconductor quantum dots in which the single-particle level-spacing $\Delta\varepsilon$ may be larger than kT[2, 3, 4]. In this case the resonances which are observed as a function of the fermi energy ε_f may be understood as *resonant* tunneling through the discrete states of the dot, which simply must be corrected for the Coulomb interaction energy of adding the Nth electron to the dot. Since this interaction energy is $U = (Ne)^2/2C$ (C is capacitance), $\delta U(N) = (N - 1/2)e^2/C$ and the resonance spacing is

$$\Delta E_N = [\delta U(N) + \epsilon_N] - [\delta U(N-1) + \epsilon_{N-1}] = e^2/C + \Delta\varepsilon_N\ , \qquad (1)$$

where $\{\epsilon_N\}$ represent effective single-particle levels of the dot(which may need to be calculated self-consistently[5]).

In the typical experimental situation $e^2/C \sim 5°K \gg \Delta\varepsilon \sim 0.5°K$ and approximately equally-spaced resonances are observed as predicted by Eq. (1). This behavior is in sharp contrast to the random conductance peaks observed in early experiments on quasi-1d MOSFETS[6], and was an indication that a new phenomenon(i.e. Coulomb blockade) was controlling the resonant tunneling in these devices[7]. Further experimental confirmation came from the observation that the resonance-spacing in different devices varied inversely with the area(capacitance) of the dot and had the correct order of magnitude for the tunneling of a single-charge [2]. In addition, the shape of the resonance peaks in gate voltage was well-fit by the derivative of the fermi function $f'(\varepsilon_f - E_N)$, the typical behavior of thermally-broadened resonances. However two other observations were initially rather puzzling. First, in all experiments the *amplitude* of the oscillations exhibited fluctuations as large as one order of magnitude on the scale of adjacent peaks (roughly $0.01V$). Second, the behavior of the peak heights as a function of temperature in the interval $0.2-2^{\circ}K$ was non-uniform; some peaks grew with decreasing T and others decreased in amplitude[2, 8]. In general the peak area was not constant as a function of T in contrast to the usual behavior of thermally-broadened resonances.

This non-uniform T-dependence was addressed by Meir et al.[8] based on a finite temperature transport theory for quantum dots developed independently by Meir et al. and Beenakker[9]. In this theory when $kT \ll \Delta\varepsilon$ only one single-particle state contributes to the Nth resonance which has a peak height

$$g_{max} = \frac{e^2}{h}\frac{\Gamma^l_N\Gamma^r_N}{4\pi kT(\Gamma^l_N + \Gamma^r_N)} \quad , \tag{2}$$

where $\Gamma_N = \Gamma^l_N + \Gamma^r_N$, the total decay width, is the sum of the decay widths into the right and left leads. Thus in this regime, in which a single thermally-broadened level controls the resonance, its width is fixed to be $\sim kT$ but its amplitude varies and reflects directly the T=0 decay widths(tunneling rates) from a given state. When $kT > \Delta\varepsilon$ however, a cluster of several levels contribute to each resonance; Meir et al. showed that if the decay widths of adjacent levels vary significantly(e.g. by a factor of four) non-uniform T-dependence similar to that observed experimentally occurs in the interval $\Delta\varepsilon < kT < e^2/C$. Its physical origin is fairly simple within this model; as T decreases the height of a given peak grows if the level nearest the chemical potential has a decay width which is larger than the mean width and shrinks if it is smaller. Hence both the non-uniform T-dependence as kT approaches $\Delta\varepsilon$ from above and the amplitude fluctuations when $kT < \Delta\varepsilon$ can be explained if the decay widths of adjacent quasi-bound levels within the dot vary substantially. We propose an explanation and quantitative description of these level-width fluctuations in this work.

2. Origin of Amplitude Fluctuations

Single semiconductor quantum dot devices of the type studied in refs. [2, 3] are quite new and the shape of the electrostatic confining potential and its gate voltage dependence has not been studied in detail either experimentally or theoretically. However a number of pieces of evidence suggest that the fluctuations in the decay widths inferred from Coulomb blockade resonances cannot be controlled by effects related to the electrostatic barriers confining the electrons within the dot. First, the scale of variation is $0.01V$ which corresponds to a change in the fermi energy too small to change the tunneling rates through the barriers substantially; in fact the expected increase in tunneling rates as ε_f increases and the tunneling barriers become more transparent is observed experimentally[2, 3], but this effect is *monotonic* and simply provides a smooth background against which the fluctuations occur. Although the barriers themselves may distort electrostatically as the gate voltage is varied, the scale of this variation may be estimated at $\sim 0.2V$ based on measurements of edge-state equilibration in very similar GaAs devices in the quantum Hall regime[10], again too large a scale to explain the Coulomb blockade fluctuations. Finally, recent experiments of McEuen et al.[11] find that the amplitude pattern is completely rearranged by a magnetic field B of order 1kG, too weak to affect substantially the tunneling rates.

Based on these observations we propose[12] that the fluctuations arise from spatial variations in the amplitude of the quasi-bound states *inside* the dot. Such apparently random variations in the electron density of adjacent levels is perhaps initially surprising in these GaAs heterostructures which are very high-mobility and exhibit ballistic transport phenomena such as quantized point contact resistance at energies above the tunneling barriers[3]. However even though these systems are clean enough that the only significant scattering within the dot comes from collisions with the boundaries, the study of *quantum chaos* demonstrates that even modest complexity of shape, exemplified e.g. by the stadium or Sinai billiards[13], will lead to effectively random behavior described by the Wigner-Dyson random-matrix ensembles. Moreover recent theoretical results[14] show that chaotic scattering in very similar GaAs "quantum wire" devices leads to conductance fluctuations similar to those of highly-disordered metals, and several recent experiments have confirmed this[15, 16]. Hence we conjecture that the confining potential in these semiconductor quantum dots is sufficiently complex in shape and topography to generate fully chaotic classical motion (at low or zero magnetic field). This implies that the energy-levels and eigenstates of the dot (in isolation) are well-described by Wigner's Gaussian random-matrix ensembles [17, 18, 13]; from this statistical ansatz we can calculate the probability density of decay widths which determine the amplitude fluctuations from Eq. (2).

3. R-Matrix Theory of Decay Widths

It is well-known that quantum decay widths are related to the spatial distribution of the quasi-bound states; a state with high probability density near the tunneling barrier will decay more rapidly than that with low density. Many years ago in nuclear physics it was understood that the scattering resonances of a complex system such as the nucleus would exhibit level-width fluctuations due to eigenstate fluctuations which for strongly-resonant (compound nuclear) scattering were shown to have a characteristic probability density known as the Porter-Thomas distribution[17]. A formal connection between eigenstates and decay widths was made in this context using *R-Matrix* theory [17, 19]. Because this approach has not been used (to our knowledge) for calculations in quasi-1d transport theory, we review it briefly here, treating only the case of non-interacting electrons moving in strictly one dimension.

Consider the 1d Schrödinger equation on the interval $[-\infty, \infty]$ with a scattering potential $V(x)$ which becomes negligible for $x < x_1$ and $x > x_2$. The formal development does not require any further assumptions about the nature of the potential at this point, however later when we make approximations leading to the Breit-Wigner resonance lineshape we will need to assume that $V(x)$ has a double-barrier shape with an internal classically-allowed region which leads to strong, narrow transmission resonances at discrete energies. There are scattering solutions $\psi_E(x)$ to this equation for all positive energies E and in order to find the behavior of the solutions near the resonance energies we must relate these to eigenstates of the potential $V(x)$ "in isolation". By eigenstates "in isolation" we mean solutions $X_\lambda(x)$ of the Schrödinger equation on the finite interval $[x_1, x_2]$ with discrete energies ε_λ determined by linear homogeneous boundary conditions $X'_\lambda(x_1) + a_1 X_\lambda(x_1) = 0, X'_\lambda(x_2) + a_2 X_\lambda(x_2) = 0$. Any two solutions of the Schrödinger equations ϕ_1, ϕ_2 with energies E_1, E_2 (regardless of boundary conditions) satisfies the condition equivalent to current conservation

$$-\frac{\hbar^2}{2m}\frac{(W_{12}(x_2) - W_{12}(x_1))}{E_1 - E_2} = \int_{x_1}^{x_2} \phi_1^*(x)\phi_2(x)dx, \tag{3}$$

where $W_{12}(x) = \phi_2^* d\phi_1/dx - \phi_1 d\phi_2^*/dx$. If we choose $\phi_1(x) = \psi_E(x), \phi_2(x) = X_\lambda(x)$ in Eq. (3) and expand $\psi_E(x) = \sum_\lambda^\infty A_\lambda X_\lambda(x)$ in the interval $[x_1, x_2]$ (using completeness of the X_λ) the right-hand side of Eq. (3) becomes simply A_λ. The left-hand side may be simplified by a specific choice of the boundary conditions determining the X_λ, $a_1 = a_2 = 0$, implying $X'_\lambda(x_1) = X'_\lambda(x_2) = 0$(our final result is insensitive to this choice). In this case one obtains

$$A_\lambda = -\frac{\hbar^2}{2m}\frac{X_\lambda^*(x_2)\psi'_E(x_2) - X_\lambda^*(x_1)\psi'_E(x_1)}{E - \varepsilon_\lambda}. \tag{4}$$

If we substitute this result into the expansion for $\psi_E(x)$ on the interval $[x_1, x_2]$ and evaluate this formula at the endpoints we obtain the basic result:

$$\psi_E(x_m) = \sum_{n=1}^{2} R_{mn}\psi'_E(x_n) \tag{5}$$

$$R_{mn} = -\frac{\hbar^2}{2m}\sum_\lambda^\infty \frac{X_\lambda(x_m)X_\lambda^*(x_n)}{E - \varepsilon_\lambda}. \tag{6}$$

Here $\psi'_E(x)$ is the *outward* derivative from the scattering region and the R-matrix relates this quantity to the wavefunction amplitude at each interface. The R-matrix is hermitian in general; it will be real-symmetric in 1d because wavefunctions may be chosen real, but we retain the more general form for ease of generalization to the quasi-1d case in the presence of a magnetic field.

4. Relation of R-matrix to S-Matrix

Since the R-matrix is expressed in terms of the eigenstates and energies of the dot "in isolation", by relating it to the S-matrix we can express scattering properties in terms of the properties of these bound states as desired. In 1d an arbitrary solution on the interval $[-\infty,\infty]$ can be written as $\psi_E(x) = Ie^{ikx} + Oe^{-ikx}$, $x \leq x_1$; $\psi_E(x) = O'e^{ikx} + I'e^{-ikx}$, $x \geq x_2$. The S-matrix relates the incoming amplitudes (I, I') to the outgoing amplitudes (O, O'). From the known asymptotic form of $\psi_E(x)$ one can write a 4×4 matrix equation relating the vector $(\psi_1, \psi_2, \psi_1', \psi_2')$ to (O, O', I, I') (where we have adopted the notation $\psi_E(x_1) \equiv \psi_1$). If one replaces the components $(O, O') = \mathbf{S}(I, I'), (\psi_1, \psi_2) = \mathbf{R}(\psi_1', \psi_2')$ in this equation one obtains the 2×2 matrix equation

$$\mathbf{S} = -\mathbf{U}(1 + ik\mathbf{R})(1 - ik\mathbf{R})^{-1}\mathbf{U}^{\dagger} \equiv -\mathbf{U}\mathbf{Q}\mathbf{U}^{\dagger} , \tag{7}$$

where $\mathbf{U}$ is a diagonal unitary matrix. Because $\mathbf{U}$ is unitary and diagonal it drops out when computing $|S_{mn}|^2$ and one finds the exact result $T = |S_{21}|^2 = |Q_{21}|^2$ relating the transmission coefficient to the R-matrix.

Although this relation is exact, it is not very useful as the exact R-matrix involves an infinite summation over the eigenstates X_λ. The crucial approximation is that near a given resonance with energy $\tilde{\varepsilon}_\lambda$ the sum in Eq. (6) is dominated by the term with $\varepsilon_\lambda \approx \tilde{\varepsilon}_\lambda$; this is reasonable if the resonances are much narrower than the energy-level spacing of the dot in isolation, a condition satisfied if the two barriers are sufficiently high compared to the incident energy. In this *single-level* approximation the sum in Eq. (6) is reduced to a single term:

$$R_{mn}(E \approx \varepsilon_\lambda) \approx -\frac{\hbar^2}{2m}\frac{X_\lambda(x_m)X_\lambda^*(x_n)}{E - \varepsilon_\lambda} \equiv -\frac{\gamma_\lambda^m \gamma_\lambda^{n*}}{E - \varepsilon_\lambda} . \tag{8}$$

The inversion of the matrix $(1-ik\mathbf{R})$ needed to evaluate $\mathbf{Q}$ is then simplified by the relation $det(\mathbf{R}) \approx 0$, one finds $Q_{21} = 2ikR_{12}^*/[1 - ik(R_{11} + R_{22})]$. Defining the partial width

$$\Gamma_\lambda^m \equiv |\gamma_\lambda^m|^2 \equiv (\hbar^2 k/m)|X_\lambda(x_m)|^2 \tag{9}$$

one finds the Breit-Wigner formula

$$T \approx \frac{\Gamma_\lambda^l \Gamma_\lambda^r}{(E - \varepsilon_\lambda)^2 + (\Gamma_\lambda/2)^2} , \tag{10}$$

where the total width $\Gamma_\lambda = \Gamma_\lambda^l + \Gamma_\lambda^r$ and we have replaced the superscripts 1, 2 by l, r denoting left and right. Note that in the strict single-level approximation we obtain no shift in the resonance energy; if we had included the smooth background arising from the other terms in Eq. (6) we would have obtained such a shift [19]. However we have obtained the crucial result (9) relating the decay width to the "resonance wavefunction" X_λ evaluated at the interfaces of the scattering region.

The generalization of Eq. (9) to the quasi-1d geometries of interest is similar to the derivation sketched above, but involves some subtle issues relating to neglect of evanescent modes which we will not discuss here. One obtains a Breit-Wigner formula for each transmission coefficient T_{ab} between propagating states a, b at energy E of the uniform perfect leads of width W; the partial decay width into open channel a in the left lead(beginning at $x = x_1$) is

$$\Gamma^l_{\lambda,a} = |\gamma^l_{\lambda,a}|^2 = (\hbar^2 k/m)|\int_0^W dy \phi_a(y) X_\lambda(x_1, y)|^2, \tag{11}$$

where $\phi_a(y)$ is the transverse wavefunction channel a in the leads. A general treatment for M open channels in each lead is straightforward, but since the leads will typically narrow smoothly near the tunneling barrier a single channel in each lead will have the shortest tunneling distance and will dominate the resonance; thus we focus on the case of a single channel in each lead and will omit the channel index a in our calculations below.

We argued above that the tunneling barriers themselves are not the source of the amplitude fluctuations in the Coulomb blockade oscillations; hence we will consider only geometrically simple barriers which are uniform across their width(y-direction). In this case the transfer matrix relating the wavefunctions outside the barrier to those inside does not mix channels and $X_\lambda(x, y)$ just inside the barriers is proportional to $X_\lambda(x, y)$ just outside. We then define a reduced width $|\tilde{\gamma}_\lambda|^2 = \Gamma_\lambda / P(\varepsilon_\lambda)$ where $P(\varepsilon_\lambda)$ is the barrier penetration factor which is a smooth function of energy. In this discussion we will only treat the case of equal barriers on each side of the dot and hence $P(E)$ will be the same on left and right. This means that the *average* decay widths to left and right are equal, $\overline{\Gamma}^l = \overline{\Gamma}^r = \overline{\Gamma}/2$ and we can define a dimensionless amplitude for the thermally-broadened resonance by

$$\alpha_\lambda = \frac{\Gamma^l_\lambda \Gamma^r_\lambda}{\overline{\Gamma}(\Gamma^l_\lambda + \Gamma^r_\lambda)} = \frac{|\tilde{\gamma}^l_\lambda|^2 |\tilde{\gamma}^r_\lambda|^2}{2 < |\tilde{\gamma}^l_\lambda|^2 > (|\tilde{\gamma}^l_\lambda|^2 + |\tilde{\gamma}^r_\lambda|^2)}, \tag{12}$$

where from Eq. (2) the measured amplitude of the resonance arising from this state is $g^\lambda_{max} = (e^2/h)(\overline{\Gamma}/4\pi kT)\alpha_\lambda$. Note for the case of symmetric barriers $P(E)$ cancels completely in α, which only depends on wavefunctions *inside* the dot through the quantity $\tilde{\gamma}^{l,r}_\lambda = (\hbar^2 k/m)^{1/2} \int_0^W dy \phi(y) X_\lambda(x_{l,r}, y)$, where $x_{l,r}$ denotes the inside edge of the barrier on the left(right).

It may appear questionable to use the results of Eqs. (9)-(12) to study the level widths in Eq. (2), which describe Coulomb blockade resonances in which interaction effects are clearly important. However, although the Breit-Wigner lineshape and its relation to the resonance wavefunctions were derived for single-electron potential scattering, the approach can be generalized to many-electron wavefunctions[19] and again a Breit-Wigner resonant lineshape is predicted with the many-body wavefunctions at the interfaces of the scattering region replacing the one-body wavefunctions[20]. If we make a random-matrix theory ansatz for these wavefunctions then the resulting theory should lead to the same *statistical* distribution of decay widths. For example in the nuclear case interactions within the compound nucleus are crucially important, nonetheless Porter-Thomas level-width fluctuations are observed [17]). Thus our theory can describe Coulomb blockade amplitude fluctuations not only including the mean interaction via the charging energy, but even if residual interactions on the dot are important.

5. The α-distribution

First we calculate the distribution of T=0 decay widths $P(\Gamma)$ from which $P(\alpha)$ follows. We assume that the hamiltonian of the isolated dot is described at $B = 0$ by the Gaussian orthogonal ensemble(GOE) and at B large enough to break time-reversal(TR) symmetry by the Gaussian unitary ensemble(GUE)[18]. To relate this statistical ansatz to the decay widths of interest we first expand $X_\lambda(x,y) = \sum_{\mu=1}^{d} a_\mu^\lambda \rho_\mu(x,y)$, where $\rho_\mu(x,y)$ are an arbitrary regular basis. Although this sum in principle should run over an infinite basis, the description of quantum-chaotic hamiltonians in terms of the classic random-matrix ensembles has only been demonstrated semi-classically[13]; i.e. it is only expected to be correct on length scales large compared to the electron wavelength, λ_f. Thus the GOE description only holds if we coarse-grain the system and truncate the basis to a finite d-dimensional set with e.g. Fourier coefficients corresponding to distances much greater than λ_f. After this truncation the statistical ansatz is very simply stated: the GOE and GUE are isotropic in Hilbert space, thus each basis vector(direction in Hilbert space) has equal probability of occurence. This implies that the joint probability density of $\{a_\mu\}$ is uniform on the surface of a d-dimensional sphere of unit radius,

$$P(\{a_\mu\}) = \frac{2}{\Omega_d}\delta[\sum_{\mu=1}^{d}|a_\mu|^2 - 1], \tag{13}$$

where Ω_d is the solid angle in d-dimensions(and we have suppressed the index λ since this distribution describes each level and their eigenstates are uncorrelated). We begin by considering the quantity $\tilde{\gamma}^l$ defined after Eq. (12); defining $\eta_\mu^l = \int_0^W dy \rho_\mu(x_l,y)\phi(y)$ we have (suppressing the index l), $\tilde{\gamma} = \sum_{\mu=1}^{d} a_\mu \eta_\mu = \mathbf{a}\cdot\boldsymbol{\eta}$. It follows that

$$P(\tilde{\gamma}) = \int \prod_{\mu=1}^{d} da_\mu \delta(\tilde{\gamma} - \mathbf{a}\cdot\boldsymbol{\eta})\frac{2}{\Omega_d}\delta[\sum_{\mu=1}^{d}|a_\mu|^2 - 1]. \tag{14}$$

Since the $\{\rho_\mu\}$ are complete as $d \to \infty$ we must have $\boldsymbol{\eta}\cdot\boldsymbol{\eta} \propto d$(to generate the δ-function in the completeness relation); hence by choosing $\boldsymbol{\eta} \parallel \mathbf{a}_d$ we obtain

$$P(\tilde{\gamma}) \propto \int \prod_{1}^{d-1} da_\mu \delta[\sum_{\mu=1}^{d-1}|a_\mu|^2 - (1 - \tilde{\gamma}^2/d)] \sim (1 - \tilde{\gamma}^2/d)^{(d-3)/2} \sim e^{-\tilde{\gamma}^2/2}, \tag{15}$$

where the final proportionality only holds in the limit $d \gg 1$. For simplicity we have sketched this calculation assuming real $\tilde{\gamma}$ describing the GOE case of real wavefunctions. The same reasoning holds for $Re[\tilde{\gamma}], Im[\tilde{\gamma}]$ separately in the GUE case, thus these variables have a joint gaussian distribution in each channel. For the case of M channels per lead it is easily shown that the scaled total width $\Gamma/\overline{\Gamma}$ has a χ^2_ν distribution with $\nu = 2\beta M$ ($\beta = 1,2$ for GOE,GUE),

$$P_\nu(\Gamma) = A_\nu \Gamma^{\nu/2-1} \exp\left(-\frac{\nu\Gamma}{2\overline{\Gamma}}\right) , \tag{16}$$

where $A_\nu = (\nu/2\overline{\Gamma})^{\nu/2}/\mathcal{G}(\nu/2)$ and $\mathcal{G}(p)$ is the gamma function. We note that $P_\nu(\Gamma)$ is peaked at small widths for $\nu = 1,2$ but as ν increases it peaks at a value approaching one and its variance decreases. The case $\nu = 1$, in which the density has a square-root divergence at small widths, corresponds to the famous Porter-Thomas distribution of compound nuclear reaction widths[17]. It only arises here if TR-symmetry is present *and* the potential of the dot has reflection symmetry; this would imply $\Gamma^l_N = \Gamma^r_N$ and the two partial widths would not fluctuate independently. For this special case $\alpha = (1/2)(\Gamma/\overline{\Gamma})$ and $\mathcal{P}_1(\alpha) = \chi^2_1(\alpha)$ also has the Porter-Thomas form. More generally $\mathcal{P}_\nu(\alpha)$ must be obtained from $P_\nu(\Gamma)$ by convolution from Eq. (12) and (16) and can only be expressed as a one-dimensional integral[21] for all ν. The most relevant cases are $\nu = 2$(for which the integral may be done analytically) and $\nu = 4$ which describe an asymmetric dot with a single decay channel per lead with and without TR symmetry. One finds

$$\mathcal{P}_2(\alpha) = \sqrt{\frac{2}{\pi\alpha}} e^{-2\alpha} \tag{17}$$

$$\mathcal{P}_4(\alpha) = 2e^{-4\alpha} \int_0^\infty dz e^{-z} \sqrt{\frac{z+4\alpha}{z}}. \tag{18}$$

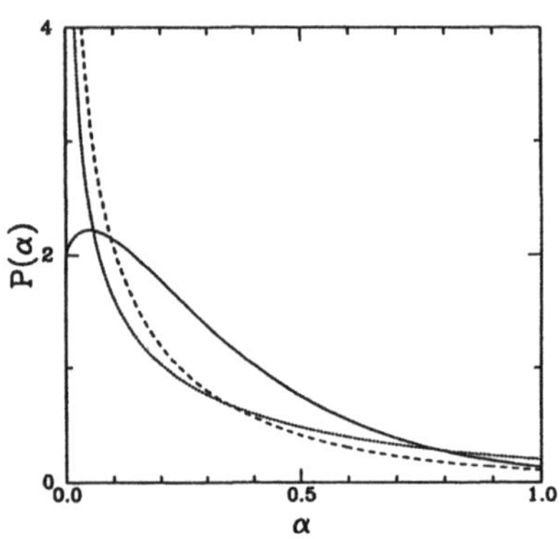

Figure 1: Probability distribution $\mathcal{P}_\nu(\alpha)$ for the amplitude α of Coulomb blockade resonances for 3 cases: $\nu = 1$(dotted) describing a reflection-symmetric dot with time-reversal symmetry(Porter-Thomas distribution); $\nu = 2$(dashed) describes an asymmetric dot with TR symmetry; $\nu = 4$ (solid) describes an asymmetric dot without TR symmetry ($B > B_c$).

$\mathcal{P}_1, \mathcal{P}_2, \mathcal{P}_4$ are plotted in Figure 1. Note the substantial suppression of small amplitudes caused by breaking TR symmetry; this leads to the simplest qualitative prediction of the theory: imposing a B-field sufficient to break TR symmetry will measurably suppress the amplitude fluctuations. The behavior of $\mathcal{P}_\nu(\alpha)$ as $\alpha \to 0$ can be understood by noting that for small amplitudes typically one of the partial widths Γ^l, Γ^r will dominate in Γ and hence will cancel in α, leading to a behavior of $\mathcal{P}_\nu(\alpha)$ similar to $\chi^2_{\nu/2}$ as $\alpha \to 0$.

6. Effects of Symmetry-Breaking

Although in general the distribution $\mathcal{P}_\nu$ cannot be expressed in closed form, all of its moments can be calculated analytically for arbitrary ν [21] and some interesting results are predicted. First, *breaking TR symmetry reduces amplitude fluctuations*: the variances satisfy $\Delta\alpha_2^2 = 1/8 = .125$, $\Delta\alpha_4^2 = 4/45 \approx .089$. This follows simply from the decrease in the variance of the underlying χ^2 distributions as ν increases. Second, *breaking TR symmetry increases the mean amplitude*: $\overline{\alpha}_2 = 1/4, \overline{\alpha}_4 = 1/3$. Because it relates to $\overline{\alpha}$, this prediction should hold *even when* $kT \gg \Delta\varepsilon$, i.e. for metallic as well as semiconducting quantum dots. The origin of this effect can be seen qualitatively by expanding each partial width in α for small variations around their mean which yields $\overline{\alpha} \approx 1/4(1 - \Delta(\Gamma/\overline{\Gamma})^2)$. Since $\Delta(\Gamma/\overline{\Gamma})^2$ decreases when TR-symmetry is broken, $\overline{\alpha}$ increases as long as the TR-symmetry-breaking field B_c has negligible effect on the mean tunneling rate $\overline{\Gamma}$.

Clearly a simple and straightforward experimental test of this theory would be to measure the statistical properties of the amplitude fluctuations in the presence and absence of a weak magnetic field. However this will only produce an unambiguous test of the theory if the symmetry-breaking field scale B_c is small enough that the magnetic field only affects the phase of the electrons and does not appreciably affect their dynamics (we will amplify on this point below). In the context of conductance fluctuations[22] it is well known that for a sample of area A with normal leads $B_c^{open} \sim (h/e)/A$; the reason for this is that a diffusive trajectory crossing a sample encloses an area of order A, and if the sample is open it escapes as soon as it reaches one of the leads. For the open chaotic billiards used to model GaAs junctions B_c is essentially the same order of magnitude, although its dependence on A^{-1} is not strictly linear[14]. On the other hand study of B_c in closed diffusive and chaotic ballistic systems[23, 24] has indicated a parametrically smaller field scale enters, $B_c^* \sim B_c^{open}/\sqrt{g}$, where g is the dimensionless conductance. This new and smaller scale arises becauses for closed systems the electronic trajectories never escape and continue to accumulate phase from the field until a time of order $\hbar/\Delta\varepsilon$(after which the semi-classical arguments break down). Since the decay time for escape from the quantum dot is by assumption longer than $\hbar/\Delta\varepsilon$ (since $\overline{\Gamma} < \Delta\varepsilon$), we should expect B_c^* to also set the scale for symmetry-breaking in our theory. For the diffusive case $g \approx \hbar/\Delta\varepsilon\tau_0$ where $\tau_0 = L^2/D$ is the traversal time; for the chaotic ballistic case considered here it is plausible to replace this by $\tau_0 = L/v_f$ where $L \sim \sqrt{A}$ is a typical linear dimension of the dot. This yields $B_c L^2 \equiv$

$\Phi_c \sim (h/e)/(hv_f/\Delta\varepsilon L)^{1/2}$. Since B_c sets the scale for altering the phase of the wavefunctions, it will also set the flux scale for sample-specific fluctuations of the amplitude pattern in a given sample via the familiar ergodic hypothesis for disordered metals[22]. Below we will show numerical results confirming that the model proposed here generates sample-specific amplitude fluctuations on a field scale consistent with the conjecture that $B_c = B_c^*$; however these results do not demonstrate that $B_c \ll B_c^{open}$ as the factor $(\hbar v_f/\Delta\varepsilon L)^{1/2}$ corresponding to $\sqrt{g}$ is of order unity. Nonetheless, for the experimental systems of ref. [2] the criterion $B_c \approx B_c^*$ predicts $B_c \sim 500G$, consistent with the findings of recent extensions of that work by McEuen et al.[11].

7. Numerical Tests of the Theory

To test the theory microscopically we have performed numerical calculations of conductance resonances(Figure 2) in a model system consisting of the stadium billiard[13] connected to leads[14] through tunnel barriers of height $15\varepsilon_1$ (ε_1 is the first sub-band threshold) and desymmetrized by replacing one quarter-circle by a cosine curve(inset, Fig 2). Although matrix element statistics for chaotic systems have been studied previously with results consistent with Eq. (16) [25], we found no previous direct tests of resonance statistics for chaotic systems. We studied $g(E)$ at fixed B for energies between ε_1 and $2\varepsilon_1$ so that only the single-channel cases ($\nu = 2, 4$) occur. From Weyl's law[13] this interval should contain roughly the 24th to 49th level of the stadium. The

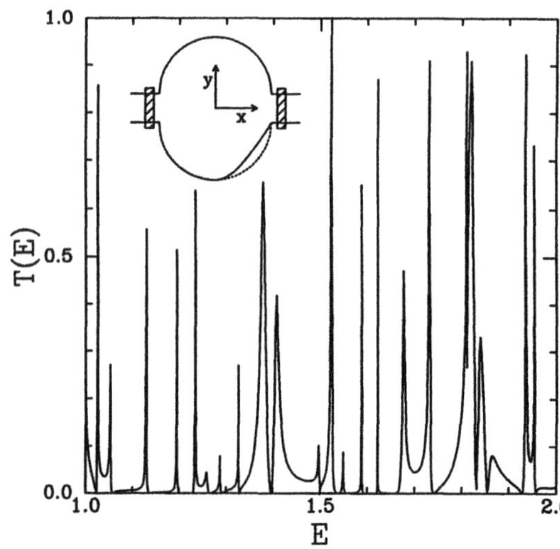

Figure 2: Resonances of the transmission coefficient T as a function of $E = \varepsilon_f$ in the first transverse sub-band of the leads for the desymmetrized stadium dot shown in the inset; cross-hatched regions indicate tunnel barriers. Resonances are fit to a sum of Breit-Wigner lineshapes and generate the data for histograms in Figure 4.

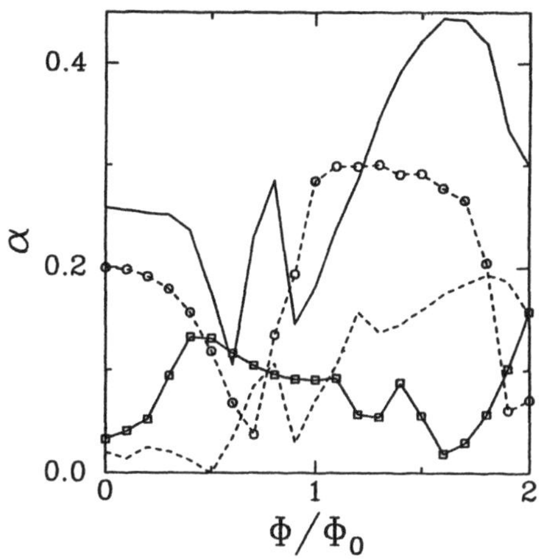

Figure 3: Sample-specific fluctuations of the Coulomb blockade amplitude $\alpha/\overline{\alpha}$ as a function of magnetic flux Φ through the dot(in units of $\Phi_0 = h/e$) for four consecutive peaks(in energy) of T(E) for desymmetrized stadium. These fluctuations have a correlation range of B_c and represent a new mesoscopic fingerprint of the dot potential.

amplitude coefficient α equal to the area under the T=0 resonance may be directly evaluated as a function of B and we obtain sample-specific fluctuations on the appropriate scale B_c as shown in Figure 3.

Quantitative statistical analysis requires more than the 25 resonances given by the billiard of Figure 2; since in the experimental systems smooth potential fluctuations on the scale of the dot are to be expected anyway, we construct a weakly-random ensemble of dots based on this billiard. We add to the "floor" of the dot three randomly-located hills (or valleys) smoothly joined to the edges with random height always less than the electron kinetic energy and uniformly distributed in the interval $[-\epsilon_1, +\epsilon_1]$(thus the hills would cause no backscattering classically). It is striking that this small perturbation is sufficient to produce an uncorrelated resonance pattern. With this procedure we generate resonances for 30 samples and fit the total half-widths to Breit-Wigner lineshapes. We find that $\overline{\Gamma}(E)$ varies smoothly by roughly a factor of two over the interval studied, so that agreement to theory is significantly improved if the resonances are rescaled by a linear $\overline{\Gamma}(E)$ instead of by a constant $\overline{\Gamma}$[12]. The fitting requires some care and details will be given elsewhere [21]. In the histograms of Figure 4a,b we compare the distribution of total widths Γ (corresponding to T=0 behavior) to the predictions of Eq. (16). This is sufficient to test Eqs. (17),(18) since the distribution $\mathcal{P}_\nu(\alpha)$ follows analytically); due to finite energy resolution the correct population of the first bin (small Γ) is not resolved and we need to compare the data to truncated χ^2 distributions[26, 21]. We find excellent agreement for the case $B = 0$ and reasonable agreement for B corresponding to flux $\Phi = 2\Phi_0$. The cross-over scale for TR-symmetry

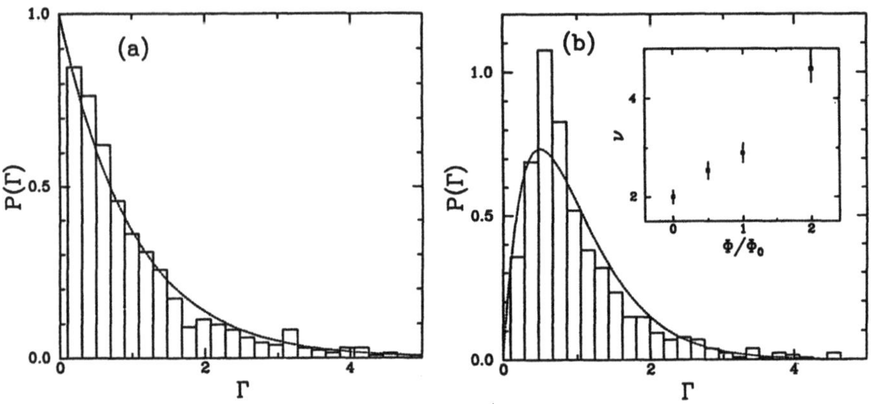

Figure 4: histograms of total widths $\Gamma(E)$ normalized by $\overline{\Gamma}(E)$ for $\Phi = 0$ (a) and $\Phi/\Phi_0 = 2$ (b). Dotted curves are expected χ^2_ν distributions of Eq. (16) with $\nu = 2(a), \nu = 4(b)$. Inset b: best value of ν obtained by fitting histograms obtained at several flux values to a truncated χ^2_ν distribution showing the field scale B_c for the $\nu = 2 \rightarrow 4$ (GOE to GUE) transition [25].

breaking is quantified by fitting the distribution at intermediate fields[25] to ν between two and four(inset Figure 4b). Unfortunately no strong saturation at $\nu = 4$(corresponding to GUE) is found.

We believe that the continued dependence of the decay width distribution on magnetic field is due to the fact that for systems of the size studied numerically the cyclotron radius r_c will approach the dot radius shortly after TR symmetry is broken. Beyond this field scale *edge-states* will be formed(or skipping orbits classically) and the classical phase-space will have regions of stability which alter the semi-classical behavior significantly and lead to a suppression of fluctuations. Numerical evidence in favor of this interpretation was obtained by performing the same calculations and analysis of the resonances of a dot with short-range disorder[21]. In this case the cyclotron effects should only become significant at a much large field scale corresponding to r_c smaller than the mean free path; as expected excellent agreement is obtained to the GUE prediction for $P(\Gamma)$ at fluxes much larger than $2\Phi_0$. Thus for ballistic dots the condition for observing the transition between the GOE and GUE behavior is more stringent than in strongly disordered dots. Nonetheless if we define B_{edge} by the field scale at which r_c equals the dot radius, it is easily estimated that $B_{edge}/B_c \sim \sqrt{N}$. Thus unlike our simulations (for which $\sqrt{N} \sim 5-7$), in typical experimental systems with $\sqrt{N} \sim 10-25$ there should be a large field interval over which our current theory will apply.

8. Summary

In summary, amplitude fluctuations of Coulomb blockade oscillations in semiconductor quantum dots arise from quantum-chaotic fluctuations in the decay widths of quasi-bound states which have universal distribution laws describable by random-matrix theory. A weak magnetic field enhances the mean amplitude while reducing fluctuations in the amplitude, an effect which should be experimentally observable.

Acknowledgements

We acknowledge useful discussions with H. Baranger, O. Bohigas, D. DiVincenzo, P. McEuen, B. Mottelson and K. Zyczkowski, and programming assistance from A. Szafer and N. Whelan. R.A.J acknowledges the support of the NORDITA workshop on Quantum Chaos. This work was partially supported by NSF grant DMR-8658135 and DOE grant DE-FG-0291ER-40608.

References

[1] D.V. Averin and K.K. Likharev, in *Mesoscopic Phenomena in Solids*, edited by B.L. Altshuler, P.A. Lee, and R.A. Webb, (Elsevier, Amsterdam, 1991).

[2] U. Meirav, M.A. Kastner and S.J. Wind, Phys. Rev. Lett. **65**, 771 (1990).

[3] L.P. Kouwenhoven, N.C. van der Vaart, A.T. Johnson, W. Kool and C.J.P.M. Harmans, Z. Phys. B, in press (Proceedings of 1991 Nato ASI on "Single-Charge Tunneling", Les Houches, France).

[4] P.L. McEuen, E.B. Foxman, U. Meirav, M.A. Kastner, Y. Meir, N.S. Wingreen and S.J. Wind, Phys. Rev. Lett. **66**, 1926 (1991).

[5] P.L. McEuen, E.B. Foxman, J. Kinaret, U. Meirav, M.A. Kastner, N.S. Wingreen and S.J. Wind, Phys. Rev. B15 **45**, in press (1992).

[6] R.A. Webb et al., Phys. Rev. Lett. **54**,1577 (1985); A.B. Fowler et al., Phys. Rev. Lett. **57**,128 (1986).

[7] H. van Houten and C.W.J. Beenakker, Phys. Rev. Lett. **63**, 1893 (1989).

[8] Y. Meir, N. Wingreen and P.A. Lee, Phys. Rev. Lett. **66**, 3048 (1991).

[9] C.W.J. Beenakker, Phys. Rev. **B44**, 1646 (1991).

[10] A. Szafer, A. Douglas Stone, P.L. McEuen and B.W. Alphenaar, Proceedings of NATO ASI on Granular Nanoelectronics, Springer-Verlag (1991).

[11] P.L. McEuen, private communication.

[12] R.A. Jalabert, A.D. Stone and Y. Alhassid, Phys. Rev. Lett. **68** 3468 (1992).

[13] O. Bohigas, Les Houches, Session L11, 1989, *Chaos and Quantum Physics*, Eds. M.-J. Giannoni, A. Voros and J. Zinn-Justin, Elsevier (1991) and refs. therein.

[14] R.A. Jalabert, H.U. Baranger and A.D. Stone, Phys. Rev. Lett. **65**, 2442 (1990).

[15] C.M. Marcus, A.J. Rimberg, R.M. Westervelt, P.F. Hopkins, and A.C. Gossard, preprint.

[16] M.W. Keller, O. Millo, D.E. Prober, and R.N. Sacks, Bull. Am. Phys. Soc., **37**, 712 (1992).

[17] C.E. Porter(Ed.), *Statistical Theory of Spectra*, Academic, New York (1965).

[18] T.A.Brody, J. Flores, J.B. French, P.A. Mello, A. Pandey and S.S.M. Wong, Rev. Mod. Phys. **53**, 385 (1981).

[19] A.M. Lane and R.G. Thomas, Rev. Mod. Phys. **30**, 257 (1958).

[20] Ng and Lee, Phys. Rev. Lett. **61**, 1768 (1988), and Kawabata(preprint, 1991) have discussed the possibility of a mixed-valence type of interaction between the dot and the fermi sea of the leads which would lead to an alteration in the resonance line-shape when $\overline{\Gamma} < kT$ and B=0. We do not discuss this regime in the present work.

[21] R.A. Jalabert, Y. Alhassid, A.D. Stone, unpublished.

[22] P.A. Lee, A.D. Stone and H. Fukuyama, Phys. Rev. B**35**, 1039 (1987).

[23] S. Oh, A. Yu. Zyuzin, and R.A. Serota, Phys. Rev. B**44**, 8858 (1991).

[24] A. Szafer and B.L. Altshuler, preprint.

[25] Y. Alhassid and R.D. Levine, Phys. Rev. Lett. **57**, 2879 (1986); Y. Alhassid and M. Feingold, Phys. Rev. A **39**, 374 (1989).

[26] J.D. Garrison, Ann. of Phys. **30**, 269 (1964).

Electron Charge Distribution and Transport in Mesoscopic Systems

A. Kawabata

Department of Physics, Gakushuin University,
1-5-1 Mejiro, Toshima-ku, Tokyo 171, Japan

We discuss the electronic transport in one-dimensional systems and show how the conductance at zero temperature is related to the electronic charge distribution, using an generalized form of Friedel sum rule. We apply the theory to the transport through a quantum dots. It is predicted that the transmission probability is unity when the dot has a magnetic moment of magnitude 1/2 due to an unpaired electron.

1. Introduction

Many years ago, Friedel[1] pointed out that the phase shift of the electrons at the Fermi level is related to the charge distribution of the electrons. With the use of Friedel sum rule combined with symmetry arguments, one can show that in some cases the conductance of the system is determined only by the charged distribution[2], [3]. Since in most cases the calculation of the transport coefficient for a given system is of no simple matter, specially when the electron-electron interaction plays an important roll, Friedel sum rule provides a very powerful tool for the calculation of the transport coefficient.

In this paper, we apply a generalized Friedel sum rule to the calculation of the conductance of mesoscopic system and apply the method to the Kondo-resonant transmission through a quantum dot.

2. Friedel sum rule and transport coefficient in one-dimensional systems

For the sake of simplicity we consider one-dimensional electron gas placed in a potential energy $v(x)$ which is symmetric around $x = 0$, i.e., $v(-x) = v(x)$. Then, in the absence of the many body interaction between the electrons, the eigenstates of the system can be taken to be symmetric or antisymmetric around $x = 0$. We assume that $v(x)$ vanishes in the regions $a < |x|$, a being a constant, then in these regions the symmetric and antisymmetric eigenfunctions can be written as

$$\phi_{k,s}(x) = \cos\{k|x| + \eta_s(k)\} , \tag{1}$$

$$\phi_{k,a}(x) = \mathrm{sign}x \sin\{k|x| + \eta_a(k)\} , \tag{2}$$

except for the normalization constants. We can easily show that the in the present case Friedel sum rule is of the form,

$$\eta_s(k_F) + \eta_a(k_F) = \frac{\pi}{2}\Delta N , \tag{3}$$

where k_F is the Fermi momentum and ΔN is the number of electrons displaced into the region around $x = 0$ by the potential $v(x)$.

Friedel sum rule becomes very powerful if it holds independently for separate channels (e.g., for each spin components, or each components of angular momentum (in three dimension)). The generalization of the Friedel sum rule in this context was discussed by Shiba[4] and by Yoshimori and Zawadowski[5]. In the present case, we can derive the relation (3) separately for symmetric and antisymmetric states, i.e.,

$$\eta_\alpha(k_F) = \frac{\pi}{2}\Delta N_\alpha , \tag{4}$$

where $\alpha = a, s$, and ΔN_s and ΔN_a are the contribution to ΔN of the electrons in the symmetric and antisymmetric states.

On the other hand, it is easy to see that the transmission probability of an electron through the region where $v(x) \neq 0$ is given by

$$T = \cos^2\{\eta_s(k_F) - \eta_a(k_F)\} , \tag{5}$$

from an appropriate linear combination of $\phi_{k,s}(x)$ and $\phi_{k,a}(x)$. Then the conductance of the system is given by Landauer formula [6],

$$G = \frac{2e^2}{h}T , \tag{6}$$

where e is the electronic charge and h is the Planck constant.

The results obtained in the above are valid even in the presence of Coulomb interaction between the electrons as long as the ground state of the system is not degenerate[7]. In this case, there is no inelastic scattering of quasi-particles at the Fermi level, and hence we can define the phase shifts of the quasi-particles without an ambiguity. The phase shifts in eqs. (3)–(5) should be regarded as those of the quasi-particles. ΔN_α is defined by

$$\Delta N_\alpha = \sum_{k,\sigma} \left\{ \langle c^{\dagger}_{\alpha k,\sigma} c_{\alpha k,\sigma} \rangle - \langle c^{\dagger}_{\alpha k,\sigma} c_{\alpha k,\sigma} \rangle_0 \right\} \tag{7}$$

where $c^{\dagger}_{\alpha k,\sigma}$'s are the creation operators for the one-electron states of which the complete set of the system is composed(α indicates the symmetry of the states), $\langle \cdot \cdot \cdot \rangle$ is the expectation value at the ground state, and $\langle \cdot \cdot \cdot \rangle_0$ is that without potential $v(x)$.

In the following we will apply the generalized Friedel sum rule to the transmission of electrons through a quantum dot when the Coulomb interaction is considerably strong.

3. Transmission of electrons through a quantum dot: Kondo-resonant transmission

As a model of a quantum dot, we consider the case when the potential $v(x)$ is of the form shown in Fig. 1. We assume also that the potential wall at $x = \pm a$ is high and thick enough so that the transmission probability of an electron through the dot is almost 0, when there is no electron in the dot. We assume that the Coulomb repulsion between the electrons are strong enough. Suppose v_0, the energy of the bottom of the dot(see Fig. 1), is lowered. Then,

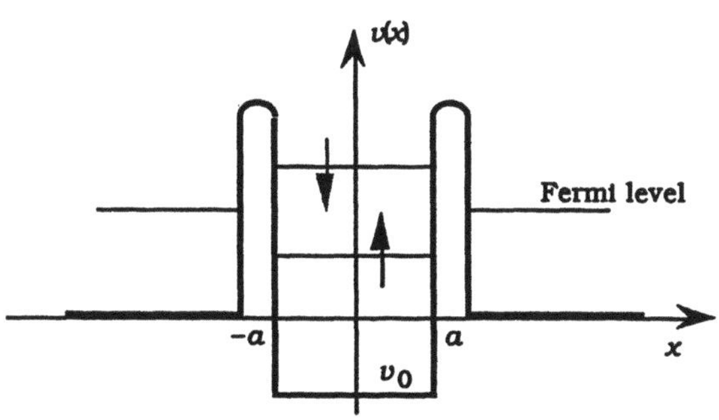

Fig. 1 Model of a quantum dot. The potential $v(x)$ is shown by the thick line. The horizontal lines in the dot indicate the Hartree-Fock energy for up and down electrons.

when the lowest virtual bound state in the dot crosses the Fermi level, only one electron can occupy that state because of the strong Coulomb repulsion. As $v(x)$ is symmetric around $x = 0$, the virtual bound states are symmetric or antisymmetric. Therefore, if an electron is put into a symmetric (antisymmetric) virtual bound state, ΔN_s (ΔN_a) changes by 1, and the argument of cosine in eq. (5) changes by $\pi/2$, as is seen from eqs. (4) . Since $T \cong 0$ when there is no electron in the dot, as was assumed in the above, we have $T \cong 1$ when there is one electron in the dot.

This transparency of the quantum dot is due to Kondo resonance[8]: When free electrons interact with a localized spin, a resonant state appears at the Fermi level and we have a pronounced resonant transmission. In order that the Friedel sum rule is valid, the ground state must be singlet. In fact, as is well known, the localized spin is compensated by the free electrons and the ground state is singlet if the interaction between them is antiferromagnetic. We can show that the interaction is antiferromagnetic in a quantum dot applying the theory of magnetic impurities in metals.

If v_0 is decreased further, the second electron is introduced in the dot. In the case when the Coulomb repulsion is not too strong, this electron occupies the same virtual bound state as the first one. As their spin states must be different, the dot has no total spin and the ground state is singlet, and we can apply the Friedel sum rule. The argument of cosine of eq. (5) changes by $\pi/2$ again, and hence T becomes very small. In this way we can show that the quantum

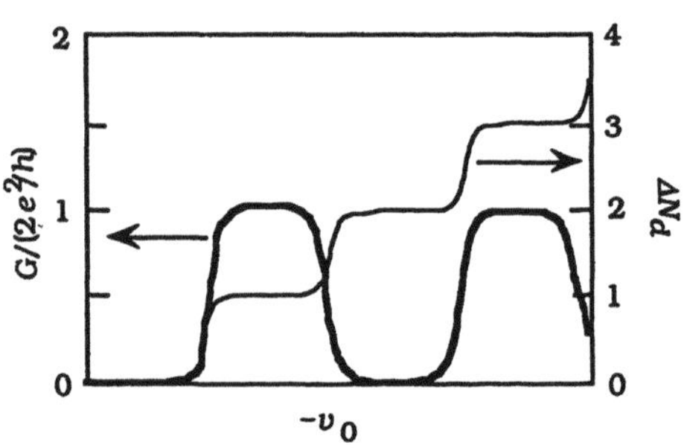

Fig. 2 The dependences of the conductance G (thick line) and the number of the electrons in the dot ΔN_d (thin line) on the energy of the bottom of the dot (schematic). This figure is collect under the conditions that the total spin of the electrons in the dot is zero when ΔN_d is even and that it is one half when ΔN_d is odd[9].

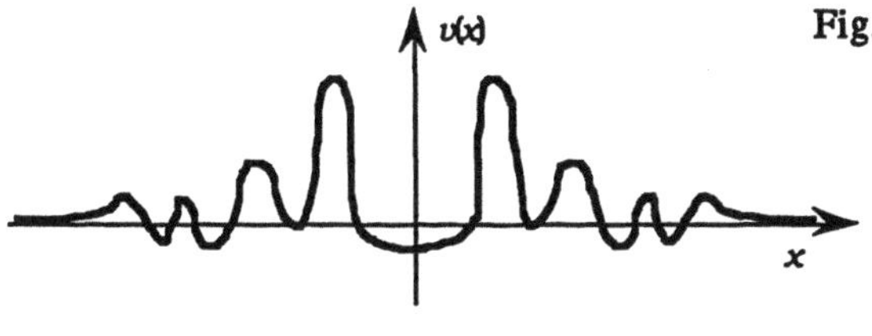

Fig. 3

dot is transparent or opaque according as the number of electrons in the dot is odd or even. Thus the conductance changes as a function of υ_0 as in Fig. 2.

The transparency of a one-dimensional system due to Kondo resonance has already been investigated by T.K. Ng and P.A. Lee[10], and L.I. Grazman and M.É. Raikh[11] on a specific model, but in this paper we dealt with a more general models. In fact, the present theory can be applied even to the case when $\upsilon(x)$ is of the form shown in Fig. 3, as long as it is symmetric. The effects of the scattering from the potential fluctuation out side the dot are cancelled by the interferences of the electron waves.

4. Cases of large localized spin

Next we consider the case when the magnitude of the total spin S in the dot is larger than 1/2. In this case, the nature of the ground state is dependent on the ratio of $2S$ to f, the number of the channels of free electron coupled to the localized spin[12].

Suppose that a dot with $S = 1$ was obtained by adding two electrons to the dot of which the total spin had been 0. When those unpaired electrons occupy a symmetric and an antisymmetric state, we have $f = 2$. When $2S \leq f$, the localized spin is compensated completely and the ground state is not degenerate. Then, since both ΔN_s and ΔN_a increase by 1, the argument of the cosine in eq. (5) does not change, and the transmission probability is the same as the case $S = 0$, i.e., $T \cong 0$.

When the two unpaired electrons occupy the symmetric states, for instance, we have the case of $2S > f = 1$, and the ground state is degenerate, for the localized spin is compensate only partly and the free spin of magnitude $S - 1/2$ remains. Fortunately, this free spin does not interact with the other part of the system and its freedom does not invalidate the Friedel sum rule. In this case, however, we need to assume the generalized Friedel sum rule of the form[13]

$$\eta_{\alpha,\sigma}(k_F) = \pi \Delta N_{\alpha,\sigma} \ , \tag{8}$$

where $\eta_{\alpha,\sigma}(k_F)$ and $\Delta N_{\alpha,\sigma}$ are the phase shift and the contribution to ΔN of the states with symmetry α and spin σ. Then eq. (6) has to be corrected as

$$G = \frac{e^2}{h} \sum_{\sigma} T_\sigma \ , \tag{9}$$

where

$$T_\sigma = \cos^2\{\eta_{s,\sigma}(k_F) - \eta_{a\sigma}(k_F)\} \ . \tag{10}$$

If the free spin is directed to the z-axis, only $\Delta N_{s,\uparrow}$ increases by 1 when the second electron is added to the dot with $S = 1/2$. Therefore, the introduction of the second electron into the dot does not change G as $\eta_{s,\sigma}(k_F)$ changes by π. The cases $S > 1$ can be treated in the same way, and we can show that $G \cong 2e^2/h$ for $f = 1$ and that $G \cong 0$ for $f = 2$.

Generally, resonances do not always give rise to a large transmission probability, even in the case when the Coulomb interaction is not effective. For instance, if a symmetric and an antisymmetric virtual bound states are degenerate for some reasons, $\eta_{s,\sigma}(k_F)$ and $\eta_{a,\sigma}(k_F)$ change by the same amount when the the virtual bound states cross the Fermi level, and the conductance does not change.

So far, the investigations have been restricted in the case of 0K. In order to observe the effects of Kondo resonance in the experiments, the temperature must be lower than the Kondo temperature. As for the estimation of the Kondo temperature, the reader is referred to the paper by P.A. Lee in this proceedings. The Kondo temperature is very sensitive to the structure of the dot, but it is not inaccessible.

In conclusion, we have shown that the conductance through a quantum dot in the presence of a localized spin is nearly equal to a universal value $2e^2/h$ due to Kondo resonance only when the localized spin is coupled with only one of the symmetric and antisymmtric channel of electrons outside the dot. In real systems, such ideal condition might not be realized because of the configuration mixing, and we need more investigation on this problem.

Acknowledgements

The author is grateful to Prof. H. Shiba, Prof. J. Kondo and Prof. Y. Ono for their valuable discussions. This work is partly supported by Grant-in-Aid for Scientific Research in Priority Area "Electron Wave Interference Effects in Mesoscopic Structures" from the Ministry of Education, Science and Culture.

References

[1] J. Friedel, Phil. Mag. **43**,153 (1952), Advance in Physics **3**, 446 (1954).
[2] A. Kawabata, J. Phys. Soc. Jpn. **60**, 3222 (1991).
[3] A. Kawabata, Proc. Int. Symp. on Science and Technology of Mesoscopic Structures, Nara, 1991, (Springer-Verlag, to be published).
[4] H. Shiba, Prog.Theor. Phys. **54**, 967 (1975).
[5] A. Yoshimori and A. Zawadowski, Phys. C (Solid State Physics) **15**, 5241 (1982).
[6] R. Landauer, Z. Phys. B **68**, 217 (1987), and references therein.
[7] J.S. Langer and V. Ambegaokar, Phys. Rev. **121**, 1090 (1961).
[8] J. Kondo, Solid state physics vol.23. , eds. F. Seitz , D. Turnbull, and H. Ehrenreich (Academic, Press, New York, 1969), p.183.
[9] Note that ΔN is not the electron number in the dot, but the difference between the total electron numbers with and without the potential $v(x)$ in the regions where $v(x) \neq 0$.
[10] T.K. Ng and P.A. Lee, Phys. Rev. Lett. **61**, 1768 (1988).
[11] L.I. Grazman and M.E. Raikh, JETP Lett. **47**, 452 (1988).
[12] Ph. Nozieres and A. Blandin, J. Physique, **41**, 193 (1980).
[13] In [4] and [5] the authors proved eq. (8) for the Anderson model, but, within the knowledge of the present author, it has not yet been proved for more general models.

Electronic Properties of Quantum Dots and Quantum Rings in Magnetic Fields

T. Chakraborty[1] *and P. Pietiläinen*[2]

[1]Institute for Microstructural Sciences, National Research Council, Montreal Road, M-50, Ottawa K1A 0R6, Canada

[2]Department of Theoretical Physics, University of Oulu, SF-90570 Oulu 57, Finland

Abstract. We discuss the electronic properties of quantum dots and quantum rings in magnetic fields. The results for the magnetization in these two systems show very different behavior reflecting the different energy level spectra in these systems. Information about the single- and many-electron energy levels are essential in understanding the recent magneto-optical and magneto-transport measurements on quantum dots.

1. Introduction

Quantum dots are systems of few electrons, formed at the interface of a semiconductor heterojunction or in a narrow quantum well, whose motions are constrained in all three spatial dimensions [1–3]. Typically, the dot sizes are about $100nm$ and they contain between 1 and 200 electrons. The experimental information about the electronic properties of quantum dots is primarily from capacitance spectroscopy [2,4,5], far–infrared (FIR) magnetospectroscopy [1,6,7], and transport measurements [8,9]. Oscillatory structure in the capacitance observed by Smith et al. [1,4] has been attributed to the discrete energy levels of a quantum dot. In the presence of a perpendicular magnetic field, Zeeman bifurcation of the energy levels of a quantum dot has also been observed [1,4]. The splitting is believed to occur due to the interplay between competing spatial and magnetic quantization. Another interesting result in capacitance spectroscopy: Hansen et al. [5] observed fractionally quantized states in quantum dots at gate voltages and magnetic fields corresponding to the filling factors $\frac{1}{3}$ and $\frac{2}{3}$, similar to the fractional quantum Hall effect in two-dimensional electron systems [10].

FIR spectroscopic measurements on quantum dot structures in InSb by Merkt et al. [6] showed that the measured resonant frequency is independent of electron number within the experimental error. These resonances are in fact related to single-particle transition energies in a bare confinement potential and have been explained as the center of mass excitations in the system [11]. Demel et al. [7] created small quantum dot structures in AlGaAs/GaAs heterostructures. The FIR resonance absorption they observed shows similar dispersion with magnetic field to that seen for the excitations in Ref. 6. However, a resonant *anticrossing* was resolved in the energy levels determined in Ref. 7. The origin of the anticrossing is still not clear.

In addition, there are very interesting transport measurements recently performed on quantum dot structures [8,9]. In one such experiment, McEuen et al. [9] found that in the quantum Hall regime the conductance as a function of the gate voltage has a periodic series of sharp peaks. The basic periodicity of the peaks is explained by the so-called *Coulomb-blockade* model of the single-electron tunneling through a quantum dot [12]. They also found that with increasing magnetic field the height and position of a particular peak show a distinct periodic structure. The oscillations were explained as due to an electron occupying alternately a state in the first or the second Landau level. From this result, and assuming a constant Coulomb energy they were able to map out the single-particle energy spectrum and found it to be in very good agreement with that for a parabolic confinement studied over half a century ago in Ref. 13.

In this paper, we present a comparative theoretical study of the electron states in quantum dots and quantum rings in magnetic fields. We focus our attention in particular on the behavior of the magnetization in these two systems. It is shown that the behavior of the electronic states is very well reflected in the magnetization, which is a measurable quantity (*e.g.* Störmer et al. measured the magnetization in a two-dimensional electron system [14]). The structure of the magnetization curves in both systems are explained in terms of the behavior of interacting or non-interacting electron energy levels.

2. Quantum Dots

The effect of the electron-electron interaction on a quantum dot in a magnetic field was first studied by Maksym and Chakraborty [11]. They found that the effect of a magnetic field on the energy spectrum of a dot is quite intricate. One consequence is that different ground states are expected to occur at different magnetic fields. From self-consistent numerical solutions of the Poisson and Schrödinger equations for a quantum dot in the Hartree approximation, Kumar et al. [15] showed that the confining potential has nearly circular symmetry and therefore angular momentum is approximately a good quantum number. Also with

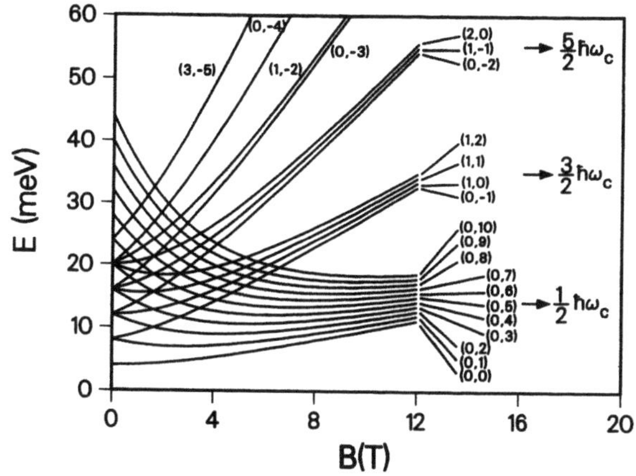

Fig. 1. Single-electron energy levels for a parabolic dot in a magnetic field. The levels are indicated by their quantum numbers (n, l). The confinement energy is $\hbar\omega_0 = 4\,meV$

increasing magnetic field, the evolution of energy levels was found to be similar to that for a parabolic potential. For a parabolic dot in a perpendicular magnetic field, B, the single-electron energies depend on the two quantum numbers n and l in the form [13]:

$$E_{nl} = (2n + |l| + 1)\hbar\Omega - \tfrac{1}{2}l\hbar\omega_c\ ,$$

where $n = 0, 1, 2, \ldots$, $l = 0, \pm1, \pm2, \ldots$, $\Omega \equiv \left(\frac{1}{4}\omega_c^2 + \omega_0^2\right)^{\frac{1}{2}}$, $\omega_c = eB/m^*c$ and $\hbar\omega_0$ is the confinement energy. In the limit of zero confinement, they however reduce to $E_{nl} = (\mathcal{N} + \frac{1}{2})\hbar\omega_c$, where $\mathcal{N} = n + (|l| - l)/2$ is the Fock-Darwin level (FDL) index. This is shown in Fig. 1 for a GaAs quantum dot. Without the confining potential the energies of positive l states are independent of l but in its presence they increase with l. For very large B ($\omega_c \gg \omega_0$) and for $n = 0, l \geq 0$, we get $E = \frac{1}{2}\hbar\omega_c$, while for $n = 1, l \geq 0$, $E = \frac{3}{2}\hbar\omega_c$, *etc.* States with $l < 0$ have much higher energies than those for $l > 0$ (Fig. 1). The single-particle wave function is[7]: $\phi_{nl} = r^{|l|}\exp(-il\theta)L_n^{|l|}(r^2/2a^2)\exp(-r^2/4a^2)$, where the effective magnetic length is $a = [\hbar/2m^*\Omega]^{\frac{1}{2}}$, $L_n^{|l|}$ is a Laguerre polynomial and $-l$ is the angular momentum quantum number.

The states of the interacting electrons in a parabolic quantum dot are calculated by the standard procedure of numerically diagonalizing the many-electron Hamiltonian of the form:

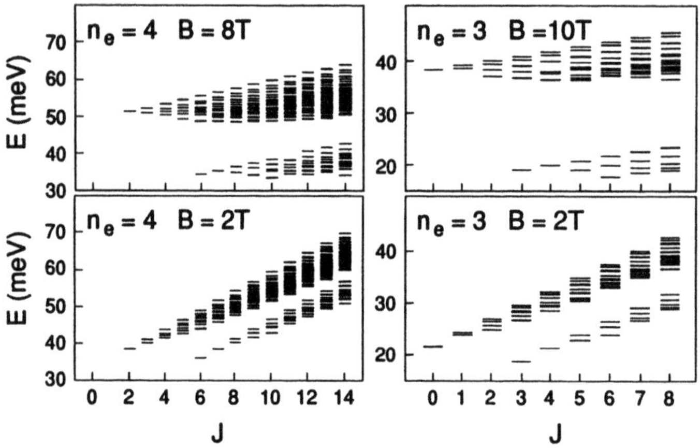

Fig. 2. Energy levels as a function of J for three and four electrons in a parabolic quantum dot [11]

$$\mathcal{H} = \sum_{nls} E_{nl} c^{\dagger}_{nls} c_{nls} + \sum_{\substack{n_1 \ldots n_4 \\ l_1 \ldots l_4 \\ ss'}} \mathcal{A}_{n_1 l_1, n_2 l_2, n_3 l_3, n_4 l_4} c^{\dagger}_{n_1 l_1 s} c^{\dagger}_{n_2 l_2 s'} c_{n_3 l_3 s'} c_{n_4 l_4 s}$$

where $\mathcal{A}$ is the Coulomb matrix element:

$$\mathcal{A} = \tfrac{1}{2} \int d\mathbf{r}_1 d\mathbf{r}_2 \phi^*_{n_1 l_1}(\mathbf{r}_1) \phi^*_{n_2 l_2}(\mathbf{r}_2) V(\mathbf{r}_1 - \mathbf{r}_2) \phi_{n_3 l_3}(\mathbf{r}_2) \phi_{n_4 l_4}(\mathbf{r}_1).$$

The eigenstates of the system are the eigenstates of the total angular momentum J (the sum of the single-electron l), which is conserved by the electron-electron interaction. The results of Maksym and Chakraborty for the energy levels of the three and four electron systems where the Coulomb interaction is used for $V(\mathbf{r}_1 - \mathbf{r}_2)$ [11] are shown in Fig. 2. We have considered the parameters appropriate to GaAs. The energies are plotted after subtracting $\hbar\Omega$ per electron. Clearly there are always two sets of broadened levels separated by a gap. In the limit of zero confinement potential these would be the lowest two Landau levels. The general trend is that energies increase with J because the single-electron energies increase with l.

The main difference between the high-field and low-field behavior is in the ground-state angular momentum. It is to be noted that the single-electron contribution increases linearly with J. In contrast, the interaction term decreases because electrons with higher angular momenta move in orbitals of larger radii, thereby decreasing the Coulomb energy.

The net result is that the total energy as a function of J has a *minimum.* At low fields, this mimimum occurs at the lowest available J, *i.e.*, the smallest angular momentum compatible with placing the electrons in $\mathcal{N} = 0$ states. At high fields, the minimum appears at a higher J value. The ground state of electrons in a magnetic field therefore occurs only at certain *special* values [11] of the total angular momentum J.

3. Magnetization

The first theoretical study of the magnetization of interacting electrons in a quantum dot was also by Maksym and Chakraborty [16]. They calculated the magnetization by differentiating the eigenvalues with respect to the magnetic field. More precisely, the numerically generated eigenvalues (including spin) are used to compute the free energy which is then differentiated to get the magnetization. In Fig. 3 we show the magnetization as a function of the magnetic field B for a three and four electron GaAs dot with $\mathcal{N} = 1$ (solid line) and $\mathcal{N} = 2$ [dotted line in the inset of Fig. 3(a)]. The ground state J and the total spin S as a function of B are

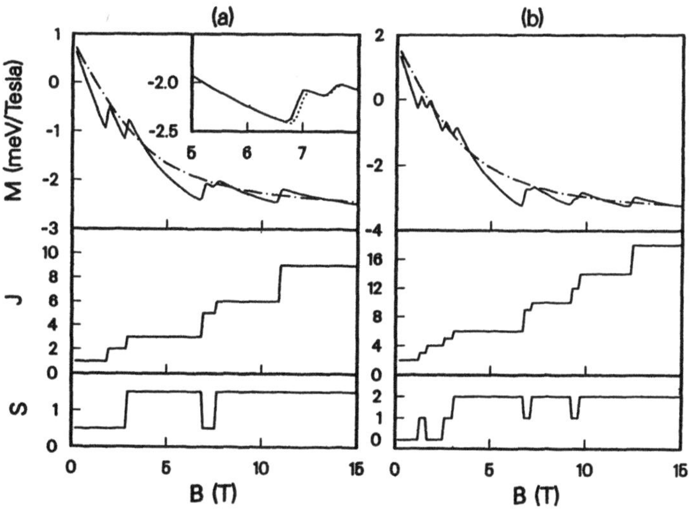

Fig. 3. Magnetization $\mathcal{M}$ $(meV/Tesla)$, the ground state angular momentum J and the ground state spin S at $T = 0.1K$ of a parabolic quantum dot containing (a) three and (b) four electrons per dot ($\mathcal{N} = 1$). The dash-dot line corresponds to the non-interacting case. A comparison of the results for the $\mathcal{N} = 2$ case (dotted line) is given in the inset of (a) [16]

also plotted for comparison. While the magnetization of noninteracting electrons (dash-dot line) changes smoothly with field the magnetization curves for the interacting system show steps with increasing magnetic field. The steps occur every time the ground-state J changes with the magnetic field and are clearly a consequence of the interaction.

The magnetization for non-interacting electrons has no steps because the lowest two single-electron levels are unaffected by level crossings as the field is increased. Hence systems of up to four non-interacting electrons in the lowest spin state stay in the same angular momentum state throughout the field range so the magnetization curve is smooth. All the steps in this case are a consequence of the interaction. For five or more non-interacting electrons the magnetization would be affected by negative l levels crossing positive l levels. However the position of these crossings would be drastically affected by the interaction. In addition, there are relatively few of them when the electron number is small (for five electrons in the lowest spin state there is only one) and they would tend to occur at low field. In contrast the steps due to the interaction occur at a regular sequence of J values throughout the field range [16].

Spin effects are important at fields $B \lesssim 10T$; the system is spin polarized at higher fields. One effect of spin is that it causes extra steps in the magnetization. Each spin state has its own sequence of special J values and each of those J values correspond to a possible ground state at that spin.

4. Quantum Rings

van Houten et al. recently suggested that [17,18] when the electrons in a disk (quantum dot) are in the lowest Landau level and when the charging energy is comparable to the cyclotron energy, the magnetoconductance oscillations are suppressed (*Coulomb blockade of the Aharonov-Bohm (AB) effect*). The blockade is lifted and the AB effect is recovered if one considers a *quantum ring* instead. In semiconductor nanostructures, quantum rings can be created from a two-dimensional electron system with an additional gate within the gates shaping the disk [8,9]. Application of a negative voltage to this additional gate would deplete the central region of the disk thereby forming a ring. In the lowest Landau level the basic difference between the two systems is the behavior of the electrochemical potential (or the energy cost associated with the addition of a single electron to the disk or the ring) as a function of the magnetic field which determines the conductance oscillations. The situation is different when the second Landau level is occupied in which case the electrochemical potential for a dot has the same *sawtooth* behavior as that for a ring and the Coulomb blockade of the AB effect is lifted – a situation exploited in the experiment of McEuen et al. [9].

The ring geometry is also very useful to study various other phenomena, *e.g.*, periodic oscillation of thermodynamic quantities like the magnetization in weak [19] and strong magnetic fields [20], as a function of the magnetic flux and observation of the persistent currents [21,22] (which is related to the periodic nature of the free energy) etc. Also in the quantum Hall regime, Halperin [23] employed it to demonstrate the importance of the *edge states* in quantizing the Hall conductivity.

The Hamiltonian for an electron confined in a parabolic ring and subject to a magnetic field is given by [24]

$$\mathcal{H} = \frac{1}{2m^*}\left(\mathbf{p} - \frac{e}{c}\mathbf{A}\right)^2 + \tfrac{1}{2}m^*\omega_0^2(r-r_0)^2\,,$$

where the vector potential is $\mathbf{A} = \frac{1}{2}(-By, Bx, 0)$ (symmetric gauge). Introducing the quantities [17]:

$$N_s = \frac{\Phi}{\Phi_0} = \frac{BeA}{hc}, \qquad \alpha = \frac{\omega_0 m^* A}{h}$$

where A is the area of the ring, $A = \pi r_0^2$ and $\Phi_0 = hc/e$ is the flux quantum, the radial part of the Schrödinger equation is written as:

$$f'' + \frac{1}{x}f' + \left[4\lambda + 2N_s l - 4\alpha^2 - (N_s^2 + 4\alpha^2)x^2 + 8\alpha^2 x - \frac{l^2}{x^2}\right]f = 0. \quad (1)$$

Here we have introduced the dimensionless quantities, $x = r/r_0$ and $\lambda = \dfrac{2m^*\pi A}{h^2}E$. As a check, let us consider the case of δ-function confinement. Eq. (1) then has the solution $\lambda = \frac{1}{4}(N_s - l)^2$. This case was first studied by Büttiker et al. [21] and shown in Fig. 4(a).

In Fig. 4(b)–(d), we have also plotted the single-electron energy levels for the parabolic ring for various values of α (α is inversely proportional to the width of the ring). The results are obtained by numerically solving Eq. (1). With decreasing α (wider ring), the *sawtooth* behavior of the narrow ring is gradually replaced by the formation of the Landau levels.

The results for the magnetization ($\mathcal{M}$) and susceptibility (χ) for a three- and four-electron system are shown in Figs. 5 – 7 at various temperatures [kT = 0.1 (dotted line) and 0.2 – 1.0 (solid lines)]. For

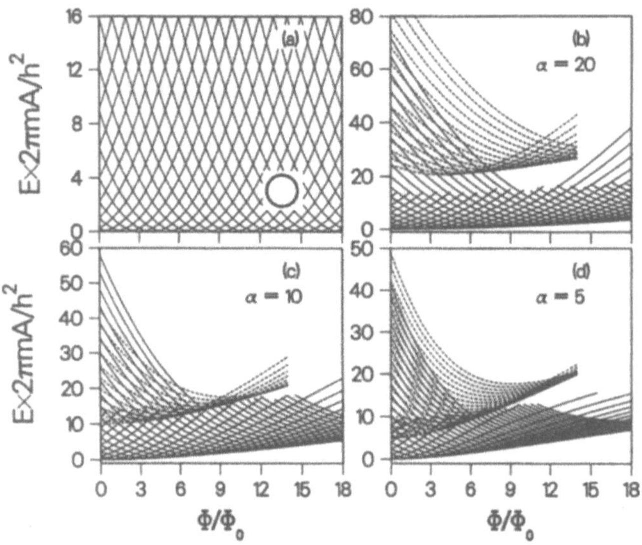

Fig. 4. (a) Single-electron energy levels as a function of B for (a) narrow ring and (b)-(d) wide parabolic quantum rings with various values of the confinement potential strength. The second FDL is plotted as dotted lines

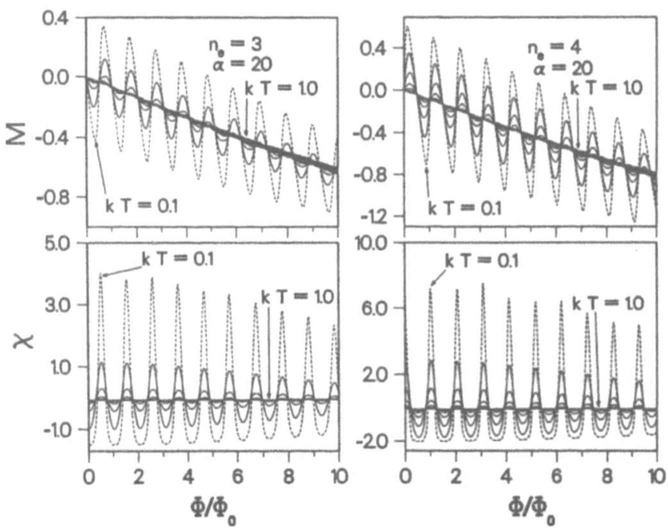

Fig. 5. Magnetization $\mathcal{M}$ and susceptibility (χ) for three and four non-interacting electrons in a parabolic quantum ring with $\alpha = 20$ at various temperatures

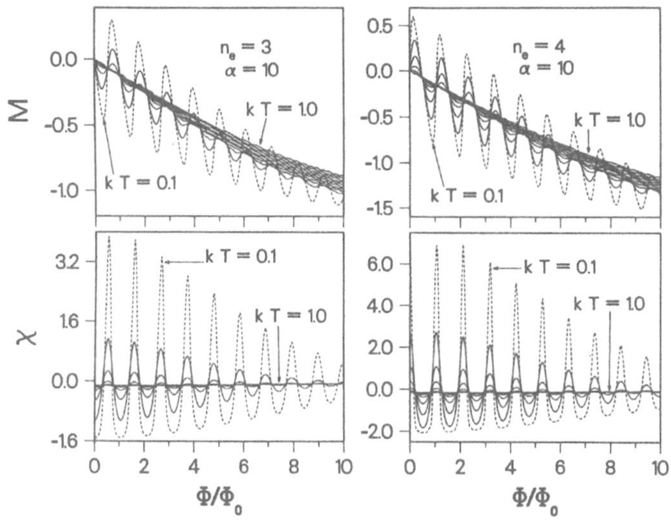

Fig. 6. Same as in Fig. 5, but for $\alpha = 10$

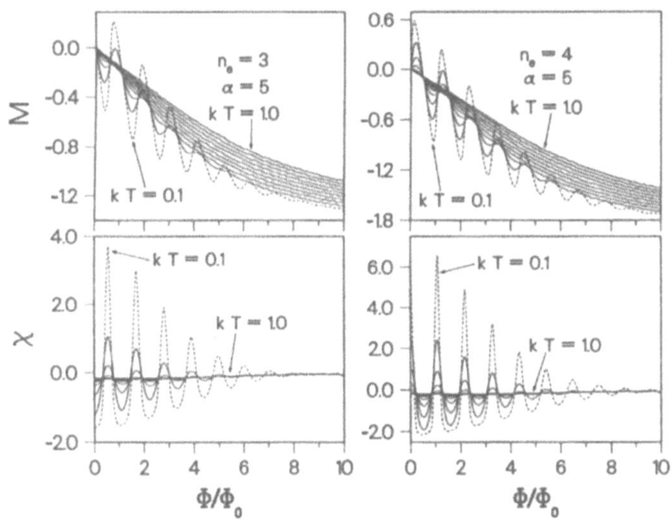

Fig. 7. Same as in Fig. 5, but for $\alpha = 5$

$\alpha = 20$, the system is still close to the ideal ring and $\mathcal{M}$ and χ show periodic oscillations as a function of the field Φ (albeit with decreasing amplitude as the magnetic field increases). The situation here is therefore completely different from that of a quantum dot where, as discussed above, the magnetization of the non-interacting three- and four-electron systems does not show any oscillation at all. For $\alpha = 10$ the oscillations are damped at the high magnetic field range, and for $\alpha = 5$, the oscillations are quickly damped with magnetic field, reflecting the behavior in the energy at this value of α.

The single-particle energy for any l behaves as a function of the magnetic field Φ roughly as

$$E_l = b + c\Phi + d(l - \Phi)^2 ,$$

where b, c and d are constants depending on the confinement and the principal quantum number but not on the field Φ nor the angular momentum l. Therefore, for any l the dispersion is a parabola with the minimum roughly at $\Phi \sim l$. When one moves slightly away from the minimum position, the lowest single-particle energy will increase until one is so far away that the next (or previous) parabola will be lower in energy. This means, of course, that the ground state energy itself oscillates as a function of Φ (or B). As a derivative of the energy the magnetization exhibits, of course, stronger oscillations and the susceptibility as a derivative of the magnetization exhibits even stronger oscillations.

In order to understand the high-magnetic-field behavior, let us consider the limit where the magnetic field is large compared to the confinement α. In that case, it can be shown [24] that Eq. (1) becomes the familiar equation for the Landau levels, *i.e.*, all the angular momentum values are degenerate. Hence there are no level crossings and as a consequence, no oscillations occur in the magnetization or susceptibility.

In our present study of the quantum ring in magnetic fields, the electron-electron interaction was not included. We have seen above that in a quantum dot, the effect is quite strong because the behavior of the magnetization as a function of magnetic field with and without interaction is quite different. Similar studies for a quantum ring are therefore of interest and will be reported elsewhere [24].

Acknowledgments

The work on quantum dots was done in collaboration with Dr. Peter Maksym (Leicester, UK).

References

1. U. Merkt, Advance in Solid State Physics **30**, 77 (1990)
2. T. P. Smith, III, Surf. Sci. **229**, 239 (1990)

3. For a review see, Tapash Chakraborty: Comm. Cond. Matt. Phys. (to be published)
4. W. Hansen, T. P. Smith, III, K. Y. Lee, J. A. Brum, C. M. Knoedler, J. M. Hong, D. P. Kern: Phys. Rev. Lett. **62**, 2168 (1989)
5. W. Hansen, T. P. Smith III, K. Y. Lee, J. M. Hong and C. M. Knoedler, Appl. Phys. Lett. **56**, 168 (1990)
6. Ch. Sikorski and U. Merkt: Phys. Rev. Lett. **62**, 2164 (1989)
7. T. Demel, D. Heitmann, P. Grambow, K. Ploog: Phys. Rev. Lett. **64**, 788 (1990)
8. B. J. van Wees, L. P. Kouwenhoven, K. J. P. M. Harmans, J. G. Williamson, C. E. Timmering, M. E. I. Broekaart, C. T. Foxon and J. J. Harris, Phys. Rev. Lett. **62**, 2523 (1989); L. P. Kouwenhoven, B. J. van Wees, K. J. P. Harmans and J. G. Williamson, Surf. Sci. **229**, 290 (1990)
9. P. L. McEuen, E. B. Foxman, U. Meirav, M. A. Kastner, Y. Meir, N. S. Wingreen, S. J. Wind: Phys. Rev. Lett. **66**, 1926 (1991)
10. D. C. Tsui, H. L. Störmer and A. C. Gossard, Phys. Rev. Lett. **48**, 1559 (1982); Tapash Chakraborty and P. Pietiläinen, *The Fractional Quantum Hall Effect* (Springer-Verlag, New York, Berlin, Heidelberg, 1988)
11. P. A. Maksym and Tapash Chakraborty: Phys. Rev. Lett. **65**, 108 (1990)
12. C. W. J. Beenakker, H. van Houten, A. A. M. Staring: in *Single Charge Tunneling*, edited by H. Grabert and M. H. Devoret (Plenum, New York, 1991)
13. V. Fock: Z. Physik **47**, 446 (1928); C. G. Darwin: Proc. Cambridge Philos. Soc. **27**, 86 (1930)
14. T. Haavasoja, H. L. Störmer, D. J. Bishop, V. Narayanamurti, A. C. Gossard, and W. Wiegmann, Surf. Sci. **142**, 294 (1984); J. P. Eisenstein, H. L. Störmer, V. Narayanamurti, A. Y. Cho, A. C. Gossard, and C. W. Tu, Phys. Rev. Lett. **55**, 875 (1985)
15. A. Kumar, S. E. Laux, F. Stern: Phys. Rev. B**42**, 5166 (1990)
16. P. A. Maksym and Tapash Chakraborty: Phys. Rev. B**45**, (15 January, 1992)
17. C. W. J. Beenakker, H. van Houten, A. A. M. Staring: Phys. Rev. B**44**, 1657 (1991)
18. C. W. J. Beenakker, H. van Houten, A. A. M. Staring: *Granular Nanoelectronics*, edited by D. K. Ferry, J. Barker, C. Jacobini, (Plenum, New York, 1991)
19. E. N. Bogachek and G. A. Gogadze: JETP **36**, 973 (1973)
20. U. Sivan and Y. Imry: Phys. Rev. Lett. **61**, 1001 (1988)

21. M. Büttiker, Y. Imry, R. Landauer: Phys. Lett. **A96**, 365 (1983)
22. V. Chandrasekhar, R. A. Webb, M. J. Brady, M. B. Ketchen, W. J. Gallagher, A. Kleinsasser: Phys. Rev. Lett. **67**, 3578 (1991)
23. B. I. Halperin: Phys. Rev. B**25**, 2185 (1982)
24. P. Pietiläinen and Tapash Chakraborty: to be published

Part III

Tunneling

Single Charges in Networks of Small Capacitance Tunnel Junctions

J.E. Mooij

Department of Applied Physics, Delft University of Technology,
P.O. Box 5046, NL-2600 GA Delft, The Netherlands

Abstract. If metallic regions are separated by tunnel junctions with resistance above the quantum resistance, and if capacitances are small, the charging energy associated with the transfer of one single electron has a strong influence on transport properties. The resultant Coulomb gap is dependent on the environmental properties. In the normal state tunneling is a stochastic process. Circuits can be realized in which the Coulomb blockade is modified by an external gate voltage. In the superconducting state the phase of the order parameter and the charge are conjugate variables; for large charging energy the phase fluctuations are strong. In two-dimensional junction arrays with small capacitances a charge inbinding transition takes place, both in the normal and in the superconducting state.

1. Introduction

It is possible with present-day fabrication techniques to obtain high quality tunnel junctions with dimensions below 100 nm. Such junctions have a capacitance C below 10^{-15} F, which has the consequence that the Coulomb energy for one single electron, $E_C=e^2/2C$, is of order 1 K. At low temperatures transport through circuits of such junctions is completely dominated by the Coulomb blockade effects. Recent reviews of the field are provided by Averin and Likharev [1] and the proceedings of the Les Houches winter school [2]. The junctions are usually made of aluminum with a shadow evaporation technique. For measurements in the normal state, superconductivity is suppressed by a magnetic field. For occurrence of a Coulomb gap, the resistance of the junctions R_t has to be high with respect to the quantum resistance as can be understood on the basis of the following simple argument: Quantum fluctuations against the Coulomb blockade with typical energy scale E_C exist for times below $h/E_C=2hC/e^2$. The circuit response time RC has to be larger, which leads to the requirement $R_t \gg h/e^2$. Typical junction resistances are 50 to 200 kΩ.

2. Coulomb gap

For Coulomb blockade effects, the circuit connecting the junction has a strong influence. The simplest system to study is a series circuit of two high resistance, small capacitance junctions. When an electron tunnels in or out of the island between the junctions, the electrostatic energy of the system increases by $e^2/4C$, when the junction capacitances are equal. At zero temperature and low bias voltage, no tunnel current runs through the system. Only for an applied voltage across the two junctions larger than $e/2C$, the free energy is decreased by a tunnel process. Figure 1 shows the current-voltage characteristic of a double junction system at different temperatures.

For the case of tunneling in a junction with resistance R_t much higher than the quantum resistance R_q (so that the tunneling can be treated as a perturbation), unshorted by capacitance of the leads or by a low source impedance, the tunneling rate can in the normal state be calculated with the golden rule. This treatment is usually referred to as the 'orthodox' theory. The expression for the rate is:

$$\Gamma = (\Delta E / e^2 R_t) \{ \exp(\Delta E/k_B T) - 1 \}^{-1} \tag{1}$$

Here ΔE is the total energy after tunneling minus the energy before the tunnel process,

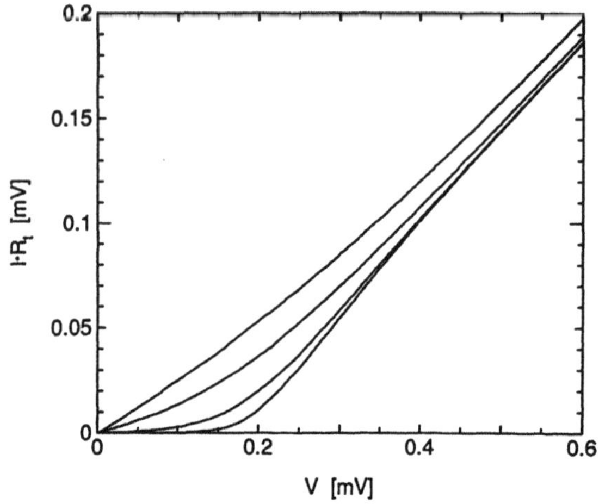

Fig.1. Current-voltage characteristic of two junctions in series with R_t=117 kΩ and C=0.3 fF at different temperatures. From top to bottom: T=700, 400, 200, 25 mK [3].

including the electrostatic energy of other junctions and capacitors in the circuit and including the energy of sources. The temperature dependence of the tunneling rate, as shown in figure 1, follows from this equation. For $|\Delta E| >> k_BT$, the rates can be approximated by:

$$\Delta E>0: \quad \Gamma = (\Delta E/e^2R_t) \exp(-\Delta E/k_BT) \tag{2a}$$

$$\Delta E<0: \quad \Gamma = |\Delta E| / e^2R_t \ . \tag{2b}$$

As can be seen, tunneling is exponentially forbidden for increasing energy, as expected, while for negative ΔE the rate is proportional to $|\Delta E|$ and to R_t^{-1}. For a single junction that is current-biased, ΔE is positive when the absolute value of the voltage across the junction is smaller than e/2C, and the charge smaller than e/2. For a voltage bias, ΔE is always negative as the source energy decreases by eV regardless of the charge on the junction. In practice it is very difficult to observe the Coulomb gap in a single junction. Unless high impedances are placed so close to the junction that the capacitances of the leads between junction and impedances are smaller than the junction capacitance, no effect is seen.

For tunneling in junctions with $R_t>>R_q$ the influence of a low impedance environment can be taken into account as shown by various authors [4-8]. The environment is characterized by the density of harmonic oscillator modes as a function of energy; these oscillators make it possible to tunnel with absorption or emission of a photon. The effect of the environment is strongest at the lower voltages and currents. In the asymptotic behavior for large bias of the current-voltage characteristic the Coulomb offset is seen much longer.

In networks, the islands between junctions contain typically 10^7 to 10^8 electrons. Consider, for example, the double junction system. Here, an electron (or hole) will tunnel into the island through one junction; the most likely next event is the tunneling out of an electron (hole) through the other junction. One might ask whether there is a direct connection between the tunneling events in the two junctions. For the processes considered so far, that connection does not exist. When the bias voltage exceeds the Coulomb gap, one electron will tunnel into the island. It will occupy an energy above the Fermi level. Although the tunneling probability in the other junction may be higher for this hot electron, it will likely have been scattered inelastically before it has diffused from one junction to the other. The first tunneling event has an effect on all electrons in the island by increasing their Coulomb energy. Because of that a different electron will tunnel out through the second junction. This type of tunneling is a stochastic process, there is no correllation between tunneling events in both junctions, other than that differences of rates lead to increased charging.

If the electron-phonon scattering rate is low, the effective electronic temperature in an island may easily be much higher than the ambient temperature. On analysis for typical present-day samples, within the uncertainties of the parameters one finds that heating might or might not be serious. It are the experimental results that show that relaxation is usually strong enough. For smaller junctions, E_C goes up and the energy of incoming tunneled electrons is higher. For heating this is advantageous, as electron-phonon scattering increases much faster with energy than the energy itself.

Apart from the previous incoherent tunneling processes in more than one junction, there are also higher order tunneling processes in which more than one junction is involved. This leads to a certain amount of tunneling within the Coulomb gap. Tunneling of the whole system occurs through a virtual state with increased energy. Two (or more for more than two junctions) different electrons are involved in the single event. Averin and Nazarov [9] have first pointedthis out, Geerligs et al. [10] have verified their predictions experimentally. For applications this so called co-tunneling is extremely important in a negative sense. The associated rates are proportional to $(R_n/R_q)^{N-1}$, where N is the number of junctions in series. To suppress co-tunneling, one needs high-ohmic junctions or more junctions in series. Given the practically available capacitance values, this leads to limitations for the frequencies at which charging circuits can be driven.

3. Gate modulation

When a gate electrode is coupled to an island between junctions by a capacitance C_g, it can draw charge to its electrode. In the calculation of tunneling rates, dE now has to include the elctro-static energy of the gate capacitor as well as the free energy of the gate voltage supply. For the purpose of the tunneling probability, the system behaves much as if the charge on the gate capacitor has been taken away from the island. If, for example, in a two-junction system the gate charge is e/2, the Coulomb gap is completely suppressed. So, a gate can be used to modify the I-V characteristic: the single electron transistor. When the induced charge is a multiple of e, the charge can disappear by tunneling. As a consequence, the modulation is periodic in the gate voltage V_g with period e/C_g. In figure 2 the suppression of the Coulomb gap is shown for a double junction system. The extremes of integer and integer-plus-a-half electrons on the island are given. In figure 3 the periodicity is illustrated. For fixed bias voltage, the current as a function of gate voltage is shown for systems of 2,3 and 5 junctions in series.

On studying figure 3 more closely, one notices that V_g=0 does not coincide with the minimum current. This is probably due to random offset charges on the islands, that are

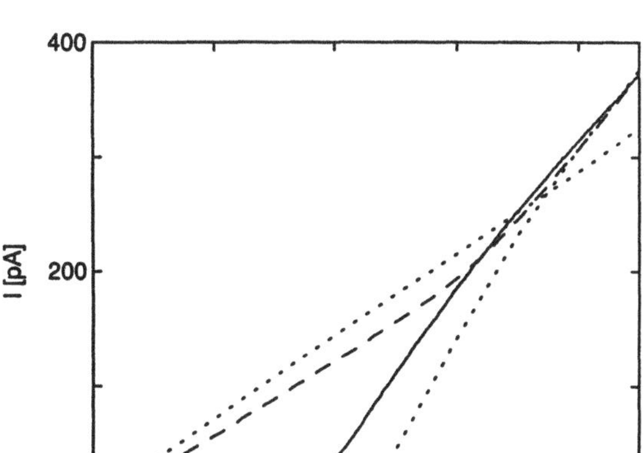

Fig. 2. Influence of gate voltage on double junction system with R_t=347 kΩ and C=0.3 fF at 15 mK [3]. The solid line gives the I-V characteristic for minimized island charge, the dashed line for induced charge e/2. Dotted lines: curves calculated with eq. (1).

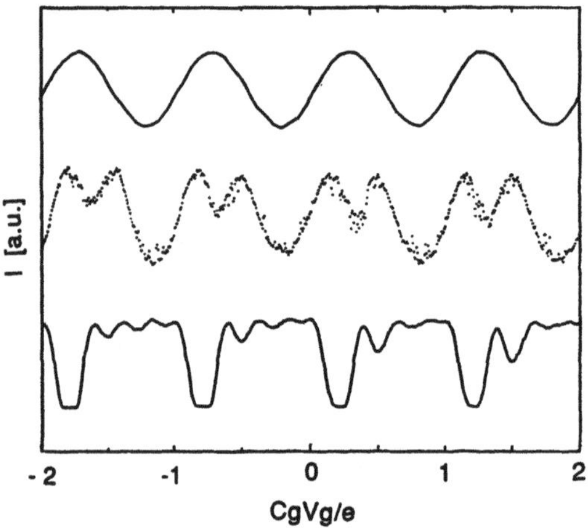

Fig. 3. Modulation of the current in series circuits of N junctions at fixed bias voltage, as a function of gate voltage. Top to bottom: N=2,3,5. The modulation period is e/C_g.

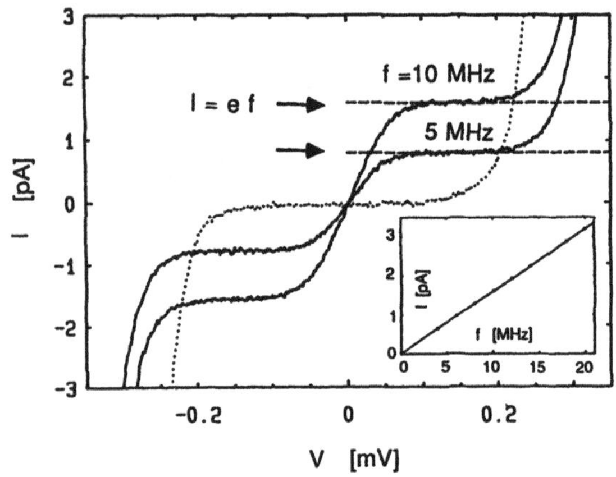

Fig. 4. Turnstile. The dotted line shows the I-V characteristic with the Coulomb gap for the system without RF signal. The drawn lines are the characteristics for applied RF with frequency 5 and 10 MHz. Current plateaus develop on which I=fe, with f the frequency. The principle may provide a fundamental standard of current.

induced by defects in the tunnel barriers or in the insulators near the islands. The random charges can be tuned away with a gate, but it is not always very practical to supply every island in a circuit with a correction gate. It may be that fabrication procedures are available where the charges adjust themselves in such a way on cooling down that the charging energies are minimized and the offset charges reduced. Even more problematic are offset charges that change in time. In metallic junction circuits one sometimes finds telegraph noise due to such variations. In semiconductor single electron circuits this type of telegraph noise seems more pronounced.

The modulation by a gate can be used to develop single electron devices. The basic element is the single electron transistor: the double junction with gate electrode. To suppress co-tunneling, the number of junctions on the two sides of the gate can be increased.

There exists a very interesting analogy between single electron devices and Josephson junction circuits. The critical current of the Josephson junctions is replaced by a critical voltage, the modulation of the critical current of two parallel junctions by a magnetic flux translates into the modulation of the series circuit by the charge. The unit of the flux quantum h/2e for the squid becomes the electron charge e for the single electron transistor. The transistor can be operated as an electrometer with a sensitivity of order 10^{-5} e $/\mathrm{Hz}^{1/2}$. Unfortunately there is no equivalent to the flux transformer, and leads to the transistor have to be very short to avoid capacitive shunting.

One of the interests in single electron devices is in developing a fundamental standard of current, in much the same way as the Josephson voltage standard. One can design circuits, where for each cycle of an applied RF signal on a gate, exactly one electron is transmitted. An example is the turnstile, consisting of two branches of multiple junctions, with a central RF gate [11]. In figure 4 the influence of the RF signal is shown. For adequate parameters, plateaus develop at I=fe, where f is the RF frequency. With only two junctions per branch, the accuracy is limited, mainly by co-tunneling, to about 0.5%. By inserting more junctions, expectations are that a very high accuracy may be achieved. For practical current standards, one needs much higher current levels. This can be achieved by making the junctions smaller (allowing a higher frequency) and in particular by putting more junctions in parallel. Other schemes for the current standard are for example the "pump" [12] and various semiconductor single electron circuits [13].

4. Superconducting circuits

In a superconductor, the phase of the order parameter and the charge are conjugate variables in the quantum mechanical sense. Usually, the phase is the relevant parameter and charge is not determined. In circuits with very small capacitance, the charge becomes the relevant observable and the phase fluctuates. In the Hamiltonian of the junction:

$$H = Q^2/2C - E_J\cos\phi - (h/4\pi e)\, I\phi \qquad (3)$$

with E_J the Josephson coupling energy, the charge is an operator $Q= (2e/i)\, \partial/\partial\phi$. The energy of

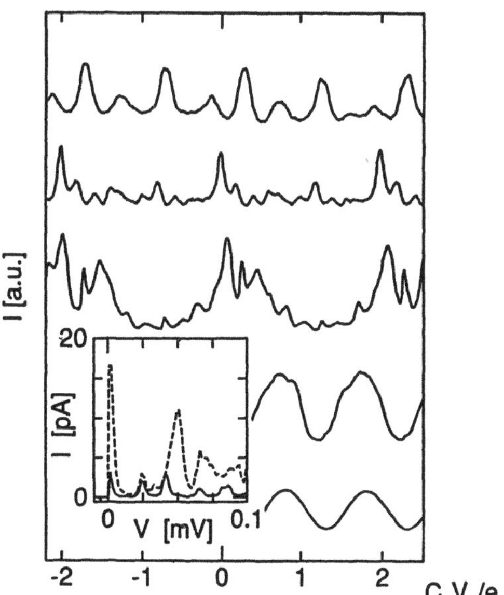

Fig. 5. Gate modulation of a superconducting double junction circuit. Inset shows I-V characteristic for two values of gate voltage. The modulation traces have been taken from top to bottom at 0, 20 and 40 μV, at the gap edge and in the normal state.

a single junction is a periodic function of the quasicharge, with bands separated by E_J. A current source drives the junction through the band, leading to Bloch oscillations if Zener tunneling to a higher band can be avoided. The Bloch oscillations can be observed by their interference with an external RF signal, giving steps at current current [14].

If circuits of more than one junction are considered, solutions must be found in an N-dimensional Hilbert space with N the number of junctions. In addition the junctions can couple to resonating modes of the environment with exchange of photons. Tunneling requires equal energies before and after the event. Unlike in the normal state where tunneling is stochastic, in the superconducting state tunneling is deterministic. There are however many processes that have to be taken into account such as coupling to the environment, the superconducting energy gap and the presence of even only one quasiparticle. At this time the superconducting single charge effects are not yet well charted and little experimental data is available [15-17].

In figure 5 as an example the modulation of the current by a gate voltage is given for a double junction circuit. The lowest trace is in the normal state, showing the e/C_g modulation period as in figure 3. The inset shows part of the I-V characteristics for two extreme values of the gate voltage. Resonances are seen at 20 and 40 μV. The modulation traces in the main figure indicate the current at 0 V (the supercurrent), on the 20 μV peak, on the 40 μV peak and at the onset of the gap from top to bottom. From the lower curve in the normal state, the period e/C_g is determined. The upper traces show periodicity with e/C_g as well as periodicity with $2e/C_g$. In particular in relatively slow measurements one expects that island charging energies will be minimized by tunneling of a quasiparticle. Why so often a quasiparticle is available even at the extremely low temperatures is a question that is not answered by standard BCS theory.

5. Two-dimensional junction arrays

The charging effects lead to a special kind of phase transition in a two-dimensional array of small capacitance tunnel junctions. Here the normal state is considered. If one of the islands in the array contains one additional charge, this charge distributes itself over the capacitances to ground and to neighboring islands. These neighboring islands are in turn polarized and the effects may extend over considerable distance. A model can be considered in which each island has a capacitance C to nearest neighbors, a capacitance C_0 to ground and where all other elements of the capacitance matrix are ignored. We consider a square array with 4 nearest neighbors [18]. The screening length for the potential, measured in cell numbers, is $\Lambda=(C/C_0)^{1/2}$. The potential falls off as $K_0(r/\Lambda)$. For distances much smaller than the screening length, the potential falls off as $-\ln r/\Lambda$.

If one electron tunnels from an island to a neighboring island, a charge-anticharge pair is formed. If the electron tunnels on, the pair separation increases. The interaction between the charge and anticharge is attractive, with logatithmic dependence on separation. Figure 6 gives an impression of the way in which the charges on islands redistribute in the presence of such a pair. Only two islands in the picture contain a net charge.

When the system is much smaller than Λ, a Kosterlitz-Thouless/Berezinskii phase transition will occur. Below the transition temperature, only bound pairs are present. Their density and their separation increase with increasing temperatures. Bound pairs do not lead to conduction.

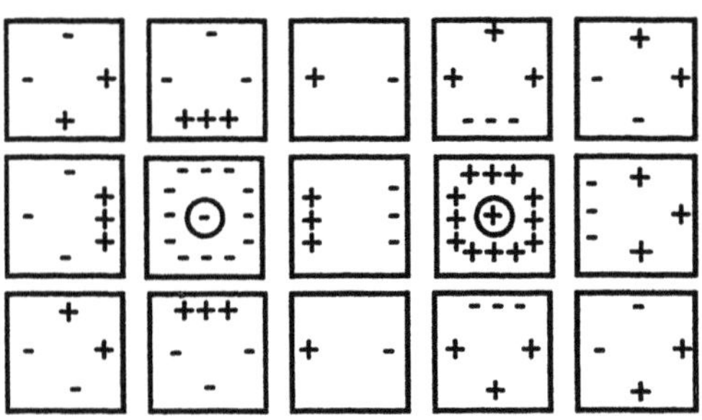

Fig. 6. Charge distribution in two-dimensional junction array that contains a charge-anticharge pair.

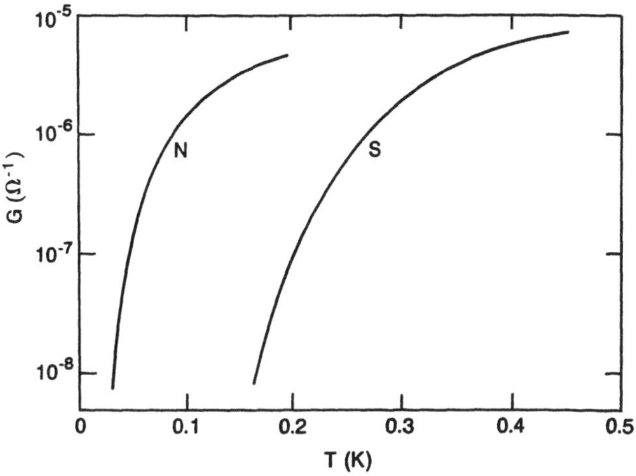

Fig. 7. Conduction of a two-dimensional array of tunnel junctions as a function of temperature [18]. N normal state; S superconducting state for the same sample with R_t=15 kΩ and C=1fF.

At the transition temperature:

$$k_B T_c = (4\pi\varepsilon_c)^{-1} E_C \quad (4)$$

the first free charges are generated. In the expression, ε_c is a non-universal constant of order one. The density of free charges (positive as well as negative) will increase slowly above T_c with temperature dependence $\exp\{-2b(T/T_c-1)^{-1/2}\}$, with b of order 1.
The influence of dissipation on the transition is considerable. It can be interpreted as if the tunneling leads to a renormalization of the capacitance, with the result that the effective E_C is reduced. The first order correction to the transition temperature is a factor $(1- 0.1R_q/R_t)$, for higher values of R_q/R_t the reduction is stronger. When R_q/R_t is 0.45, T_c is zero.

It is not easy to observe the charge unbinding transition in a real sample. The values of the capacitances C and C_0 yield a screening length of 10 to 20 with the cell dimensions of the presently available arrays. This means that the maximum scale is limited to the logarithm of that screening length and is certainly not much larger than 1. As a result, the transition will be rounded off and over most of the temperature range show thermally activated behavior rather than the specific 2D dependence. If the screening length is increased by increasing the nearest-neighbor capacitance, the transition temperature goes down below the practically accessable range.

In figure 7 the conduction is shown of a two-dimensional array of 15 kΩ junctions with a capacitance of 1 fF. Without correction for dissipation the transition should be at 60 mK in the normal state. Both the dissipation and the finite size reduce the transition. Still, it is clear that at zero temperature there is no conduction and that a sharp rise occurs in a narrow temperature range. In the same figure the transition is also shown for the same sample in the superconducting state. In spite of the name the superconducting system has a significantly lower conduction over a certain range of temperatures. The ratio of the transition temperatures is about 4, as expected from the ratio of the charging energies for Cooper pairs and single electrons.

In the superconducting state, single charges and vortices exhibit a strong duality. Flow of vortices leads to resistance, flow of charges to conductance. Currents and electric fields lead to forces, the elementary units are the flux quantum h/2e and the charge 2e. Array behavior will be more dominated by vortex dynamics when E_J is larger than E_C, more by the charge dynamics in the opposite case. The vortex system has a vortex-antivortex unbinding transition, from a superconducting to a metallic state. The charge system crosses from insulating to metallic as derscribed. At zero temperature there is a phase transition between the superconducting and the insulating regimes, for E_J/E_C about 1. Arrays close to that transition show very peculiar behavior, including reentrance. Near the phase boundary, a weak magnetic field has a very strong influence. A small frustration can drive a system from purely superconducting to purely insulating at the lowest temperatures. Such effects are described in reference 19.

References

1. D.V. Averin and K.K. Likharev, in: "Quantum effects in small disordered systems", B.L. Al'tshuler, P.A. Lee and R.A. Webb, editors; Elsevier 1991.
2. Proceedings of the NATO Advanced Study Institute on Single Charge Tunneling, M.H. Devoret and H. Grabert, editors; Plenum 1992.
3. L.J. Geerligs, Ph.D. thesis, Delft University, 1990.
4. Yu.V. Nazarov, Pis'ma Eksp. Teor. Fiz. **49**, 105 (1989) [JETP Lett. **49** , 126 (1989)].
5. M.H. Devoret, D. Esteve, H. Grabert, G.-L. Ingold, H. Pothier and C. Urbina, Phys. Rev. Lett. **64** , 1824 (1990).
6. H. Grabert, G.-L. Ingold, M.H. Devoret, D. Esteve, H. Pothier and C. Urbina, Z. Phys. **84** , 143 (1991).
7. S.M. Girvin, L.I. Glazman, M. Jonson, D.R. Penn and M.D. Stiles, Phys. Rev. Lett. **64**, 3183 (1990).
8. G. Falci, V. Bubanja and G. Schön, Z. Phys.B **85**, 451 (1991).
9. D.V. Averin and A.A. Odintsov, Phys. Lett. A **140**,251 (1989).
10. L.J. Geerligs, D.V. Averin and J.E. Mooij, Phys. Rev. Lett. **65**,3037 (1991).
11. L.J. Geerligs, V.F. Anderegg, P.A.M. Holweg, J.E. Mooij, H. Pothier, D. Esteve and M.H. Devoret, Phys. Rev. Lett. **64**, 2691 (1990).
12. H. Pothier, P. Lafarge, C. Urbina, D. Esteve and M.H. Devoret, Europhys. Lett. **17**, 249 (1992).
13. L.P. Kouwenhoven, A.T. Johnson, N.C. van der Vaart, C.J.P.M. Harmans, C.T.Foxon, Phys. Rev. Lett. **67**, 1626 (1991).
14. L.S. Kuzmin, Yu.V. Nazarov, D.B. Haviland, P. Delsing and T. Claeson, Phys. Rev. Lett. **67**, 1161 (1991).
15. T.A. Fulton, P.L. Gammel, D.J. Bishop, L.N. Dunkleberger and G.J. Dolan, Phys. Rev. Lett. **63**, 1307 (1989).
16. L.J. Geerligs, V.F. Anderegg, J. Romijn and J.E. Mooij, Phys. Rev. Lett. **65**, 377 (1990).
17. A.G. Maassen van de Brink, G. Schön, L.J. Geerligs, Phys. Rev. Lett. **67**, 3030 (1991).
18. J.E. Mooij, B.J. van Wees, L.J. Geerligs, M.Peters, R. Fazio and G. Schön, Phys.Rev. Lett. **65**, 645 (1990).
19. J.E. Mooij and G. Schön, in reference 2.

Fermi-Surface Effects in Small-Capacitance Tunnel Junctions

M. Ueda

NTT Basic Research Laboratories, Musashino-shi,
Tokyo 180, Japan

Abstract. In a small-capacitance tunnel junction, even a single-electron tunneling almost completely inverts the junction voltage and causes significant electrical and thermal relaxation. This paper uses a simple model to study the effects of the associated nonequilibrium shake-up of electron-hole pairs on Coulomb blockade. Possible ways to observe these effects are discussed. The Fermi-surface environment and the electromagnetic environment are compared.

1. Introduction

Tunneling is a unique quantum phenomenon in that the transition of states depends strongly on the junction environment. In macroscopic junctions, tunneling occurs exclusively as a result of the wave nature of an electron and the background thermal noise is large enough to ensure the thermal equilibrium of the electrodes. The dynamic variables characterizing the junction environment can therefore be eliminated using the fluctuation-dissipation theorem.

However, in a small-capacitance tunnel junction, where the electrostatic energy of a single electron is comparable to or even greater than the masking thermal energy, such a picture breaks down because a tunneling electron excites the environment and the back reaction of the excitation can no longer be disregarded. A tunneling electron excites both the electromagnetic modes along the transmission line and the electron-hole pairs in the junction electrodes. Recent studies have revealed that Coulomb blockade emerges as a result of exciting the electromagnetic environment [1]. The main purpose of this paper is to study the effect of electron-hole pair excitation on Coulomb blockade.

Suppose that a normal tunnel junction is initially in thermal equilibrium. The orthodox Coulomb-blockade theory assumes that the junction is also in thermal equilibrium immediately after tunneling. That is, the theory neglects the transient behavior between the two equilibrium states before and after tunneling. This assumption is be justified for large junctions because the perturbation brought about by tunneling is usually negligible. For a tunnel junction whose capacitance is less than 10^{-15} F, however, even a single-electron tunneling causes a voltage change on the order of hundreds of microelectron volts. The ensuing electrical and thermal relaxation should significantly alter the density of final states available for tunneling because conduction electrons in the junction electrodes constitute Fermi liquids that display the *orthogonality catastrophe* [2].

It has recently been pointed out that if a tunneling event causes a local change in the Coulomb potential, infinitely many electron-hole pairs should be excited near the Fermi surface [3]. This leads to a power-law anomaly in the tunneling conductance. This paper extends the theory developed in Ref. 3 to finite temperatures, and discusses several phenomena, comparing the Fermi-surface environment and the electromagnetic environment.

2. Effects of the Fermi-surface environment on tunneling

Most analyses of tunneling assume that tunneling is a small perturbation and does not disturb the equilibrium of the electrodes. For small-capacitance tunnel junctions, however, even a single electron tunneling almost completely inverts the junction voltage and causes significant electrical and thermal relaxation. The electrical relaxation has two stages occurring on different time scales. In the first stage, the surface charge is formed within a time scale of the inverse plasma frequency (the screening process). Following this, the electronic configuration undergoes fine rearrangement as low-energy electron-hole pairs are excited by the sudden change in the screened Coulomb potential. If this change occurs locally in space, Fermi liquid theory predicts that infinitely many electron-hole pairs should be excited near the Fermi surface. When the shake-up energy associated with this final-state interaction becomes comparable to the charging energy, Coulomb blockade will be significantly different from the orthodox result.

This is the physical picture of what we shall refer to as the Fermi-surface environment. The Fermi-surface environment causes the electrical relaxation in the second stage. On the other hand, the excited electronic states are coupled with the thermal reservoir, and this coupling is in the long run responsible for the thermal relaxation. This process will also be incorporated into the theory using real-time Green's functions.

3. Formulation of the problem

As described above, the orthodox Coulomb-blockade theory neglects the transient electronic response to a sudden potential change due to tunneling. Here, in contrast, we take into account the electrical relaxation and the ensuing thermal relaxation. Since the electrical relaxation is basically the rearrangement of the electronic configuration by the screened Coulomb potential, it can be described by the time-dependent scattering potentials $\sum_{k\neq k} V^L_{kk'}(t)c^\dagger_k c_{k'}$ and $\sum_{q\neq q'} V^R_{qq'}(t)c^\dagger_q c_{q'}$, where L and R represent the left and right electrodes. The strengths of the scattering potentials $V^L_{kk'}$ and $V^R_{qq'}$ are zero before tunneling, and are suddenly switched on after tunneling. The Hamiltonian of the model is therefore given by

$$H = H_L + H_R + H_T + H_C, \tag{1}$$

where

$$H_L = \sum_k \varepsilon_k c^\dagger_k c_k + \sum_{k\neq k'} V^L_{kk'}(t) c^\dagger_k c_{k'}, \tag{2}$$

$$H_R = \sum_q \varepsilon_q c^\dagger_q c_q + \sum_{q\neq q'} V^R_{qq'}(t) c^\dagger_q c_{q'}, \tag{3}$$

$$H_T = \sum_{k,q} \left(T_{kq} c^\dagger_k c_q + T_{qk} c^\dagger_q c_k \right), \tag{4}$$

$$H_C = \frac{Q^2}{2C}, \quad Q = -\frac{e}{2}(N_L - N_R) + \text{const.} \tag{5}$$

Here N_L (or N_R) is the number operator of electrons in the left (or right) electrode. Equation (1) is the standard Hamiltonian of a normal tunnel junction except for the scattering terms. These terms rearrange conduction electrons, performing fine adjustment to a new equilibrium.

The tunneling characteristic is determined by the response function of the tunneling Hamiltonian:

$$iR(t-t') = \langle 0|T\{H_T(t)H_T(t')\}|0\rangle, \tag{6}$$

where $|0\rangle$ describes the equilibrium state of the junction before the first tunneling event and T represents the Dyson time-ordering operation. Because the Coulomb term can be expressed in terms of operators of the right electrode only, the probability amplitude can be factorized, giving

$$R(t-t') = 2i|T|^2 \mathcal{G}^R(t-t')^* \mathcal{G}^L(t-t') + \text{c.c.}, \tag{7}$$

where c.c. denotes the complex conjugate of the preceding term. The problem thus reduces to obtaining the transient Green's functions $\mathcal{G}^R$ and $\mathcal{G}^L$.

4. Transient Green's functions and response function

A. Transient Green's function for the total system

The transient Green's function of the right (or left) electrode describes the propagation of a tunneling electron (or a hole created by it), the collective propagation of the localized potential, and associated many-body interactions such as vertex corrections and self-energy renormalizations [see Fig. 1 (a)]. Since the localized potential is assumed to be time-independent in the second stage, its propagation contributes to the transient Green's function only as an exponential factor. The propagation of the localized potential can therefore be eliminated, and the distinction between self-energy renormalizations and vertex corrections is no longer relevant [Fig. 1 (b)]. Thus the transient Green's function of the right electrode is given by the product of a dressed propagator of a tunneling electron and the contributions of loops:

$$i\mathcal{G}^R(t) \;=\; L^R(t)e^{C^R(t)}. \tag{8}$$

Note that the many-body graph in Fig.1(a) is reduced to a time-dependent one-body graph, where the time-dependence enters only as boundary conditions. Since both the self-energy renormalizations and vertex corrections become single loops, their contributions can be summed using the linked-cluster theorem [4].

B. Dressed propagators of a tunneling electron

Let us assume that the junction electrodes are initially in thermal equilibrium. The propagator of conduction electrons is given by

$$\bar{G}_0^R(t) = -\frac{\pi\rho_0^R}{\beta}\left[P\left(\frac{1}{\sinh\frac{\pi t}{\beta\hbar}}\right) + \pi\tan\theta\;\delta\left(\sinh\frac{\pi t}{\beta\hbar}\right)\right] e^{-i(\mu_R+\Delta)t/\hbar}, \tag{9}$$

where where ρ_0^R, μ_R, and P respectively represent the density of energy states at the Fermi level, the chemical potential of the right electrode, and the Cauchy principal value. $\Delta = \frac{e}{C}(Q-\frac{e}{2})$ and the delta-function term is introduced to take into account the effect of the conduction band having a finite width. The equilibrium propagator after tunneling is given by

$$\bar{G}^R(t) = -\frac{\pi\rho_o^R}{\beta}\frac{\cos^2(\theta-\delta_R)}{\cos^2\theta}\left[P\left(\frac{1}{\sinh\frac{\pi t}{\beta\hbar}}\right) + \pi\tan(\theta-\delta_R)\delta\left(\sinh\frac{\pi t}{\beta\hbar}\right)\right] e^{-i(\mu_R+\Delta)t/\hbar}. \tag{10}$$

Here the scattering of conduction electrons is characterized by their phase shift δ_R.

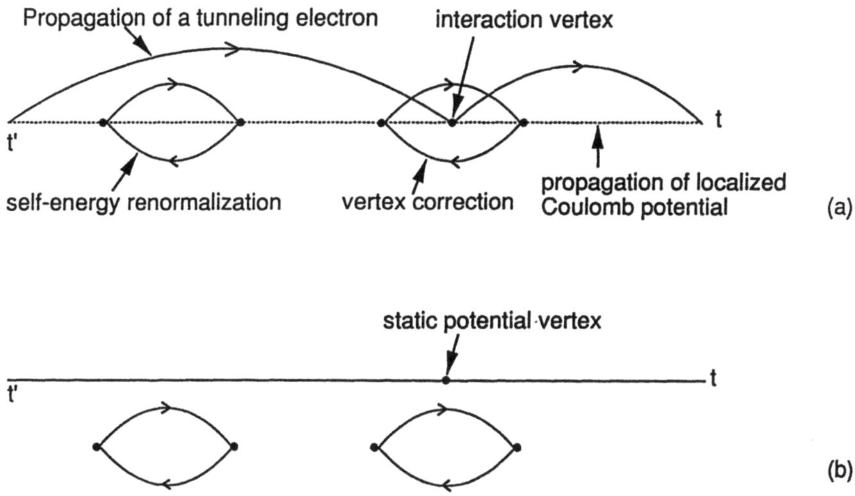

Fig.1

Disentanglement of many-body time-independent graphs to one-body time-dependent graphs: (a) Typical many-body graph represented by the Hamiltonian (1); (b) Reduced one-body graph, where time-dependence appears only as boundary conditions.

We are interested in the transient Green's function $\varphi^R_{qq'}(\tau,\tau';t_0,t)$, which describes the transient propagation from one thermal equilibrium before tunneling [described by Eq. (9)] to another after tunneling [described by (10)]. Because the scattering terms causes the fine adjustment, the transient Green's function should obey the following Dyson equation:

$$\varphi^R_{qq'}(\tau,\tau';\, t_0,t) = \bar{G}^R_{0qq'}(\tau-\tau') + i\int_{t_0}^{t} d\tau'' \sum_{pp'} \bar{G}^R_{0qp}(\tau-\tau'')V^R_{pp'}\varphi^R_{p'q'}(\tau'',\tau';t_0,t). \quad (11)$$

This equation can be solved to give [4]

$$\varphi^R(\tau,\tau';\, t_0,t) = \bar{G}^R(\tau-\tau')\left[\frac{\sinh\frac{\pi}{\beta\hbar}(t-\tau')\sinh\frac{\pi}{\beta\hbar}(\tau-t_0)}{\sinh\frac{\pi}{\beta\hbar}(t-\tau)\sinh\frac{\pi}{\beta\hbar}(\tau'-t_0)}\right]^{\frac{\delta_R}{\pi}}. \quad (12)$$

This Green's function gives the transient response to a sudden potential change due to tunneling. As time progresses, the transient Green's function tends to describe the new equilibrium state represented by Eq. (10).

The propagator of a tunneling electron and that of a hole created by it are renormalized by the static vertex correction [see Fig. 1(b)]. The dressed propagator of the tunneling electron is given by letting $\tau \to t$ and $\tau' \to t_0$ in Eq. (12):

$$L^R(t) = \lim -\varphi^R(\tau,\tau';t_0,t) = \frac{\alpha^R \frac{\pi}{\beta\hbar}\rho_0^R\hbar}{\sinh\frac{\pi}{\beta\hbar}(t-t_0)}\left[\frac{\sinh\frac{\pi}{\beta\hbar}(t-t_0)}{\sinh\frac{\pi}{\beta\hbar}\tau_{\rm cut}^R}\right]^{2\frac{\delta_R}{\pi}}, \tag{13}$$

where $\alpha^R = \cos^2(\theta-\delta_R)/\cos^2\theta$, and $\tau_{\rm cut}^R$ is introduced to avoid the spurious divergence of the algebraic factor.

C. Contribution of closed loops

The contribution of closed loops is to carry information on the vacuum-field fluctuations induced by tunneling. The fluctuation-dissipation theorem tells us that the closed loops bear a close relationship to the decay of the transient Green's function for conduction electrons. This decay is controlled by the algebraic factor, which can be decomposed into unity and a smaller factor:

$$\frac{\sinh\frac{\pi}{\beta\hbar}(t-\tau')\sinh\frac{\pi}{\beta\hbar}(\tau-t_0)}{\sinh\frac{\pi}{\beta\hbar}(t-\tau)\sinh\frac{\pi}{\beta\hbar}(\tau'-t_0)} = 1 + \frac{\sinh\frac{\pi}{\beta\hbar}(t-t_0)\sinh\frac{\pi}{\beta\hbar}(\tau-\tau')}{\sinh\frac{\pi}{\beta\hbar}(t-\tau)\sinh\frac{\pi}{\beta\hbar}(\tau'-t_0)}. \tag{14}$$

This decomposition allows us to expand φ^R near $\tau=\tau'$ as

$$\varphi^R(\tau,\tau';t_0,t) = \bar{G}_0^R(\tau-\tau') - \alpha^R\frac{\pi}{\beta\hbar}\rho_0^R\frac{\delta_R}{\pi}\left(\coth\frac{\pi}{\beta\hbar}(\tau'-t_0) + \coth\frac{\pi}{\beta\hbar}(t-\tau)\right). \tag{15}$$

The contribution of closed loops C^R satisfies

$$g\frac{\partial}{\partial g}C^R(t-t_0) = -V\int_{t_0}^{t}\varphi^R(\tau,\tau;t_0,t)d\tau, \tag{16}$$

where V is the strength of the scattering potential and g is the renormalized coupling constant. Substituting Eq. (15) into Eq. (16) yields

$$\frac{\partial}{\partial g}C^R(t-t_0) = -2\alpha^R\frac{\delta_R}{\pi}\ln\left(\frac{\sinh\frac{\pi}{\beta\hbar}(t-t_0)}{\sinh\frac{\pi}{\beta\hbar}\tau_{\rm cut}^R}\right). \tag{17}$$

Since the (relative) amplitude α^R of the propagator can be expressed in terms of δ_R and g as

$$\alpha^R = \frac{1}{\pi}\frac{d\delta_R}{dg}, \tag{18}$$

Eq. (16) can be solved to give

$$C^R(t-t_0) = -\left(\frac{\delta_R}{\pi}\right)^2\ln\left(\frac{\sinh\frac{\pi}{\beta\hbar}(t-t_0)}{\sinh\frac{\pi}{\beta\hbar}\tau_{\rm cut}^R}\right). \tag{19}$$

Thus we obtain the transient Green's functions of the right electrode,

$$
\mathcal{G}^R(t) = -i\frac{\pi}{\beta\hbar}\frac{\alpha^R\rho_0^R\hbar}{\sinh\frac{\pi}{\beta\hbar}t}\left[\frac{\sinh\frac{\pi}{\beta\hbar}t}{\sinh\frac{\pi}{\beta\hbar}\tau^R}\right]^{\frac{\delta_R}{\pi}-\left(\frac{\delta_R}{\pi}\right)^2} e^{-i(\mu^R+\Delta)t/\hbar}, \tag{20}
$$

and we get a similar result for the left electrode. Substituting these results into Eq. (7) gives the desired transient response function:

$$
\begin{aligned}
iR(t) \ &= \ 2|T|^2\left(\frac{\pi}{\beta}\right)^2\frac{\alpha^R\rho_0^R}{\sinh\frac{\pi}{\beta\hbar}\tau_{\mathrm{cut}}^R}\frac{\alpha^L\rho_0^L}{\sinh\frac{\pi}{\beta\hbar}\tau_{\mathrm{out}}^L}\left[\frac{\sinh\frac{\pi}{\beta\hbar}t}{\sinh\frac{\pi}{\beta\hbar}\tau_{\mathrm{cut}}^R}\right]^{2\frac{\delta_R}{\pi}-\left(\frac{\delta_R}{\pi}\right)^2-1} \\
&\times\left[\frac{\sinh\frac{\pi}{\beta\hbar}t}{\sinh\frac{\pi}{\beta\hbar}\tau_{\mathrm{cut}}^L}\right]^{2\frac{\delta_L}{\pi}-\left(\frac{\delta_L}{\pi}\right)^2-1} e^{i\{(\mu^R-\mu^L)+\Delta\}t/\hbar}\theta(t).
\end{aligned} \tag{21}
$$

5. Tunneling rate at finite temperatures

For junctions that have the same electrodes, the strengths of the scattering potential, V^L and V^R, are equal in magnitude but opposite in sign because conduction electrons in each electrode are scattered by mirror charges on opposite sides of the barrier. The phase shifts δ_L and δ_R are therefore equal in magnitude but opposite in sign: $\delta_R = -\delta_L \equiv \delta$. To simplify matters we assume that the electrochemical potential and the cut-off threshold for conduction electrons are the same in both electrodes, so $\mu^R = \mu^L \equiv \mu$ and $\tau_{\mathrm{cut}}^R = \tau_{\mathrm{cut}}^L \equiv \tau_{\mathrm{cut}}$. Equation (21) then reduces to

$$
R(t) \ = \ -iA(\sinh\frac{\pi}{\beta\hbar}t)^{-g-2}e^{i\Delta t/\hbar}\theta(t), \tag{22}
$$

where $A = (\pi/\beta)^2(2\alpha^L\alpha^R\hbar/e^2R_T)(\sinh(\pi\tau_{\mathrm{cut}}/\beta\hbar)^g$ and R_T is the tunnel resistance $R_T = \hbar/(e^2|T|^2\rho_0^L\rho_0^R)$. The tunneling rate is given by the imaginary part of the response function:

$$
\mathrm{Im}R(\omega) \ = \ A\frac{2g\cosh\pi y}{\frac{\pi}{\beta\hbar}\cos\frac{\pi}{2}g}B\left(iy+\frac{g}{2}+1,-iy+\frac{g}{2}+1\right), \tag{23}
$$

where $y \equiv \frac{\beta}{2\pi}(\hbar\omega+\Delta)$. The general frequency-dependent forward tunneling rate is therefore given by

$$
r(Q,\omega) = \frac{1}{\pi\hbar^2}\mathrm{Im}R(\omega) = \frac{\alpha^L\alpha^R}{e^2R_T}\frac{4g}{\beta}\ B\left(iy+\frac{g}{2}+1,-iy+\frac{g}{2}+1\right). \tag{24}
$$

This is the main result of this paper. Now let us examine two specific cases to check this result. At zero temperature, we obtain

$$
r(Q,\omega) = \frac{\alpha^R\alpha^L}{(\hbar/\tau_{\mathrm{cut}})^g}\frac{1}{\cos\frac{\pi}{2}g}\frac{1}{\Gamma(g+2)}\frac{(\hbar\omega+\Delta)^{1+g}}{e^2R_T}. \tag{25}
$$

This result is identical to the one obtained earlier [3]. Note that the contribution from electron-hole pairs appears as an anomalous power exponent. In the orthodox limit, $g = 0$, we obtain a familiar result for the forward tunneling rate:

$$r(Q,\omega) = \frac{\hbar\omega + \frac{e}{C}\left(Q - \frac{e}{2}\right)}{e^2 R_T} \frac{1}{1 - \exp\left\{-\frac{1}{k_B T}\left[\hbar\omega + \frac{e}{C}\left(Q - \frac{e}{2}\right)\right]\right\}}. \tag{26}$$

6. Possible experimental situations

The predicted effect of infrared divergence on Coulomb blockade has not yet been observed. The two main obstacles to observing it are the high mobility of the charge disturbance in normal electrodes and the electromagnetic environment, which is inevitable in single junctions. A mobile charge disturbance smears out the singularities due to electron-hole pairs [5], and the electromagnetic environment causes spontaneous charge fluctuations that mask the change in the junction charge caused by tunneling [1].

A possible configuration to circumvent these difficulties is a microcluster double junction which consists of an STM tip/vacuum barrier/microcluster junction and a microcluster/oxide barrier/substrate junction. When the STM tip is negatively biased, an electron tunnels from the tip to the microcluster. Because the traversal time for tunneling is negligible for normal metal junctions, conduction electrons in the substrate see the charging of the cluster as a sudden perturbation. When the perturbation is local and static, it causes the infrared-divergent creation of electron-hole pairs in the substrate, and the spectrum of the subsequent tunneling from the microcluster to the substrate is expected to be renormalized by this shake-up. According to the orthodox theory, the current-voltage characteristic of a voltage-biased double junction exhibits, under appropriate conditions, a Coulomb staircase with uniform current and voltage steps. The shake-up, on the other hand, would make the current steps irregular [6].

In single junctions, the electron-hole-pair shake-up and the electromagnetic field fluctuations should combine to produce conductance anomalies inside the Coulomb gap. It has been shown [7] that an asymmetric junction, whose electrodes have different electronic affinities, should act as a diode: the tunneling conductance is enhanced when the current flows in one direction, whereas it is suppressed in the other. In particular, when the tunneling resistance is much smaller than the resistance quantum and the current flows in the favorable direction, a strong tendency towards establishing phase coherence emerges, resulting in an infinite differential conductance at the zero bias voltage.

7. Two kinds of infrared divergence in single-electron tunneling

An electron tunneling suddenly changes the charge on the junction capacitor, which excites two different environments: the electromagnetic environment and the Fermi-surface environment. These environments should be regarded as complementary mechanisms of dissipation in tunneling, but they share a common feature — the *orthogonality catastrophe.*

The phenomena presented in the preceding sections result from infrared-divergent excitation of electron-hole pairs. Coulomb suppression of tunneling, in contrast, is caused by exciting the electromagnetic modes [1]. A tunneling event suddenly excites the electromagnetic modes whose excitation spectrum is determined by the impedance of the external circuit. The tunneling rate is proportional to the square of the absolute overlap integral of the wavefunctions before and after tunneling. It is roughly given by

$$\text{tunneling rate} \sim \exp\left(-N_{\text{excited modes}}\right), \tag{27}$$

where

$$N_{\text{excited modes}} = \frac{e^2/2C}{\hbar/CZ(\omega_{\text{SET}})} \tag{28}$$

is the average number of excited electromagnetic modes and $Z(\omega_{\text{SET}})$ is the environmental impedance at a typical frequency of SET operation. When the environmental impedance is infinite, so is the number of excited modes, and therefore tunneling is prohibited. A full Coulomb blockade is thus caused by infrared divergent-excitation of the electromagnetic modes [1].

Both the electromagnetic effect and the Fermi-surface effect result from the same tunneling Hamiltonian

$$H_T = \sum_{k,q} \left(T_{kq} e^{-i\phi} c_k^\dagger c_q + T_{kq}^* e^{i\phi} c_q^\dagger c_k\right). \tag{29}$$

The former can be analyzed by calculating the autocorrelation function of the electromagnetic phase without regard to the many-body final-state interactions of conduction electrons, whereas the latter can be characterized by the autocorrelation function of conduction electrons which is calculated here without regard to the electromagnetic phase. The influence of the electromagnetic environment can be evaluated using equilibrium thermodynamics of quantized harmonic oscillators by assuming that the environmental modes are in thermal equilibrium. However, the electron-hole pair creation is due to the nonequilibrium properties of conduction electrons. This sharply distinguishes the Fermi-surface environment from the electromagnetic environment.

Acknowledgments

Fruitful discussions with H. Fukuyama, F. Guinea, and S. Kurihara are gratefully acknowledged.

References

1. Yu. V. Nazarov, Sov. Phys. JETP **68**, 561 (1989); M. H. Devoret, D. Esteve, H. Grabert, G.-L. Ingold, H. Pothier, and C. Urbina, Phys. Rev. Lett. **64**, 1824 (1990); S. M. Girvin, L. I. Glazman, M. Jonson, D. R. Penn, and M. D. Stiles, Phys. Rev. Lett. **64**, 3183 (1990).
2. P. W. Anderson, Phys. Rev. Lett. **18**, 1049 (1967); Phys. Rev. **164**, 352 (1967).
3. M. Ueda and S. Kurihara, in *Macroscopic Quantum Phenomena*, edited by T. D. Clark, H. Prance, R. J. Prance, and T. P. Spiller (World Scientific, Singapore, 1991), pp. 143-146.
4. A general mathematical technique to cope with the infrared catastrophe was developed for the x-ray edge problem [P. Nozieres and C. T. De Dominicis, Phys. Rev. **178**, 1097 (1969)], and was extended to finite temperatures for the Kondo problem [G. Yuval and P. W. Anderson, Phys. Rev. B **1**, 1522 (1970)].
5. E. Müller-Hartmann, T. V. Ramakrishnan, and G. Toulouse, Phys. Rev. **B3**, 1102 (1971).
6. M. Ueda and F. Guinea, Proc. Int. Symp. on *Science and Technology of Mesoscopic Structures*, edited by S. Namba (Springer, Tokyo, 1992).
7. M. Ueda and F. Guinea, Z. Phys. B — Condensed Matter **85**, 413 (1991).

Tunneling Processes in Double Barrier Structures

Y. Isawa

Faculty of Engineering, Hiroshima University,
Higashihiroshima, Kagamiyama 724, Japan

Abstract. The electron tunneling through the double barrier separated by a quantum well is studied. We describe following four subjects. (1) Microscopic derivation of the coherent resonant current and the sequential tunneling current in terms of the non-equilibrium Green's function. (2) The effects of the phase randomizing scattering on the tunneling current. (3) The dc *V-I* characteristic of a double barrier structure. (4) Three tunneling processes resulting in the sequential tunneling current.

1. Introduction

The resonant tunneling phenomena in double barrier structures are typical mesoscopic phenomena in tunneling conduction regime. The coherence and the incoherence of the electron wave in this regime, and the nonlinear *V-I* characteristics are important not only for practical applications but also for fundamental physics.

It is known that there are two tunneling processes in double barrier structures: the coherent resonant tunneling process and the sequential tunneling process.

The former one is the tunneling process from an emitter to a collector electrode in which the phase coherence is kept between two electrodes. In 1984, Ricco and Azbel[1] gave an interesting discussion on this process analogous to the resonant transmission of light through a Fabry-Perot etalon.

The latter one was proposed by Luryi[2] in 1985, in which the transmission is regarded as two successive transitions from the emitter electrode into a resonant state in the quantum well, and then from the well into the collector electrode. Between two steps, the electron loses its phase memory.

At the present time, two processes are discussed separately and phenomenologically. The microscopic basis of two processes and the mutual relation between them are not fully understood.

In this paper, we propose a microscopic theory of tunneling processes in double barrier structures in which both processes are derived in a unified way. This theory will be useful to make clear not only the tunneling processes in double barrier structures but also the Coulomb blockade in point contacts in series.

2.1 Model and Hamiltonian

We use a tight binding model along the tunneling direction proposed by Caroli et al.[3] for one-dimensional Metal-Insulator-Metal system. It avoids the difficult matching problem at the boundaries between the barrier and the electrodes. We assume that the potential distribution is symmetric along tunneling direction in the absence of the applied bias(Fig.1) and also assume the translational invariance in lateral plane.

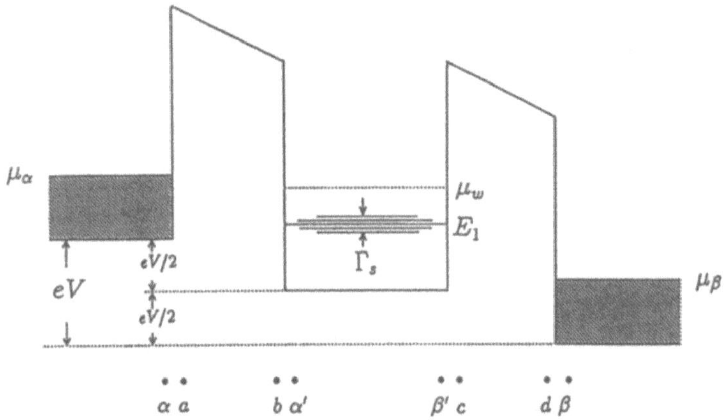

Fig.1 The shematic energy diagram of a double barrier structures in the presence of the applied voltage V. We assume that the voltage variation exists only within the potential barriers, and the structures are symmetric when $V=0$. The chemical potential on the left electrode and the right one are μ_α and μ_β, respectively. The lowest resonant energy level in the quantum well is E_1 and its width is $\Gamma=\Gamma_s+\Gamma_{es}$ where Γ_s is the phase randomizing scattering rate, and Γ_{es} the tunneling(escape) rate of the electron from the quantum well into both electrodes. The chemical potential in the well μ_w is determined from the continuity of the sequential current through the tunneling structures.

$$H=H_0+T, \tag{1}$$

where H_0 is the unperturbed Hamiltonian and the T is the transfer Hamiltoninan.

The former one is given by the sum of Hamiltonian for each of five segments in double barrier structures:

$$H_0=H_l+H_r+H_w+H_{bl}+H_{br}, \tag{2}$$

where $H_{l(r)}$ is the Hamiltonian for the left(right) electrode, H_w the well, H_{bl} the left barrier, and H_{br} the right barrier. The voltage bias is taken into account in each one of eq.(2).

The latter one, which induces the electron transfer between two neighboring disconnected segments, is given by

$$T=\sum_n\Big[T_{\alpha a}c^+_{\alpha n}c_{an}+(h.c)+T_{\alpha' b}c^+_{\alpha' n}c_{bn}+(h.c)+T_{\beta' c}c^+_{\beta' n}c_{cn}+(h.c)\\ +T_{\beta d}c^+_{\beta n}c_{dn}+(h.c)\Big], \tag{3}$$

where $c^+_{\gamma n}$ and c_{sn}($\gamma=\alpha,\alpha',\beta'$ and β, and s=a,b,c and d) are the creation and the annihilation operators at γn and sn sites respectively. The γ and s denote the location of site along the tunneling direction, while n denotes the one parallel to the layers. We assume that electrons in both electrodes propagate freely and the barriers are insulating.

The coherence of electron waves in double barrier structures is deeply affected by phase randomizing scatterings. In order to make clear the coherence and the incoherence in tunneling conduction regime, it is appropriate to investigate how the tunneling properties are modified by these scatterings. The quantum well in our model is macroscopic in lateral direction so that not only the inelastic scattering but also the elastic scattering randomize the phase of electrons.

Considering the case where an inelastic scattering exists in the quantum well, the chemical potential μ_w in the well is well defined. The value of μ_w is determined from the current continuity through the tunneling structures. Since the resonant current automatically satisfies the continuity, the voltage distribution is determined only from the continuity of the sequential current.

When the charging effect in the quantum well is considered, the energy levels of the resonant states shift upward. Thus both current components due to the sequential process and the resonant process are decreased in the low voltage region with the positive differential resistance but are increased in the region with the negative differential resistance.

In the following analysis, we neglect all complications such as band structure effects, the charge pileup in the quantum well, and the possible nonconservation of lateral momentum.

2.2 Non-equilibrium Green's Function

As a method of analysis, we use the non-equilibrium Green's function developed by Keldysh[4] since it allows to calculate the nonlinear response in non-equilibrium state. This method gives us two set of equations.

The first one is given by

$$G^{r(a)}=g^{r(a)}+g^{r(a)}TG^{r(a)}, \tag{4}$$

where $G^{r(a)}$ and $g^{r(a)}$ are the retarded(advanced) Green's functions defined as $G^r= 1/(\varepsilon-H+i\delta)$ $(G^a=(G^r)^\dagger)$ and $g^r=1/(\varepsilon-H_0+i\delta)$ $(g^a=(g^r)^\dagger)$.

The second one is

$$F=(1+G^rT)f(1+TG^a), \tag{5}$$

where the Green's functions F and f carry information on an electron distribution.

The effect of phase randomizing scattering on the Green's functions is easily introduced phenomenologically by adding an imaginary part $i\Gamma_s$ to the diagonal elements of H_0 with respect to the site in the quantum well. The Γ_s corresponds to the optical potential. The resonant level width Γ is given by the sum of the tunneling(escape) rate Γ_{es} of the electron from the quantum well into both electrodes and the phase randomizing scattering rate Γ_s, i.e. $\Gamma=\Gamma_{es}+\Gamma_s$. Not only the inelastic scattering but also the elastic scattering contribute to Γ_s as described above.

2.3 Electrical Current

Within the theoretical framework of the tight binding model and the non-equilibrium Green's function, the current at the site

αn along the tunneling direction[3] is given by,

$$J_{\alpha n}=(e/\hbar)[T_{\alpha a}F_{an,\alpha n}(t,t_+)-T_{a\alpha}F_{\alpha n,an}(t,t_+)] \quad (t_+=t+i0). \tag{6}$$

The current at the site β is obtained by replacing a and α by d and β, respectively.

By substituting the expression of $G^{r(a)}$[5] into Eq.(5) and by using eq.(6), we find that the total current at the site α perdicular to the layer, which is defined by $J_\alpha=\sum_n J_{\alpha n}$, is given by

$$J_\alpha=J_{\alpha\beta}+J_{\alpha\alpha'}+J_{\alpha\beta'}+J_{\alpha\alpha'\beta'}, \tag{7}$$

where

$$J_{\alpha\beta}=\frac{2e}{h}\sum_k\int_{\mu_\beta}^{\mu_\alpha}d\epsilon T_{\alpha\beta}(\epsilon-\epsilon_k), \tag{8}$$

$$J_{\alpha\delta'}=\frac{2e}{h}\sum_k\int_{\mu_w}^{\mu_\alpha}d\epsilon T_{\alpha\delta'}(\epsilon-\epsilon_k) \quad (\delta'=\alpha' \text{ and } \beta'), \tag{9}$$

$$J_{\alpha\alpha'\beta'}=\frac{2e}{h}\sum_k\int_{\mu_w}^{\mu_\alpha}d\epsilon T_{\alpha\alpha'\beta'}(\epsilon-\epsilon_k), \tag{10}$$

Here μ_α ,μ_β and μ_w are the chemical potentials of the left electrode, the right electrode and the quantum well,respectively. The difference of the chemical potential between two electrodes is the applied voltage V,i.e. $\mu_\alpha-\mu_\beta=eV$. The ϵ_k is the kinetic energy in lateral plane and the summation of k is taken over lateral momentum.

The coefficient $T_{\alpha\beta}$ in eq.(8) given by,

$$T_{\alpha\beta}(\epsilon-\epsilon_k)=(2\pi)^2(T_{\alpha a}T_{d\beta})^2|G^r_{ad}(\epsilon)|^2\rho_\alpha(\epsilon)\rho_\beta(\epsilon), \tag{11}$$

is the transmission coefficient from the left to the right electrode in terms of the direct tunneling process since it satisies $0\le T_{\alpha\beta}\le 1$. Thus $J_{\alpha\beta}$ corresponds to the coherent resonant tunneling current J_{res}.

Other "transmission coefficients"[5] in Eqs.(9)-(10),$T_{\alpha\alpha'}$, $T_{\alpha\beta'}$ and $T_{\alpha\alpha'\beta'}$ describe the sequential tunneling process from the well into the left electrode, but are not the true transmission coefficients, since $T_{\alpha\alpha'}$ and $T_{\alpha\beta'}$ take a value larger than 1, and $T_{\alpha\alpha'\beta'}$ is negative. However, the sum of them, $T_{seq}\equiv T_{\alpha\alpha'}+T_{\alpha\beta'}+T_{\alpha\alpha'\beta'}$ satisfies $0\le T_{seq}\le 1$, which implies that T_{seq} can be interpreted as the transmission coefficient of the sequential tunneling. Both transmission coefficients $T_{\alpha\beta}$ and T_{seq} have a single resonant peak around the lowest resonant energy, and the peak width of T_{seq} is larger than that of $T_{\alpha\beta}$ when Γ_s is small.

Thus the sequential current J_{seq} is defined as

$$J_{seq}=J_{\alpha\alpha'}+J_{\alpha\beta'}+J_{\alpha\alpha'\beta'}, \tag{12}$$

and the total current J_α at the site α is the sum of the resonant current and the sequential current:

$$J_\alpha=J_{coh}+J_{seq}. \tag{13}$$

2.4 Microscopic Tunneling Processes

Let us discuss the microscopic tunneling proceses which end at the site α. The tunneling probability due to the coherent resonant process is proportional to the square of the transmission amplitude from β to α, i.e., $|\Phi(\beta\to\alpha)|^2$, while the sequential one is proportional to the square of the sum of transmission amplitudes from α' to α and β' to α, i.e.,

$$|\Phi(\alpha'\to\alpha)+\Phi(\beta'\to\alpha)|^2. \tag{14}$$

Thus the sequential tunneling process includes an interference process, which is the origin of $J_{\alpha\alpha'\beta'}$. All the currents resulting from interference processes does not appear because we have no interference current $J_{\alpha\alpha'\beta}$ and $J_{\alpha\beta'\beta}$ in eq.(13). The condition for the interference current to appear is that the electron waves emitted from two spatially separated sources are correlated(coherent) or nonlocally coupled. The electron waves in the well are coherent, while the mutual coherence of electron waves between the well and the electrodes is broken by the potential barrier between them. This is the reason why $J_{\alpha\alpha'\beta'}$ appears and $J_{\alpha\alpha'\beta}$ and $J_{\alpha\beta'\beta}$ does not appear in eq.(13).

The interference process plays an essential role to ensure that J_{seq} vanishes when $\Gamma_S=0$. Each one of $J_{\alpha\alpha'}$, $J_{\alpha\beta'}$ and $J_{\alpha\alpha'\beta'}$ does not vanish but only the sum of them vanishes[6]. The vanishing of the sequential current at $\Gamma_S=0$ is a result of the complete destructive interference among sequential tunneling processes.

The interference effect described above will also be observed in many terminal elements of mesoscopic samples if the electron waves incident from different terminals are coherent each other, and will be observed in other waves such as light waves, acoustic waves etc.,(Fig.2).

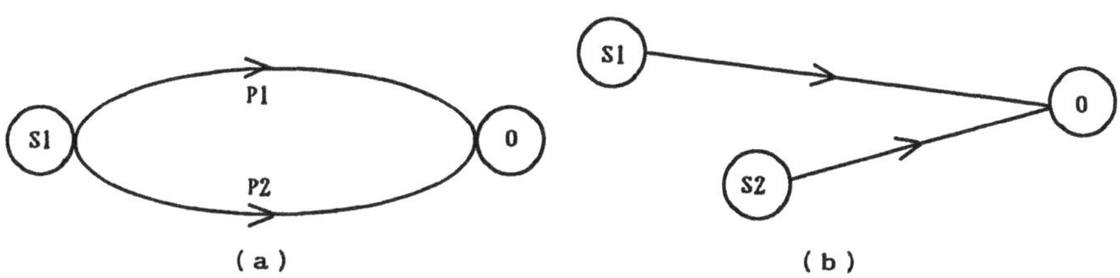

Fig.2 (a) The lowest Young's interference process. The interference effect is observed at an observation point O when waves, emitted from a single source S1, propagate along two paths P1 and P2.
(b) The interfernce process between waves emitted from two spatially separated sources, S1 and S2. The interference effect is observed at O when waves emitted from two sources are correlated or nonlocally coupled, but it is not observed when they are not correlated.

2.5 *V-I* Curve

In Fig.3, we show the tunneling current as a function of applied bias V[5]. The parameters chosen correspond to the case where the sequential current is nearly equal to the coherent resonant tunnel current. The sequential and the resonant currents behave similarly as a function of V, which differs from our previous

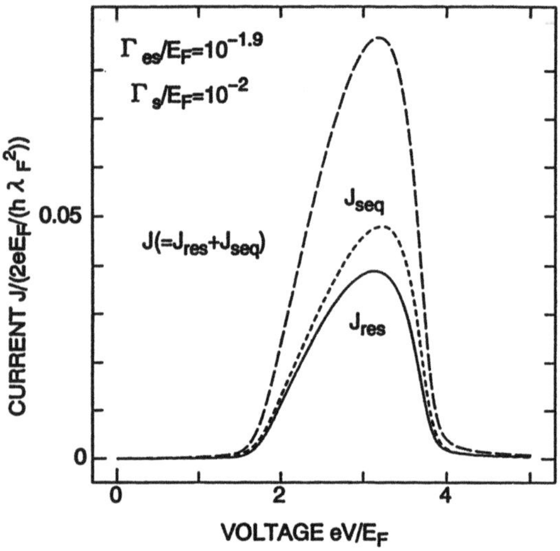

Fig.3 The coherent resonant tunneling current and the sequential one are shown as a function of the applied voltage. The sum of them, i.e., the total current, is also shown. The current is normalized by $2eE_F/(h\lambda_F^2)$, which corresponds to a current density where an electron is emitted at the mean rate $\hbar/E_F$ per an area of $\pi\lambda_F^2$.

result[5] in which we showed that the peak of the coherent resonant current is sharper than the sequential current when Γ_s is small. The difference is attributed to that of the values of parameters used: the diagonal and the off-diagonal elements in the tight binding model, and the value of Γ_s. The current-volt age curve is sensitive to these parameters.

When $\Gamma_s \leq \Gamma_{es}$, the coherent resonant current is dominant while the sequential one is dominant when $\Gamma_s \geq \Gamma_{es}$. The total current is almost independent of Γ_s/Γ_{es}[5]. This is consistent with Stone and Lee[7], and Büttiker[8].

The resonant tunneling current decreases with increasing the strength of the phase randomizing scattering in the well. The reason is that the Green's function $g_{\alpha'\beta'}$ across the quantum well included in $T_{\alpha\beta}$ becomes smaller with increasing Γ_s, i.e., the coherence length of electron in the well becomes smaller.

The chemical potential in the well μ_w is determined in such a way that the sequential current is equal at α and β. Then the current at α' and β' is identically equal to the one at α and β since we assume that the barriers are insulating.

Let us comment on the Coulomb blockade of the point contacts in series from the point of view of the coherent resonant tunneling and the sequential tunneling. We think that two effects are combined. The first one is the suppression of the sequential tunneling current due to the charging effect in the well. This is the effect of the electron charge within the well which interrupts the tunneling of another electron from the emitter into the well. The second one is the change of the resonant tunneling current induced by the energy shift of the resonant level due to the charging energy. This change might be positive or negative. Thus the first one is dominant when the sequential current is larger than the resonant current, i.e., Γ_S is large, while the second one is dominant when Γ_S is small.

2.6 Summary

We summarize our work:
(1) We show that the exact expression of the tunnleing current in double barrier structures is given by the sum of the coherent tunneling current and the sequential tunneling current: $J=J_{res}+J_{seq}$, and two tunneling currents come from two distinct tunneling processes.
(2) It is made clear that the sequential current is not caused by a single incoherent tunnel process but by a mixture of two tunnel processes and an interference process between them. The interference process plays an essential role to ensure $J_{seq}=0$ when the phase randomizing scattering vanishes.
(3) We show that the interference process is a new one observed between waves emitted from several spatially separated sources as far as the sources are nonlocally coupled or coherent.
(4) We show that the non-equilibrium Green's function developed by Keldysh is a powerful tool for the study of resonant tunneling phenomena. We think that the Green's function is applicable not only to the nonlinear properties of mesoscopic samples[9] but also to the microscopic study of the Coulomb blockade in dots. It will also make possible to extend the Landauer formula to nonlinear regime.

The author thanks Professor Y.Osaka for helpful discussions and comments. He also thanks Mr.Kanechika for numerical calculation of Fig.3. This work was partially supported by the "Project of the Mesoscopic Electronics", and by a Science Research Grant-in-Aid for "Specific Projects on the Interference Effects of Electron Waves" from the Ministry of Science Education and Culture.

2.7 References

[1] B.Ricco and M.Ya.Azbel, Phys.Rev.B29,1970(1984).
[2] S.Luryi, Appl.Phys.Lett.**47**,490(1985).
[3] C.Caroli, R.Combescot, P.Nozieres and D.Saint-James, J.Phys.C4,916(1971).
[4] L.V.Keldysh, JETP **20**,63(1965).
[5] Y.Isawa and M.Kanechika, J.Phys.Soc.Jpn,**60**,4013(1991).
[6] Y.Isawa and M.Kanechika, to be published.
[7] A.D.Stone and P.A.Lee, Phys.Rev.Lett,**54**,1196(1985).
[8] M.Büttiker, IBM J.Res.Develop,**32**,63(1988).
[9] D.E.Khmelnitskii, presented at Taniguchi Symposium(1991).

Part IV

Conductance Fluctuation

Quantum Interference in Ballistic Cavities: Conductance Fluctuations and Weak Localization

H.U. Baranger[1], *R.A. Jalabert*[2;*], *and A.D. Stone*[2]

[1]AT&T Bell Laboratories, 1D-230, Murray Hill, NJ 07974-0636, USA
[2]Applied Physics, Yale University, New Haven, CT 06520, USA
*Present Address: Service de Physique de l'Etat Condense, CEA Saclay, F-91191 Gif-sur-Yvette Cédex, France

Abstract. We summarize recent work showing that transport through ballistic cavities is strongly influenced by quantum interference effects. We find both large conductance fluctuations as a function of either magnetic field or wavevector and a substantial weak-localization correction. Both of these effects result from the interference among the complicated trapped paths in the cavity. Using semiclassical arguments, we demonstrate a connection between the classical dynamics in these systems and the quantum conductance properties: the mean area enclosed by a classical trajectory characterizes both the magnetic-field correlation length and the width in field of the weak-localization magnetoconductance. Further, we note a qualitative difference between regular and chaotic geometries in their weak-localization behavior.

1. Introduction

Quantum interference effects on transport in microstructures have been intensively investigated during the last decade [1]. The regime of diffusive transport, where the mean-free-path of the electrons is much smaller than the sample size, has received the most attention since this regime is most readily accessible experimentally. The two most studied quantum effects in this regime are weak localization— the suppression of the conductance due to interference of time-reversed paths— and conductance fluctuations— the sensitivity of the conductance to changes in magnetic field or Fermi energy. By using the high-mobility two-dimensional electron gas formed at the interface between GaAs and AlGaAs as a basis, one can now make microstructures whose size is less than the mean-free-path [1]. Transport through such structures is then ballistic. If in addition the phase coherence length is larger than the size of the structure, which is possible at low temperature, the electrons are coherent and ballistic, and the microstructures act like electron waveguides. In this paper we discuss the effect of quantum interference on conductance for structures in this coherent ballistic regime.

Extensive previous work on transport through ballistic microstructures has concentrated on either classical size effects or quantum effects associated with the mode structure in the leads. Examples of the former include the quenching of the Hall resistance [2] and the bend resistance [3, 4]; examples of modal effects include the quantization of the conductance in a constriction [5, 6] and peaks in the bend resistance [7]. For reviews of these topics see Refs. [1] and [8].

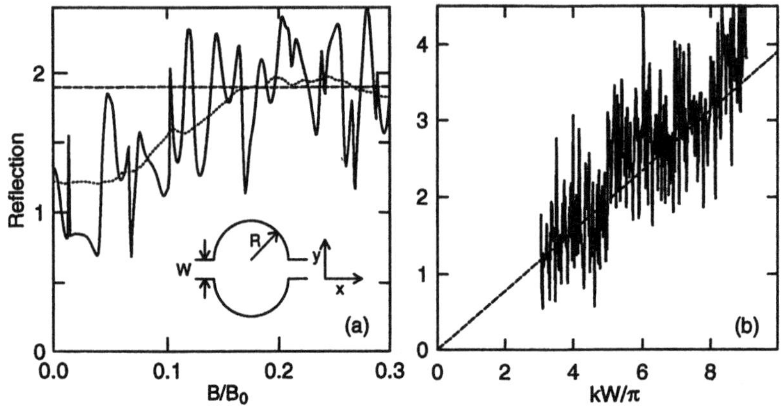

Figure 1. (a) $R(B)$ for open stadium with $R/W = 2$ (inset). The quantum result (solid, $kW/\pi = 4.5$), the smoothed $< R(B) >$ used in computing $C(\Delta B)$ (dotted), and the classical result (dashed) are shown. $B_0 = mcv/eW$. (From Ref. [9]) (b) $R(k)$ for the same structure at $B = 0$, quantum (solid) and classical (dashed).

Here we are concerned with quantum interference effects involving the whole sample and, therefore, on a longer length scale than simply the width of the incoming leads.

We start by noting the striking difference between the fluctuating quantum resistance and the smooth classical resistance for the cavity shown in Fig. 1. Following the treatment of Ref. [9], we argue that these fluctuations arise due to the complex scattering dynamics from geometric features which can trap the electrons for times much longer than the ballistic transit time. Long trapping times are essential because, e.g., transmission through a square junction which cannot trap particles shows no such behavior [10]. We will show below that the characteristic scale in magnetic field for these fluctuations in the *quantum* transport coefficients can be predicted from knowledge of the irregular *classical* scattering dynamics of the same system [9]. The results in Fig. 1 show that fluctuation effects in the classical transport coefficients are completely absent in these structures at these fields. The fluctuations of Fig. 1 are a quantum interference effect, not obtainable from the classical transmission coefficient.

The connection between the quantum resistance fluctuations and the classical dynamics suggests a connection between this mesoscopic effect and "quantum chaos" [11]. A signature of chaos in a classical dynamical system is "random" (unpredictable) behavior due to exponential sensitivity to initial conditions. It is now understood that quantum systems whose classical analogs are chaotic also exhibit "randomness" in a more subtle manner. For example, the energy levels of a Hamiltonian begin to exhibit [12] the long-range level repulsion characteristic of random matrices [13] at the classical transition to chaos. Recent studies of "quantum chaos" have turned attention to open systems [9, 14, 15, 16, 17] whose classical scattering dynamics is irregular or chaotic [18, 19, 20, 21, 22, 23] in the sense that the final scattering angle varies strongly with the incident angle on an arbitrarily fine scale. In these systems, the signature of quantum chaos is fluctuations of the quantum scattering matrix with properties described by random-matrix theory [15]. Since the transport properties of small phase-coherent conductors may be expressed completely in terms of the transmission matrix, this raises the possibility of an *experimental* probe of this random behavior characteristic of quantum chaos.

While the fluctuations in the conductance are the most striking difference between the quantum and classical conductances, we also address the possibility of a difference in average conductance between the two [8, 24]. Indeed, in the diffusive regime time-reversal symmetry at zero magnetic field leads to a suppression of the conductance due to the interference of time-reversed paths, and hence a positive magnetoconductance [1]. By using an average over wavevector, we find a weak-localization effect in the ballistic cavities that we study. We show that the form of the magnetoconductance curve is distinctly different for regular cavities as opposed to chaotic shapes. This, then, is a second connection between a quantum transport effect and the classical dynamics. A semiclassical argument shows that the weak-localization correction should have a Lorentzian form in the chaotic case, a result consistent with our numerical work. Curiously, the magnitude of the weak-localization effect that we find is smaller than in the diffusive regime while the magnitude of the conductance fluctuations is comparable.

In order to describe transport in the ballistic regime, it seems reasonable to neglect scattering from impurities completely and to view the transport properties in terms of scattering from the geometry of the junction. Thus it is natural to use the approach advocated and developed by Landauer [25] and Büttiker [26] in which one expresses the resistance of the system in terms of its transmission properties between reservoirs attached to leads. For the simple two-probe structures that we discuss here, this amounts to a proportionality between the conductance and transmission intensity: $G = (e^2/h)T$. In the quantum numerical calculations presented here, we find the transmission intensity by discretizing the problem and using the recursive Green function method as described previously [8]. Our classical conductances result from tracing ballistic trajectories between the two leads in order to evaluate the probability of transmission from one lead to another, as introduced by Beenakker and van Houten [27, 8].

2. Semiclassical Approach

The analytic tool for studying quantum interference effects in the diffusive regime is the ensemble average over short-range impurities which leads to a well-developed diagrammatic perturbation theory. In the ballistic regime, one is interested in the properties of a single specific structure and therefore this tool is not available. In order to characterize the quantum interference effects, we will introduce an average over wavevector. And in order to make progress analytically, we consider the limit of large wavevector, developing a semiclassical approximation to the transmission coefficient valid asymptotically as $N \to \infty$, where N is the number of modes in the leads, or $\hbar \to 0$ [9].

Scattering theory shows that there is a very simple relation between the transmission amplitude for a quantum scattering problem and the Green function in the absence of a magnetic field: the transmission amplitude is simply the projection of the Green function onto the transverse wave functions in the asymptotic part of the lead, $\chi_a(y)$ [28],

$$|t_{mn}| = \hbar\sqrt{v_m v_n}\left|\int dy \int dy' \chi_m(y')\chi_n(y)G^+(\mathbf{x}',\mathbf{x})\right| , \tag{1}$$

where the transmission is from mode n to mode m and v_n is the longitudinal velocity of mode n (at the Fermi energy). In fact this projection can be done at any point in the lead outside the scattering region since the population of the various transverse states will not change in the ideal lead.

In this exact expression for the transmission amplitude, we approximate the Green function by its semiclassical path-integral expression [29],

$$G(y', y, E) = \frac{2\pi}{(2\pi i\hbar)^{(3/2)}} \sum_{s(y,y')} \sqrt{D_s} \exp\left(\frac{i}{\hbar}S_s(y', y, E) - i\frac{\pi}{2}\mu_s\right) , \tag{2}$$

where S_s is the action integral along *classical* path s, at energy E, $D_s = (v\cos\theta'/m)^{-1} \mid (\partial\theta/\partial y')_y \mid$, θ and θ' are the incoming and outgoing angles, and μ is the Maslov [29] index given by the number of constant-energy conjugate points. The restriction to paths at energy E arises from a stationary-phase approximation in the time Fourier transform of the WKB propagator [29].

We evaluate t_{mn} from Eqs. (1) and (2) in an analogous manner. Assuming hard walls in the leads, the integration over the transverse (sine) wavefunctions can be performed by stationary phase approximation. This is valid in the semiclassical $\hbar \to 0$ limit, that is, for large mode index m and n. For example, the stationary-phase condition on the y-integration is $(\partial S/\partial y)_{y'} = -p_y = -\bar{m}\hbar\pi/W$ for $y = y_0$ ($\bar{m} = -m, m$); i.e. the classical trajectories that contribute are those for which the incident transverse momentum is equal to the appropriate quantized value. We obtain

$$t_{nm} = -\frac{(2\pi i\hbar)^{(1/2)}}{2W} \sum_{s(\bar{m},\bar{n})} sgn(\bar{m})sgn(\bar{n})\sqrt{\tilde{D}_s} \exp\left(\frac{i}{\hbar}\tilde{S}_s(\bar{n},\bar{m},E) - i\frac{\pi}{2}\nu_s\right). \tag{3}$$

The sum is now over trajectories between the cross sections x and x' at angles $\sin\theta = \bar{m}\pi/kW$ and $\sin\theta' = \bar{n}\pi/kW$. The action is $\tilde{S}(\bar{n},\bar{m},E) = S(y_0', y_0, E) + \hbar\pi\bar{m}y_0/W - \hbar\pi\bar{n}y_0'/W$ and the prefactor is $\tilde{D}_s = (mv\cos\theta')^{-1} \mid (\partial y/\partial\theta')_\theta) \mid$, $\nu = \mu + H(-(\partial\theta/\partial y)_{y'}) + H(-(\partial\theta'/\partial y')_\theta)$ (H is the Heaviside step function) [30].

This semiclassical expression for the transmission amplitude can now be used to study the quantum conductance properties of ballistic systems. It will be useful later to have an expression for the total transmission intensity, summed over all modes N, in which the wave-vector dependence is written explicitly. For a structure with infinite hard walls, called a "billiard", the paths do not depend on k, and we find

$$T = \frac{1}{2}\frac{\pi}{kW}\sum_{n=1}^{N}\sum_{m=1}^{N}\sum_{s}\sum_{u}(-1)^{n(s,u)}\sqrt{\tilde{A}_s\tilde{A}_u}\exp\left[ik(\tilde{L}_s-\tilde{L}_u)-i\pi(\nu_s-\nu_u)/2\right], \tag{4}$$

where $\tilde{A}_s = mv\tilde{D}_s/W$, $k\tilde{L}_s = \tilde{S}_s/\hbar$, and $n(s,u) = 0$ if the number of positive angles is even while $n(s,u) = 1$ otherwise.

3. Conductance Fluctuations

Since the fluctuation of the quantum conductance is the most notable deviation from the smooth classical conductance (see Fig. 1), we start by analyzing these fluctuations. Though the details of each bump or wiggle in the quantum trace are specific to each geometry, we look for general features of the fluctuations by considering their statistical properties, averaging over either magnetic field or Fermi wavevector. Closely following Ref. [9], we outline a semiclassical derivation of the magnetic field correlation function, $C(\Delta B) \equiv \langle \delta g(B+\Delta B)\delta g(B)\rangle$, since B is the most convenient experimental control parameter.

The magnetic field correlation function $C(\Delta B)$ can be calculated using Eq. (3). We use arguments similar to those used in Ref. [15] for the energy correlation function, $C(\Delta E)$, which in fact characterizes the fluctuations in Fig. 1(b). First, in the multiple sum over paths, only the interference of a path s at field B with the same path at $B+\Delta B$ is retained. Second, the difference of the action is expanded to find the relative phase accumulated:

$$\frac{1}{\hbar}\left(\frac{\partial\tilde{S}_s}{\partial B}\right)\Delta B = \frac{e}{\hbar c}\int \Delta\vec{A}\cdot d\vec{l} \equiv \frac{\Theta_s \Delta B}{\phi_0} \tag{5}$$

where $|\nabla\times\Delta\vec{A}| = \Delta B$ and $\phi_0 = hc/e$. Θ can be positive or negative (see Fig. 3), and we find that for weak fields (except near the origin) the distribution of Θ over classical trajectories, $N(\Theta)$, has an approximately symmetric exponential form, $N(\Theta) \propto \exp(-\alpha_{cl}\mid\Theta\mid)$. We emphasize that the entire analysis is only valid for weak fields, where the cyclotron radius is much larger than the device dimensions. For closed orbits, Θ is the area enclosed (times 2π). For typical open orbits this is true to a good approximation, and the inverse decay constant α_{cl}^{-1} will then give the typical area enclosed by trajectories traversing the structure(although Θ is not itself gauge-invariant, we have checked that α_{cl} is independent of gauge [24]). Finally, making the approximation $N(\Theta) \propto \exp(-\alpha_{cl}\mid\Theta\mid)$ for *all* Θ we can replace the sum over trajectories by an integral over Θ, giving

$$C(\Delta B) = \frac{C(0)}{[1+(\Delta B/\alpha_{cl}\phi_0)^2]^2}. \tag{6}$$

We have calculated by classical simulations α_{cl} for the structures shown, and hence can compare the predictions of Eq. (6) to exact numerical results

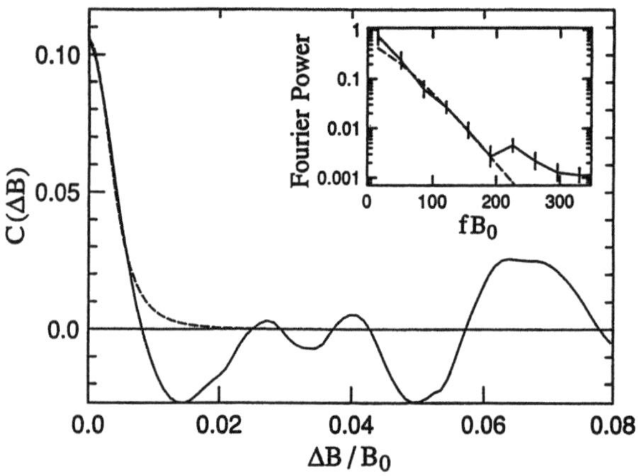

Figure 2. Magnetic field correlation function (for $W = 1000Å$, half-width $\Delta B_c = 40G$) for data in Fig. 1 (a). Dashed line is the semiclassical prediction of Eq. (6). Inset: smoothed power spectrum of $R(B)$. Error bars indicate rms variation of raw data. The dashed line is the best fit to the Fourier transform of Eq. (6) in the interval [50,200]. (From Ref. [9])

for the correlation function with no free parameters. We consider systems with 4-15 modes at E_F. It is important to note that the semiclassical approximation leading to Eqs. (2) or (3) is *not* well-justified in the few-mode limit[17] typical of experiments, since the difference between the actions of classical trajectories with the same end-points is not always much greater than $\hbar$. Thus our numerical results provide a crucial test of the accuracy of the semiclassical approximation in such systems.

First we show (Fig. 2) a typical correlation function $C(\Delta B)$ and the semiclassical prediction of Eq. (6) [32]. The agreement is excellent except in the tail which corresponds to the non-universal short trajectory behavior. Because the behavior in the tail affects the half-width, it is necessary to calculate δT from a smoothed $< T(B) >$ curve (Fig. 1); which introduces some arbitrariness in the determination of α_{qm}. To eliminate this freedom, we instead extract α_{qm} from fitting the Fourier power spectrum of the data to the Fourier transform of Eq. (6) (inset to Fig. 2); the value thus extracted is independent of the non-universal "low-frequency" behavior.

Fig. 3 shows that α_{qm} is indeed given by the classical quantity α_{cl} to high accuracy while it is varied over roughly two orders of magnitude by changing R/W in two and four-probe structures. *Thus it is possible to predict quantitatively measurable properties of these ballistic quantum conductors from a knowledge of the chaotic classical scattering dynamics.*

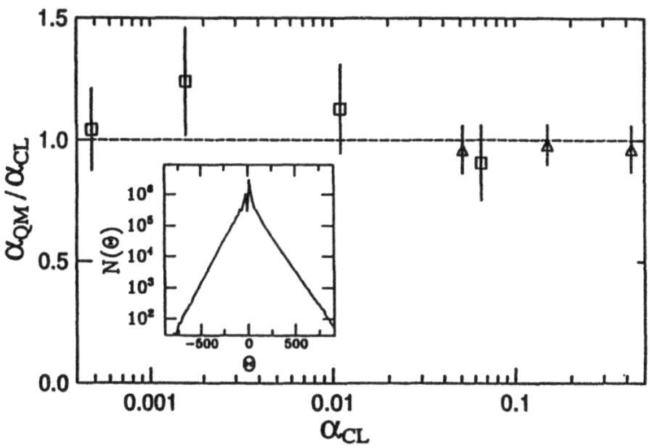

Figure 3. Ratio of magnetic field correlation length (obtained as in inset of Fig. 2) to α_{cl}, the exponent of the distribution of effective areas, as a function of α_{cl} for both four-disk structure (triangles) with $R/W = 1,2,4$, and open stadium (squares) with $R/W = 1,2,4,6$. α_{cl} (in units W^{-2}) is obtained from the classical $N(\Theta)$ curve. Inset: $N(\Theta)$ for structure of Fig. 1. The correlation lengths of the quantum fluctuations agree with the semiclassical prediction over two decades. (From Ref. [9])

4. Weak Localization: Numerics

While the most striking difference between the classical and quantum conductance (see Fig. 1) is the conductance fluctuations discussed in the last section, deviations of the average quantum conductance from the classical are also possible. The most well-known example is weak-localization in the diffusive regime: the interference of time reversed paths causes a suppression of the ensemble averaged conductance at $B = 0$ [1]. The magnitude of this suppression is independent of either the mean-free-path or the Fermi wavevector in the quasi-one-dimensional limit. Experimentally, the ensemble average can be carried out by considering a system larger than the phase-coherence length, thus producing effectively an average over independent coherent regions. In fact, because the experimental constraints for observing the average effect are much less severe than for the fluctuations, weak-localization was studied experimentally well before conductance fluctuations.

In contrast, in the ballistic regime one is interested in the behavior of a *particular* sample, so one is led to consider a single coherent region. In this case, the weak-localization is obscured by the conductance fluctuations, since in any particular sample the phase of the fluctuations at $B = 0$ is random and hence the low-field magnetoconductance has no particular sign. In order to investigate average conductance properties, one needs to average over some other quantity. In the context of ballistic cavities, it seems natural to use an average over wavevector [33] to reveal the magnetoconductance behavior. Referring to Fig. 1, we see that the general trend of the quantum conductance

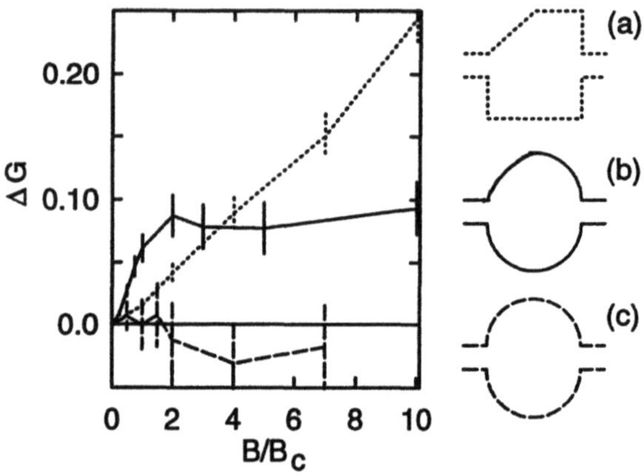

Figure 4. The weak-localization correction to the conductance as a function of magnetic field for the three structures shown. ΔG is obtained by averaging $G(B) - G(0)$ over wavevector; the statistical error is indicated. For a symmetric structure, the magnetoconductance is zero. For the asymmetric structures, note the difference between the regular case (dotted, linear magnetoconductance) and the chaotic case (solid, saturated magnetoconductance).

is a linear increase with wavevector as in the classical conductance. An average deviation from the classical result would correspond to a deviation in either the slope or the intercept which could then be studied as a function of magnetic field.

We have carried out a numerical study of the field-sensitive part of the conductance. First, we fit the trace $G(k)$ to a line for each value of B. The slope of this line is independent of B for small B and equal to the classical slope, to within the statistical error. However, the intercept of each line differs from the classical value (zero) and changes as a function of B. This corresponds to the familiar weak-localization effect: a positive magnetoconductance whose value is constant independent of k. Second, in order to extract the change in intercept more carefully, we compute $\Delta G(B) \equiv \langle G(B,k) - G(0,k)\rangle_k$. The results in Fig. 4 show that weak localization does occur in a variety of strictly ballistic cavities.

There are a number of intriguing differences between the three structures studied in Fig. 4. First, structure (a) shows a linear magnetoconductance while the magnetoconductance of structure (b) has a Lorentzian form. The important difference between these two structures is that the classical dynamics of structure (a) is regular— the straight sides lead to conservation of both momentum components— while structure (b) is classically chaotic. *Thus, it seems that the qualitative difference in the character of the classical dynamics— regular versus chaotic— is reflected in the quantum interference properties.*

Second, the magnetoconductance for the chaotic structure (c) is zero (within statistical error). Structure (c) has several mirror symmetries which are absent in the purposely asymmetrized structure (b). The mirror symmetries in structure (c) leads to the interference of parity-related trajectories. Our result suggests that the interference of both the parity related and time-reversed paths leads to a net magnetoconductance of zero.

Finally, the saturation magnitude of the magnetoconductance in the asymmetrized chaotic case is substantially smaller than for a diffusive quasi-one-dimensional strip for which $| \langle \Delta G \rangle | = 1/3$. This is surprising in view of the fact that the magnitude of the conductance fluctuations is comparable to the quasi-one-dimensional value [24, 31]. Because we have not calculated the weak-localization correction in a diffusive cavity at this time, we cannot say whether this quantitative difference is mainly due to the ballistic nature of the transport or to the cavity geometry.

5. Weak Localization: Semiclassics

To gain insight into the numerical results on weak-localization, we turn again to the semiclassical approach used in analyzing the conductance fluctuations. The numerical results for $T(k)$ (Fig. 1) suggest that an expansion of the average transmission in terms of $1/k$ or the inverse number of modes, N^{-1}, is reasonable:

$$\overline{T}(k) = \mathcal{T}\frac{kW}{\pi} + b + \mathcal{O}\left(\frac{\pi}{kW}\right). \tag{7}$$

Our goal, then, is to find the coefficients in this expansion.

We start by finding the leading order contribution to the average conductance, namely $\mathcal{T}$ in Eq. 7, summarizing results from Ref. [8]. The coefficient of the linear term in $\overline{T}(k)$ can be found by taking the average

$$\mathcal{T} = \left\langle \frac{T(k)}{kW/\pi} \right\rangle \equiv \lim_{q\to\infty} (1/q) \int_0^q dk \frac{T(k)}{kW/\pi}. \tag{8}$$

Inserting the semiclassical approximation for $T(k)$ (Eq. (4)) into this expression, we note two simplifications. First, to find the leading order behavior in kW/π, we can approximate the sums over modes by an integral over angle, $(\pi/kW)\sum_n \to \int d\theta \cos\theta$. Similarly, the sum over modes m introduces an integral over angles θ'. Second, note that the only remaining k-dependence is in the exponent. The average of this oscillating term will be zero unless the two phases are exactly equal, $\tilde{L}_s = \tilde{L}_u$. Because both paths s and u satisfy the same boundary condition, namely incoming at angle θ and outgoing at angle θ', they are chosen from a discrete set of paths. In the absence of symmetries, $\tilde{L}_s = \tilde{L}_u$ requires the paths to be identical, $s = u$. Using these two simplifications, we find that

$$\mathcal{T} = \frac{1}{2}\frac{kW}{\pi} \int_{-\pi/2}^{\pi/2} d\theta \cos\theta \int_{-\pi/2}^{\pi/2} d\theta' \cos\theta' \sum_{s(\theta,\theta')} \tilde{A}_s. \tag{9}$$

Using the definition of $\tilde{A}_s$, one can replace the integral over outgoing angle by an integral over incoming position,

$$\mathcal{T} = \int_{-\pi/2}^{\pi/2} d\theta \frac{1}{2}\cos\theta \int_0^W \frac{dy}{W} f(y,\theta) \quad , \tag{10}$$

where $f(y,\theta) = 1$ if the path starting at (y,θ) is transmitted and zero otherwise. This is exactly the prescription for the classical probability of transmission and shows that the leading order term in the average quantum conductance is the classical conductance. An analogous expression holds for the reflection coefficient and its average, $\overline{R}(k)$, in which the paths contributing are those which reflect.

Now we turn to corrections to the classical conductance which come in at the next order in kW/π, namely those which are independent of k and thus contribute to the coefficient b in Eq. 7. From results in the diffusive regime, one expects that coherent backscattering— the enhancement of the backscattered intensity by interference of time-reversed paths— is an essential ingredient of weak-localization [1]. We therefore consider the diagonal part of the reflection coefficient, $R_D(B) \equiv \sum_{n=1}^{N} R_{nn}$. Since R_D is approximately $1/N$ of the total reflection, its contribution to $\overline{R}(k)$ is independent of kW/π to leading order. The contribution of R_D to $\mathcal{R}$ is, of course, just given by pairing of the paths, $s = u$. We now show that there is an additional contribution to R_D if there is an exact symmetry in the problem. This additional contribution, which we denote δR_D, is also independent of kW/π, but because it exists only for R_D and not for the off-diagonal parts of R_{nm}, δR_D contributes to the constant term b in $\overline{R}(k)$ and not $\mathcal{R}$ (see Eq. 7). δR_D thus contributes to the weak-localization correction.

Using the semiclassical reflection amplitudes Eq. (3) in δR_D, we explicitly exclude the possibility of the two paths being the same, $s \neq u$, since this contributes to the classical conductance. We thus find that δR_D is given by Eq. (4) above with $m = n$ and $s \neq u$. Now the same two simplifications are possible that were used in the case of the leading order behavior. First, convert the sum over modes to an integral over angle. Then, note that the only k-dependence is in the exponent, so that one keeps only those paths for which $\tilde{L}_s = \tilde{L}_u$ exactly and $s \neq u$. In the absence of any exact symmetries, there are no such pairs of paths. However, at $B = 0$ the system is time-reversal invariant, and one can find such a pair by taking path u to be path s time-reversed. It is crucial in obtaining this pair that we are considering the diagonal part of the reflection coefficient since it is only for the diagonal part that a path and its time-reversed partner satisfy the same boundary conditions on the angles. For a general path injected at θ and exiting at θ', the time-reversed partner enters at angle θ' which is only equal to $\pm\theta$ for the diagonal terms. While these time-reversed pairs of paths also exist in the calculation of $\mathcal{R}$, they give no contribution to $\mathcal{R}$ [8]: because they exist only for the diagonal reflection coefficients, they are higher order in N^{-1}.

At this point one can include the influence of the magnetic field on the interference between the time-reversed paths using the same technique as for the magnetic field correlation function above. From Eq. (5), the additional phase introduced by the magnetic field is $(\tilde{S}_s - \tilde{S}_u)/\hbar = 2\Theta_s B/\phi_0$. Because this additional phase is independent of wavevector, the averaging argument in the last paragraph is unchanged. Using the results of the last two paragraphs in the expression for the correction to the diagonal reflection coefficient, we arrive at the simple and general expression

$$\langle \delta R_D(B) \rangle = \frac{1}{2} \int_{-\pi/2}^{\pi/2} d\theta \cos\theta \left(\sum_{s(\theta,\theta)} + \sum_{s(\theta,-\theta)} \right) \tilde{A}_s \exp\left(i 2\Theta_s B/\phi_0\right). \tag{11}$$

In the case of a chaotic system, some further analytic progress can be made through certain "uniformity" assumptions [34]. First, it is plausible that in a chaotic system no preference is shown to scattering through any particular angle. If one takes a distribution of incoming trajectories, the probability of coming out at a particular angle is uniform. To be precise, $\sin\theta$ is distributed uniformly in order to preserve area in phase space. Thus we can replace the sum over the backscattered trajectories in Eq. (11) by an average over all outgoing angles

$$\langle \delta R_D(B=0) \rangle = \frac{1}{2} \int_{-\pi/2}^{\pi/2} d\theta \cos\theta \int_{-1}^{1} d(\sin\theta') \sum_{s(\theta,\theta')} \tilde{A}_s = \mathcal{R}. \tag{12}$$

The resulting expression is the same as that for the slope of the average reflection in Eq. 9. Recall that the diagonal reflection contributes approximately a fraction π/kW of the total average reflection. Thus the diagonal reflection is enhanced at $B=0$ by exactly the amount which it contributes to the classical reflection. This corresponds to a "diagonal enhancement factor" of exactly two [17]. Second, as noted in our treatment of the correlation function, the distribution of Θ_s takes a very simple form in a chaotic system: $N(\Theta) \propto \exp(-\alpha_{cl} \mid \Theta \mid)$. Using this form for all Θ and replacing the sum over trajectories by an integral over the areas Θ, we find

$$\langle \delta R_D(B) \rangle = \frac{\mathcal{R}}{1 + (2B/\alpha_{cl}\phi_0)^2}. \tag{13}$$

Thus the enhancement of the diagonal part of the reflection coefficient is Lorentzian with a characteristic field scale which is half that for the magnetic field correlation function and a magnitude equal to its contribution to the classical reflection. The form and field scale of this result are in good agreement with the numerical results in Fig. 4.

Although the presence of a coherent backscattering peak is closely connected to a weak-localization correction, experience in the diffusive regime indicates that these two are not equivalent. In particular, the quantitative theory of weak-localization requires corrections to the off-diagonal reflection coefficients. For instance, the random matrix theory of the diffusive regime yields [35]

$$\langle R_{nm} \rangle = (1 + \delta_{nm}) \frac{\langle R \rangle}{N(N+1)} \tag{14}$$

for the average of the reflection intensity between two modes where the average here is over disorder configurations. The coherent backscattering peak is the δ-function term. If one uses the *classical* result for $\langle R \rangle$ in Eq. 14 one gets $| \langle \Delta R \rangle | = 1$ rather than the correct weak-localization value of 1/3. In the random matrix theory, and other theories of the diffusive regime, the average reflection is not the classical value but rather includes corrections to $\langle R_{nm} \rangle$ with $n \neq m$ such that the total correction is 1/3.

Turning back to the ballistic cavity case of concern here, we see that the diagonal enhancement derived from our semiclassical technique is of the same form as that in the random matrix theory. Unfortunately, the technique used here yields zero for the off-diagonal corrections through an argument very similar to that used for δR_D above, as well as zero correction to all transmission coefficients [24]. We find, then, an incorrect value for the weak-localization correction, $\langle R(B=0,k) - \mathcal{R}kW/\pi \rangle = 1$, showing that the theory outlined here is incomplete. The ingredients for completing the theory— whether it be the inclusion of non-classical paths or a more sophisticated treatment of the off-diagonal terms— are not known at this time and remain the subject of ongoing study.

6. Conclusions

In this paper we have summarized our recent work [8, 9, 24] concerning quantum interference in ballistic microstructures. We find that quantum interference affects the conductance of simple ballistic cavities in ways which are similar in many respects to diffusive systems. In particular, we find both large conductance fluctuations and a substantial weak-localization correction in these systems.

Further, we have demonstrated a connection between the classical dynamics in these systems and the quantum conductance properties. First, we argued semiclassically that a property of the classical dynamics characteristic of chaotic systems, namely the mean area enclosed by a trajectory, characterizes both the magnetic-field correlation length and the width in field of the weak-localization magnetoconductance. Second, we noted the qualitative difference in the weak-localization behavior of chaotic and regular structures. These examples show that some aspects of the quantum conductance properties can be characterized by the classical dynamics. It remains for future work to further elucidate the connections between the various properties of the quantum conductance and the classical dynamics.

Acknowledgements

We thank D. P. DiVincenzo, M. Gutzwiller, R. Jensen, U. Smilansky, and B. van Wees for many helpful discussions. Research at Yale University was supported in part by NSF Grant No. DMR-8658135, an IBM postdoctoral fellowship (R.A.J.), and a Sloan Foundation fellowship (A.D.S.).

References

[1] For reviews see C. W. J. Beenakker and H. van Houten in *Solid State Physics*, Vol. 44, edited by H. Ehrenreich and D. Turnbull (Academic Press, New York, 1991) pp. 1-228 and B. L. Altshuler, P. A. Lee, and R. A. Webb, eds., *Mesoscopic Phenomena in Solids* (North-Holland, New York, 1991).

[2] M. L. Roukes, A. Scherer, S. J. Allen, Jr., H. G. Craighead, R. M. Ruthen, E. D. Beebe, and J. P. Harbison, Phys. Rev. Lett. **59**, 3011 (1987).

[3] G. Timp, H. U. Baranger, P. deVegvar, J. E. Cunningham, R. E. Howard, R. Behringer, and P. M. Mankiewich, Phys. Rev. Lett. **60**, 2081 (1988).

[4] Y. Takagaki, K. Gamo, S. Namba, S. Ishida, S. Takaoka, K. Murase, K. Ishibashi, and Y. Aoyagi, Solid State Commun. **68**, 1051 (1988).

[5] B. J. van Wees, H. van Houten, C. W. J. Beenakker, J. G. Williamson, L. P. Kouwenhoven, D. van der Marel, and C. T. Foxon, Phys. Rev. Lett. **60**, 848 (1988).

[6] D. A. Wharam, T. J. Thornton, R. Newbury, M. Pepper, H. Ajmed, J. E. F. Frost, D. G. Hasko, D. C. Peacock, D. A. Ritchie, and G. A. C. Jones, J. Phys. C **21**, L209 (1988).

[7] R. Behringer, G. Timp, H. U. Baranger, and J. E. Cunningham, Phys. Rev. Lett. **66**, 930 (1991).

[8] H. U. Baranger, D. P. DiVincenzo, R. A. Jalabert, and A. D. Stone, Phys. Rev. B **44**, 10637 (1991).

[9] R. A. Jalabert, H. U. Baranger and A. D. Stone, Phys. Rev. Lett. **65**, 2442 (1990).

[10] H. U. Baranger and A. D. Stone, Phys. Rev. Lett. **63**, 414 (1989).

[11] For a recent review of issues in "quantum chaos" see M. Gutzwiller, *Chaos in Classical and Quantum Mechanics* (Springer Verlag, New York, 1991).

[12] O. Bohigas, M.-J. Giannoni, and C. Schmit, Phys. Rev. Lett. **52**, 1 (1984); M. V. Berry, Proc. Roy. Soc. Lond. A **400**, 229 (1985).

[13] T. A. Brody, J. Flores, J. B. French, P. A. Mello, A. Pandey and S. S. M. Wong, Rev. Mod. Phys. 53, 385 (1981) and refs. therein.

[14] M. C. Gutzwiller, Physica D **7**, 341 (1983).

[15] R. Blümel and U. Smilansky, Phys. Rev. Lett. **60**, 477 (1988); **64**, 241 (1990); Physica D **36**, 111 (1989).

[16] E. Doron, U. Smilansky, and A. Frenkel, Phys. Rev. Lett. **65**, 3072 (1990); Physica D **50**, 367 (1991).

[17] C. H. Lewenkopf and H. A. Weidenmüller, Annals of Phys. **212**, 53 (1991) .

[18] C. Jung and H.-J. Scholz, J. Phys. A **20**, 3007 (1987).

[19] B. Eckhardt, J. Phys. A **20**, 5971 (1987).

[20] T. Tél, J. Phys. A. **22**, L691 (1989).

[21] P. Gaspard and S. A. Rice, J. Chem. Phys. **90**, 2225 (1989);**90**, 2242 (1989); **90**, 2255 (1989).

[22] S. Bleher, E. Ott, and C. Grebogi, Phys. Rev. Lett. **63**, 919 (1989).

[23] M. L. Roukes and O. L. Alerhand, Phys. Rev. Lett. **65**, 1651 (1990).

[24] H. U. Baranger, R. A. Jalabert, and A. D. Stone, unpublished.

[25] R. Landauer, IBM J. Res. Dev. **1**, 233 (1957); Z. Phys. B **68**, 217 (1987).

[26] M. Büttiker, Phys. Rev. Lett. **57**, 1761 (1986).

[27] C. W. J. Beenakker and H. van Houten, Phys. Rev. Lett. **63**, 1857 (1989).

[28] H. U. Baranger and A. D. Stone, Phys. Rev. B **40**, 8169 (1989) and references therein.

[29] M. C. Gutzwiller, Ref. [11], pp. 184-190, 283-287.

[30] The direct paths are not properly treated in Eq. (3) since $\tilde{D}_s$ becomes infinite in the second stationary phase integration; a correct treatment will be given in Ref. [24].

[31] R. A. Jalabert, Bull. Am. Phys. Soc. **36**, 743 (1991).

[32] See related work by R. B. S. Oakeshott and A. MacKinnon, Superlat. and Microstruc. **11**, 145 (1992).

[33] J. H. Hannay, in *Chaotic Behavior in Quantum Systems: Theory and Applications*, edited by G. Casati (Plenum, New York, 1983) p. 141.

[34] For the use of uniformity assumptions in the case of closed billiards see [33] and J. H. Hannay and A. M. Ozorio De Almeida, J. Phys. A. **17**, 3429 (1984).

[35] P. A. Mello and A. D. Stone, Phys. Rev. B **44**, 3559 (1991), see Eq. (4.14).

Numerical and Analytical Studies of Quantum Transport in Quantum Wires

H. Tamura and T. Ando

Institute for Solid State Physics, University of Tokyo,
7-22-1 Roppongi, Minato-ku, Tokyo 106, Japan

Transport in quantum wires, such as conductance, conductance fluctuation and localization are studied both numerically and analytically. Numerical calculation by a scattering matrix formalism with Landauer's conductance formula shows that the conductance fluctuation is not universal for a small number of occupied subbands, but when many subbands are occupied, there appears the length region (called the universal region) longer than the mean free path and shorter than the localization length where the amplitude of fluctuation is almost independent of the length. Crossover from one to two dimensions occurs when the broadening of one-dimensional subbands exceeds their separations. In the presence of a weak magnetic field the universal region becomes wider and the fluctuations are reduced. Analytical calculation of approximate distribution function of conductance shows that the amplitude of fluctuation in the universal region is exactly the same as those in one-dimensional metallic wires.

1. Introduction

Many works on electrical transport in mesoscopic systems have revealed new quantum-mechanical effects at low temperatures [1]. One of the most interesting effects is the conductance fluctuation which was first observed in magnetoresistance of a small normal-metal conductor at low temperatures and is now understood to be a non-averaging, sample-dependent fluctuation in mesoscopic systems. Diagrammatic perturbation calculations have shown that the amplitude of fluctuation is universal (i.e. independent of sample size, degree of disorder) at zero temperature, when the sample dimensions are always much larger than the mean free path and the Fermi wavelength but are much smaller than the localization length ("Universal Conductance Fluctuations") [2,3].

Recent developments in fabrication technology of semiconductor structures enable us to obtain quantum wires, whose width is much smaller than the mean free path and is larger but comparable to the Fermi wavelength, so that the electron motion is no longer diffusive in the lateral direction. In addition, the localization effect can be essential when only several subbands are occupied. Then interesting questions are whether conductance fluctuations are also universal in a quantum wire, and if universal, whether the amplitude of fluctuation is the same as that in a metallic wire. In the following sections of this paper, we will try to answer these questions.

2. Numerical Study

In this section a numerical study is reviewed [4]. We consider a two-dimensional system of non-interacting electrons confined within the width W in the lateral (y) direction in the presence of a perpendicular magnetic field $\mathbf{B}$. The disordered region, in which impurities with potential of δ-function type are distributed randomly, has length L in the x direction and is connected on both ends to long perfect leads which themselves are connected to reservoirs. We solved a scattering problem for this many-impurity system and obtain the scattering matrix by successive composition of the scattering matrix for a single impurity and calculated the conductance between two reservoirs with the two-terminal and multi channel version [5-9] of Landauer's formula [10]. This technique has been used by many authors [11-14].

In the following calculation, three dimensionless parameters are relevant, $W/\lambda_F, l/\lambda_F = E_F\tau/\pi\hbar$ and $\tilde{\gamma} \equiv 2m\gamma/\pi\hbar^2$ (λ_F, E_F: the Fermi wavelength and energy), where τ is the relaxation time in 2D systems given by $\tau^{-1} = n_i|\gamma|^2 m/\hbar^3$ (n_i: the impurity concentration; γ: the strength of an impurity; m: the effective mass of an electron) and is related to the mean free path via $l = v_F\tau$ (v_F: the Fermi velocity). The parameter $2W/\lambda_F$ represents the number of occupied subbands at the Fermi energy. The parameter γ represents the importance of the higher Born scatterings from a single impurity relative to the first-order Born effect. We choose $\tilde{\gamma} = 0.2$, small enough that the higher-order Born scatterings can be neglected as long as the Fermi level lies well between two adjacent subband offsets. And evanescent waves also have little influence on the calculation, and in fact complete neglect of evanescent waves causes little effect on the calculated results.

2.1. Conductance Fluctuations

In Fig. 1 we show the length dependence of the averaged conductance in the absence of a magnetic field. The narrowest wire has 5 occupied subbands and the widest 20 below the Fermi energy. We have assumed $l/\lambda_F = 51.25$, for which the broadening is always smaller than the smallest subband-separation. The Fermi level lies in the middle of the two subband bottoms E_N and E_{N+1}. We have used the two different ways of averaging, the arithmetic average $\langle G\rangle$ and the geometric average $\exp[\langle \ln G\rangle]$. With the increase of the sample length, the conductance decreases starting from the quantized value $G = 2Ne^2/h$. As the system becomes much longer, the conductance begins to decrease exponentially. The localization length ξ can be estimated by fitting to $\xi = -(\partial\langle \ln G\rangle/\partial L)^{-1}$ for the region $25 < L/l < 40$. We have $\xi/l = 4.9\pm0.2$ for $N = 5$, 10.0 ± 0.3 for $N = 10$, and 19.6 ± 1.0 for $N = 20$, which are denoted in the figure by the vertical arrows. The calculation shows that ξ increases in proportion to N ($\xi \simeq Nl$ within the numerical accuracy).

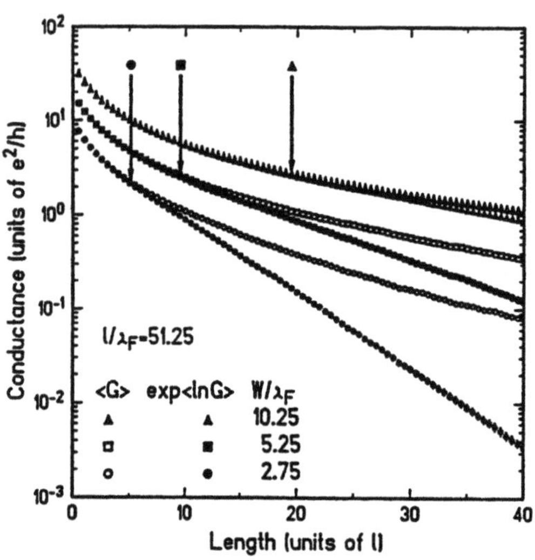

Fig. 1 Conductance as a function of the length L for different wire widths. The estimated localization lengths are indicated by the vertical arrows.

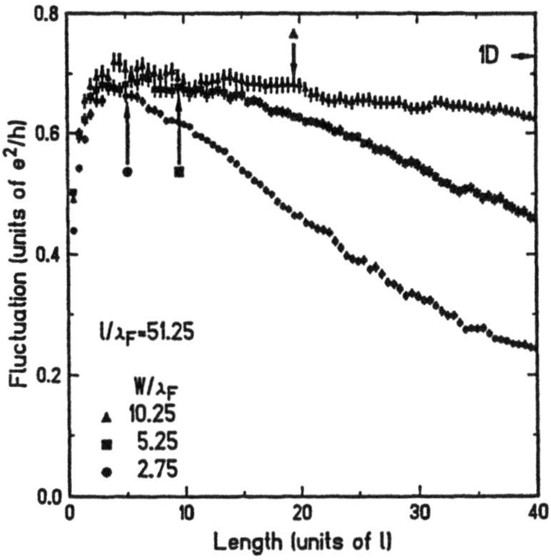

Fig. 2 Conductance fluctuations as a function of the length L for systems corresponding to those shown in Fig. 1. The universal value obtained by the perturbation for quasi-one-dimensional wires is indicated by a horizontal arrow. The localization lengths are indicated by the vertical arrows.

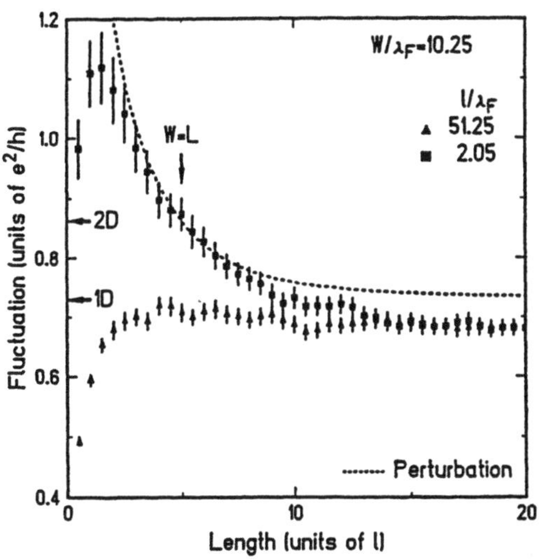

Fig. 3 Crossover of conductance fluctuations from 1D to 2D when the number of the occupied subbands is 20. The solid squares correspond to a wire where the broadening is larger than the level separation and the subband structure is smeared out. The triangles are the same data as those in Fig. 1. The vertical arrow indicates the length corresponding to the 2D square. The universal values obtained by the perturbation for quasi-one-dimensional wires and 2D squares are indicated by the horizontal arrows.

Figure 2 shows the corresponding result for the conductance fluctuation $\delta G \equiv \langle (G - \langle G \rangle)^2 \rangle^{1/2}$. With increasing length, the fluctuation first increases, takes a maximum value, and then begins to decrease. The first increase in the nearly ballistic regime is a reflection of the fact that the conductance is quantized when $L \lesssim l$, and the decrease for $L \gtrsim \xi$ is a result of the reduction of the conductance due to the localization effect. When the channel number is not so large ($W/\lambda_F = 2.75$), the nearly ballistic regime and the localized regime overlap with each other and the fluctuation is always strongly dependent on the length. As the channel number increases and the localization length becomes larger, the maximum fluctuation increases and there appears a length region where the fluctuation stays independent of length (the universal region). The maximum fluctuation for $W/\lambda_F = 10.25$ is close to $0.73e^2/h$ derived by the perturbational method for the quasi-one-dimensional metallic wires [3].

A crossover occurs from 1D to 2D when the broadening of the subbands exceeds their separation and the subband structure is smeared out. Figure 3 shows calculated δG for $l/\lambda_F = 2.05$ together with that for $l/\lambda_F = 51.25$ in the case that the number of occupied subbands is 20. For $l/\lambda_F = 2.05$ the situation is closer to that of metallic wires, i.e., the broadening is larger than the largest subband separation. The numerical result is in good agreement with that by the perturbational calculation represented by the dotted line in the region $2l < L < 10l$ including the length corresponding to the 2D square $W = L$. When the length is larger than $\sim 10l$, the fluctuation becomes smaller due to the localization effect.

2.2. Magnetic Field Dependence

An example of the calculated conductance as a function of the magnetic field $\hbar\omega_c/E_F$ ($\omega_c = eB/m$) is shown in Fig. 4. We consider a system characterized by $W/\lambda_F = 2.75$ and $l/\lambda_F = 51.25$ with fixed length $L = 3l$, where the fluctuation becomes nearly maximum in the absence of magnetic fields, and vary the magnetic field at a fixed value of the Fermi energy. In the presence of very weak magnetic fields ($\hbar\omega_c/E_F \lesssim 1 \times 10^{-2}$), the conductance increases slightly from the value at zero field probably due to a reduction in the weak localization effect. At higher magnetic fields, the conductance exhibits an oscillatory dependence on the field, which can be understood simply in terms of the magnetic-field dependence of the mean free path decreasing drastically due to the divergence in the 1D density of states whenever the Fermi level is close to a subband bottom. At the magnetic field ($\hbar\omega_c/E_F \approx 0.4$) where the channel number changes from 3 to 2, the conductance rapidly approaches the quantized value $G = 2N(e^2/h)$. In such a strong field well-defined edge states are formed and the backscatterings between the edge states are completely suppressed.

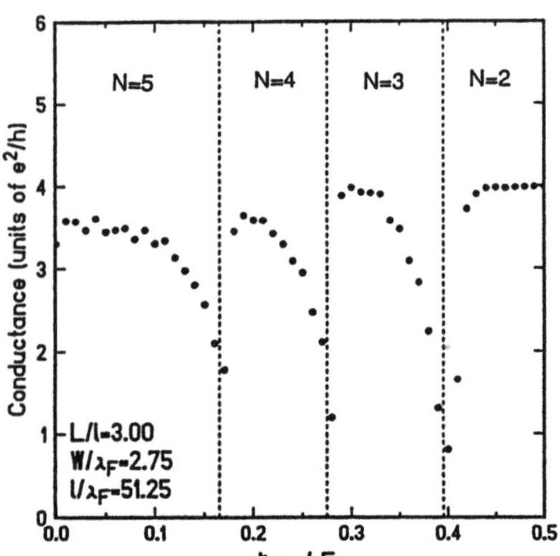

Fig. 4 Conductance as a function of the magnetic field for $L/l = 3$ and $W/\lambda_F = 2.75$. The number of occupied subbands N changes at the magnetic fields indicated by the vertical dotted lines.

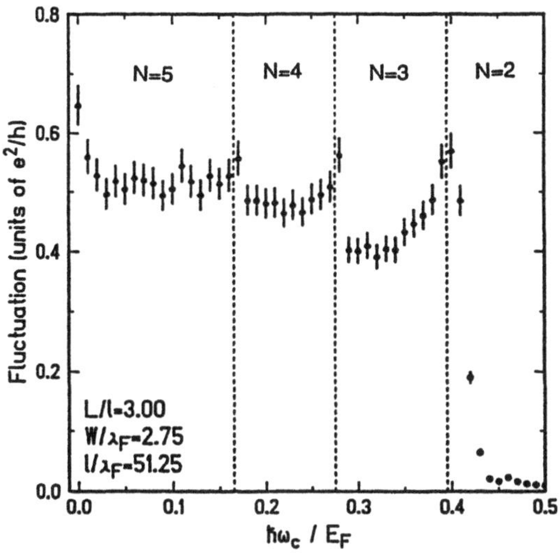

Fig. 5 The fluctuations as a function of the magnetic field for the same system as that in Fig. 4.

Figure 5 shows the corresponding result of the fluctuation. The reduction in the fluctuation which can be seen in weak magnetic fields ($\hbar\omega_c/E_F \lesssim 1\times10^{-2}$) corresponds to the suppression of the contribution from the so-called "particle-particle channel" predicted by perturbational calculations [2, 3]. In fact, the flux penetrating through the sample is $BLW \sim 13\times(h/e)$ for $\hbar\omega_c/E_F \sim 1\times10^{-2}$. When the magnetic field is increased further, the fluctuation gradually decreases except near the fields where the subband depopulation occurs. This can be explained by the gradual approach of the system to the ballistic regime due to the increase in the mean free path. In strong magnetic fields ($\hbar\omega_c/E_F \gtrsim 0.45$) the fluctuation becomes negligibly small corresponding to the quantization of the conductance itself.

Figure 6 shows G as a function of L for $W/\lambda_F = 5.25$ and $l/\lambda_F = 51.25$ in the weak field $\hbar\omega_c/E_F = 3\times10^{-3}$ together with G at $B=0$. In terms of the magnetic flux, each area normal to the field with length l and width W is penetrated by $BlW/(h/e) = 2.5$ flux. In the magnetic field the conductance becomes larger than that at $B=0$ at every length due to the reduction of the localization effect. The localization length, denoted by the vertical arrow, is 16.7 ± 0.6 in units of l smaller than twice of the value at $B=0$. This decrease of the localization effect is a result of the change in the universal class of the system from orthogonal to unitary, i.e., of the breaking of the time reversal symmetry by a magnetic field.

Figure 7 shows the corresponding results for the fluctuation. The universal region can now be clearly seen as well as the reduction of δG in comparison with δG at $B=0$, when the sample is shorter than the localization length ($L < 15l$). The actual amount of the reduction is slightly smaller than that predicted by perturbational calculations [$\delta G(B)/\delta G(B=0) \simeq 0.8$ in contrast to $1/\sqrt{2} = 0.71$]. When the sample becomes longer than the localization length, δG somewhat increases and then begins to decrease in the region $L > 35l$. It is quite interesting that the magnetic field enhances δG in the localized region, while it reduces δG in the universal region.

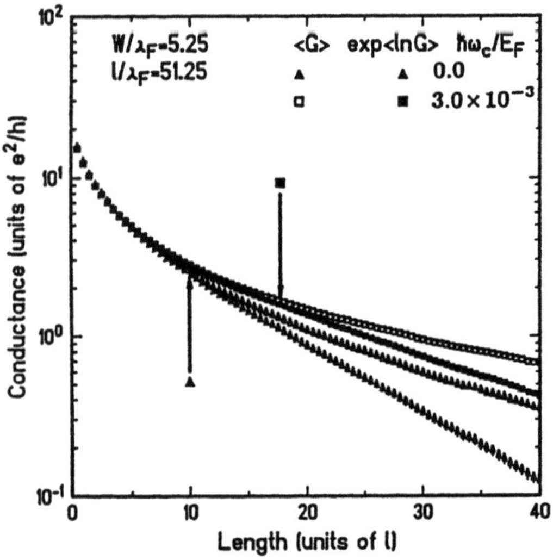

Fig. 6 Conductance as a function of the length L for $W/\lambda_F = 5.25$ in the magnetic field corresponding to $\hbar\omega_c/E_F = 3\times10^{-3}$ compared with that at $B=0$. The localization lengths are indicated by the vertical arrows.

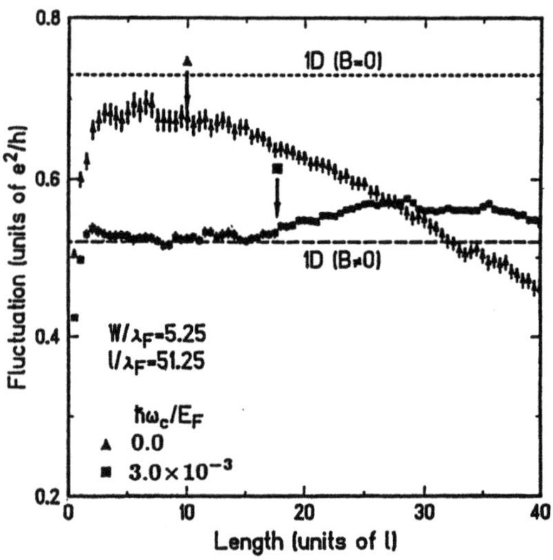

Fig. 7 The fluctuation as a function of L in the presence of the magnetic field compared with those at $B=0$ for the same system as that in Fig. 6. The localization lengths are indicated by the vertical arrows. The universal values obtained by the perturbation calculation in quasi-one-dimensional wires are indicated by the horizontal dotted line ($B=0$) and the dashed line ($BLW \gg h/e$).

3. Distribution Function of Conductance

3.1. Differential Equation of Distribution Function

In this section we describe an analytical method used by Dorokhov [15] to derive the Fokker-Planck equation of the distribution function. First we define a transfer matrix T as

$$\begin{bmatrix} O' \\ I' \end{bmatrix} = T \begin{bmatrix} I \\ O \end{bmatrix}, \tag{3.1}$$

where I, I' and O, O' are incoming and outgoing fluxes. The transfer matrix can be parametrized as [15, 16]

$$T = \begin{bmatrix} u & 0 \\ 0 & v \end{bmatrix} \begin{bmatrix} \cosh(\Gamma/2) & \sinh(\Gamma/2) \\ \sinh(\Gamma/2) & \cosh(\Gamma/2) \end{bmatrix} \begin{bmatrix} x & 0 \\ 0 & y \end{bmatrix}, \tag{3.2}$$

where Γ is a real diagonal matrix and u, v, x and y are all unitary matrices. Under this parametrization, transmission matrix t and the Landauer formula can be expressed as

$$t = u \frac{1}{\cosh(\Gamma/2)} x, \quad G = \frac{e^2}{h} \mathrm{Tr} \left(\frac{2}{1 + \cosh \Gamma} \right). \tag{3.3}$$

We add an infinitely small segment of length ΔL to the system. The transfer matrix for this small segment is calculated as

$$\begin{aligned} T'_{\Delta L} &= 1 + i \begin{bmatrix} -\gamma_{++} & \gamma_{+-} \\ -\gamma_{-+} & \gamma_{--} \end{bmatrix} - \frac{1}{2} \begin{bmatrix} -\gamma_{++} & \gamma_{+-} \\ -\gamma_{-+} & \gamma_{--} \end{bmatrix}^2, \\ (\gamma_{\pm\pm})_{nn'} &= -\frac{1}{\hbar} \frac{1}{\sqrt{v_n v_{n'}}} \iint dx dy \; \xi^*_{n\pm}(x, y) \; V(x, y) \; \xi_{n'\pm}(x, y), \end{aligned} \tag{3.4}$$

where $\xi_{n\pm}(x, y)$ and v_n are the wave function and the velocity of electron in the channel n at the Fermi level. $V(x, y)$ is the impurity potential in the segment $L < x < L + \Delta L$. The tranfer matrix of the total system with length $L + \Delta L$ is also written as Eq. (3.2) except that Γ, u, v, x, y are replaced with $\Gamma + \Delta\Gamma, u + \Delta u, v + \Delta v, x + \Delta x, y + \Delta y$ respectively, and satisfies the relation

$$T(\Gamma + \Delta\Gamma, u + \Delta u, v + \Delta v, x + \Delta x, y + \Delta y) = T'_{\Delta L} T(\Gamma, u, v, x, y). \tag{3.5}$$

Expanding Eq. (3.5), we can obtain $\Delta\Gamma, \Delta u, \Delta v, \Delta x$ and Δy in terms of Γ, u, v, x, y and γ. To obtain the Fokker-Planck equation, we need the first and the second moments of these quantities averaged over impurity configurations. For example we obtain

$$\begin{aligned} \frac{\overline{\Delta\Gamma_i}}{\Delta L} &= \coth \Gamma_i \overline{|(R_{+-})_{ii}|^2} + \sinh \Gamma_i \sum_{j \neq i} \frac{2 \overline{\left|(R_{+-})_{ij}\right|^2}}{\cosh \Gamma_i - \cosh \Gamma_j}, \\ \frac{\overline{\Delta\Gamma_i \Delta\Gamma_j}}{\Delta L} &= 2 \overline{\left|(R_{+-})_{ij}\right|^2} \delta_{ij}, \quad R_{+-} \equiv u^\dagger \gamma_{+-} v, \end{aligned} \tag{3.6}$$

where the overline denotes the average over impurity configurations and we have taken into account only two backscattering processes $\overline{(\gamma_{+-})_{nn'} (\gamma_{-+})_{n'n}}$ and $\overline{(\gamma_{+-})_{nn'} (\gamma_{-+})_{nn'}}$ giving nonzero contributions because of the momentum conservation of electrons in a quantum wire.

Using these moments, we can formally write down a Fokker-Planck equation. Since this equation contains too many parameters and is complicated, we make two assumptions: (A) Γ, u, v, x and y are statistically independent of each other, and (B) all unitary matrices u, v, x and y are equally probable, i.e., are distributed according to the invariant measure of the unitary group [17, 18]. For example, $\langle |u_{nn'}|^2\rangle = 1/N$ independent of n, n'. Using this assumption (B), we obtain the relation

$$\left\langle \overline{\left|(R_{+-})_{ij}\right|^2} \right\rangle = \frac{2-\delta_{ij}\delta_{\beta 1}}{(\beta(N-1)+2)l_H}, \tag{3.7}$$

where the symmetry factor β is defined for a system with ($\beta = 1$) or without ($\beta = 2$) time-reversal symmetry or with strong spin-orbit scatterings ($\beta = 4$), and l_H is the harmonic mean of the mean free paths of each channel given by

$$l_H^{-1} = \frac{1}{N}\sum_{n=1}^{N} l_n^{-1},\ l_n^{-1} \equiv \frac{n_i\gamma^2}{\hbar^2}\sum_{n'=1}^{N}\frac{1}{v_n v_{n'}}\int dy\ |\phi_{n+}(y)|^2\,|\phi_{n'-}(y)|^2, \tag{3.8}$$

where $\phi_{n\pm}(y)$ is the wave function in the y direction in the channel n at the Fermi level. l_H goes to $1.27l$ in a large N limit. The validity of these assumptions will be discussed in Sec. 3.2.

The expectation value of an arbitrary function of Γ is defined by

$$\langle F(\Gamma)\rangle_L \equiv \int d\Gamma\, F(\Gamma)W(L;\Gamma). \tag{3.9}$$

where the average over unitary matrices is already taken. Expanding $\langle F(\Gamma)\rangle_{L+\Delta L}$ with respect to $\Delta\Gamma$,

$$\frac{\langle F\rangle_{L+\Delta L}-\langle F\rangle_L}{\Delta L} = \sum_i \frac{\overline{\Delta\Gamma_i}}{\Delta L}\frac{\partial F}{\partial\Gamma_i} + \frac{1}{2}\sum_{i,j}\frac{\overline{\Delta\Gamma_i\Delta\Gamma_j}}{\Delta L}\frac{\partial^2 F}{\partial\Gamma_i\partial\Gamma_j}. \tag{3.10}$$

Substituting Eq. (3.6) and integrating by parts, we finally obtain the desired Fokker-Planck equation for the distribution function W as

$$\begin{aligned}\frac{\partial W(s;\Lambda)}{\partial s} = \frac{2}{\beta(N-1)+2}\Bigg[&\sum_{i=1}^{N}\frac{\partial}{\partial\Lambda_i}(1-\Lambda_i)\frac{\partial}{\partial\Lambda_i}\Lambda_i^2 \\ &-\frac{\beta}{2}\sum_{i=1}^{N}\sum_{j\neq i}^{N}\frac{\partial}{\partial\Lambda_i}(1-\Lambda_i)\frac{\Lambda_i\Lambda_j}{\Lambda_i-\Lambda_j}\Bigg]W(s;\Lambda),\end{aligned} \tag{3.11}$$

where $s = L/l_H$, $\Lambda_i = 2/(1+\cosh\Gamma_i)$ (we order Λ_i as $\Lambda_1 > \Lambda_2 > \cdots > \Lambda_N$). Exactly the same equation has been obtained in the case of a quasi-one-dimensional metallic wire for $\beta = 1$ and 2 by Mello et al. [19-22] by use of the maximum entropy method. However the mean free path has been introduced as just a parameter in this method.

3.2. Some Calculated Examples

It is difficult to obtain the general solution of $W(s;\Lambda)$ from Eq. (3.11), but several important transport quantities can be calculated.

When the system is not in the strongly localized regime ($L \ll \xi$) and the channel number is sufficiently large, we can use the $1/N$ expansion method proposed by Mello et al. [20, 22]. (For $\beta=1$ and 2, we can just replace the mean free path used in Refs. 20 and 22 with l_H given by Eq. (3.8).) For example, the averaged dimensionless conductance becomes

$$\langle g\rangle \equiv \frac{\langle G\rangle}{(e^2/h)} = \frac{N}{1+s} + \left(-\frac{\delta_{\beta 1}}{3}+\frac{\delta_{\beta 4}}{6}\right)\frac{s^3}{(1+s)^3} + O\left(\frac{1}{N}\right) \overset{L\gg l_H}{\longrightarrow} \frac{Nl_H}{L} - \frac{\delta_{\beta 1}}{3} + \frac{\delta_{\beta 4}}{6}, \tag{3.12}$$

where the first term corresponding to the Ohmic conductance and the second term to the correction according as the symmetry of the system. The variance of the conductance is given by

$$\mathrm{Var}(g) \equiv \langle g^2 \rangle - \langle g \rangle^2 = \frac{2}{\beta}\left(\frac{1}{15} - \frac{3}{5(1+s)^5} + \frac{1}{3(1+s)^6}\right) + O\left(\frac{1}{N}\right), \tag{3.13}$$

and the root-mean-squared fluctuation, including the factor 2 for spin degeneracy and with length $L \gg l_H$, is given by

$$\delta g = 2\sqrt{\mathrm{Var}(g)} \stackrel{L \gg l_H}{\longrightarrow} \sqrt{\frac{8}{15\beta}} = \begin{cases} 0.730 & \text{for } \beta = 1 \\ 0.516 & \text{for } \beta = 2 \\ 0.365 & \text{for } \beta = 4 \end{cases}. \tag{3.14}$$

This result shows that the conductance fluctuation is also universal in a quantum wire and moreover the universal values are exactly the same as those obtained in a metallic wire by the perturbation treatment.

In the localized region $\Lambda_1 \gg \Lambda_2 \gg, \cdots \gg \Lambda_N$, on the other hand, and only one of Λ_i's is dominant in the conductance ($g \simeq \Lambda_1$). In this case the differential equation for the distribution function Eq. (3.11) is reduced to the one-channel problem

$$\frac{\partial W(\tilde{s};g)}{\partial \tilde{s}} = \left[\frac{\partial}{\partial g}(1-g)\frac{\partial}{\partial g}g^2\right] W(\tilde{s};g) \quad (\tilde{s} \equiv s/\xi), \tag{3.15}$$

where ξ is the localization length given by

$$\xi = \left(-\frac{\partial}{\partial L}\langle \ln g \rangle\right)^{-1}\Bigg|_{L\to\infty} = \begin{cases} (N+1)l_H/2 & (\beta=1) \\ Nl_H & (\beta=2) \\ (2N-1)l_H & (\beta=4) \end{cases} \stackrel{N \gg 1}{\longrightarrow} \beta N l_H/2. \tag{3.16}$$

Equation (3.15) has been solved exactly by Abrikosov [23], and the result is ($\rho \equiv 1/g$)

$$W(\tilde{s};\rho) = \frac{2}{\sqrt{\pi \tilde{s}^3}} \int_{\mathrm{arch}\sqrt{\rho}}^{\infty} du \, \frac{u \exp\left[-\left(u^2/\tilde{s} + \tilde{s}/4\right)\right]}{\left(\cosh^2 u - \rho\right)^{1/2}}. \tag{3.17}$$

In a large channel-number limit the localization length ξ is proportional to the channel number and the symmetry factor, which is the same result given by several authors [15, 24, 25]. It is also noted that for a strictly one-dimensional wire ($N = 1$), $\xi = l_H$ regardless of β. This is the result of the fact that the one-dimensional system always has the time-reversal and the spin-rotation symmetry.

The above analytical results are compared with numerical results shown in Sec. 2 in Figs. 8-10. Generally speaking, the analytically calculated conductance is in good agreement with, but always a little bit smaller than, the numerically calculated conductance. The analytically calculated localization length in Fig. 8 is considerably smaller than the numerical result.

This deviation arises because the unitary matrices parametrizing the transfer matrix in Eq. (3.2) are not always distributed according to the invariant measure of U(N) like the assumption (B). Actually, under this assumption, the transmission probabilities T_n for the channel n is obtained as

$$T_n \equiv \langle (tt^\dagger)_{nn} \rangle = \sum_i \langle \Lambda_i \rangle \langle |u_{ni}|^2 \rangle = \frac{1}{N}\sum_i \langle \Lambda_i \rangle = \frac{\langle g \rangle}{N}, \tag{3.18}$$

showing that each channel gives always the equal contribution to the conductance. However, the numerical results seem to show that the distribution function of the unitary

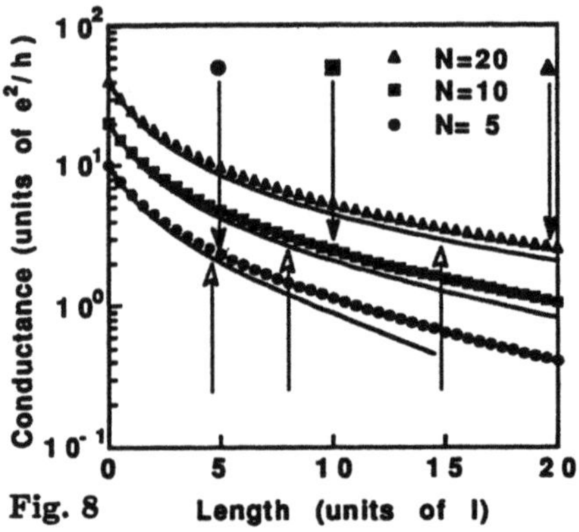

Fig. 8

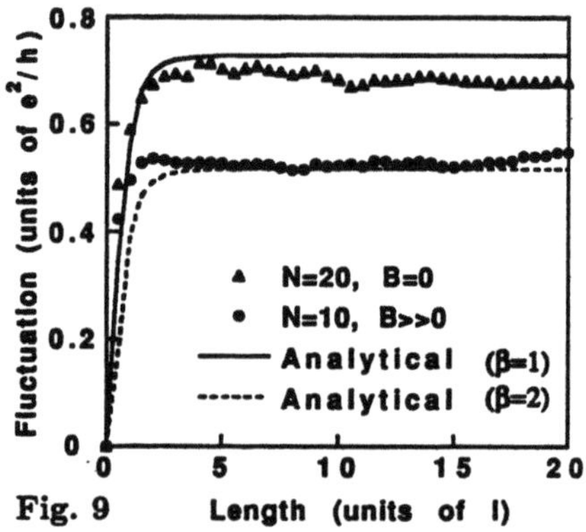

Fig. 9

Fig. 8 Length dependence of conductance for various occupied channel numbers. Symbols are the same numerical results in Fig. 1, and solid lines and upward arrows correspond to the analytically calculated conductance and localization length.

Fig. 9 Length dependence of fluctuation without and with a magnetic field. Solid line represents the analytical result with $\beta = 1$ and dotted line with $\beta = 2$.

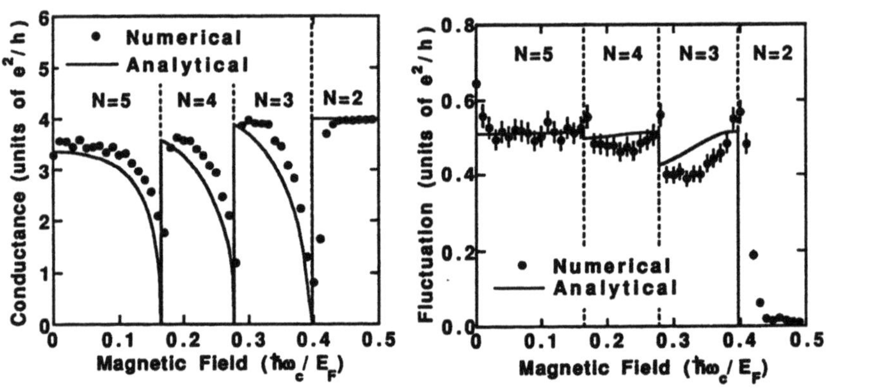

Fig. 10 Magnetic field dependence of conductance and fluctuation. Solid line represents the analytical result with $\beta = 2$

matrix deviates from the assumption (B) in such a way that the channel with a larger mean free path l_n tends to give a larger current.

As seen in Eq. (3.11), the wire length is scaled by the harmonic average l_H. This becomes quite unreasonable in the case that some channels have mean free paths much larger than others (like in the presence of strong boundary-roughness scattering for example [26]). In this case, both conductance and localization length are expected to be determined by channels with a larger mean free path. In the present result, however, they are determined by the harmonic average dominated by channels with a shorter mean free path. In the semiclassical Boltzmann transport theory [26], the mean free path appearing in conductance is given by the arithmetic average $l_A = \sum l_n/N$ ($\sim 1.6l$ in a large N limit). This arithmetic average may be more appropriate as the length scale, since it does not seem to suffer from such difficulties. Further study, including going beyond our assumptions, is highly needed to understand this interesting and important problem.

4. Summary

We have studied the electrical transport in quantum wires both numerically and analytically. It has been shown that with the increase of the length L the conductance fluctuation increases in the region $L \lesssim l$ and starts to decrease when L exceeds the localization length ξ. When only a few subbands are occupied, ξ is comparable to l and there is no universal region where the fluctuation stays independent of L. When many subbands are occupied, ξ is much larger than l ($\xi \simeq Nl$ with N the number of occupied subbands) and there appears a universal region for $l \lesssim L \lesssim \xi$. We have also demonstrated that the dimensional crossover from 1D to 2D system or from a quantum wire to a metallic wire occurs when the sample width exceeds the mean free path. In the presence of a weak magnetic field, the universal region becomes wider due to the reduction in the localization effect and the fluctuation itself is reduced. In strong magnetic fields where edge states with an extremely long mean free path are formed, the conductance is almost quantized and the fluctuation nearly vanishes. The analytical approach by use of the distribution function of conductance, under some assumptions, can show that the conductance fluctuation is universal in a quantum wire and that the universal value is exactly the same as that in a metallic wire.

Acknowledgments

We would like to thank H. Akera for useful discussion. This work is supported in part by Industry-University Joint Research Program "Mesoscopic Electronics" and by Grant-in-Aid for Scientific Research on Priority Area "Electron Wave Interference Effects in Mesoscopic Structures" from Ministry of Education, Science and Culture, Japan.

References

1. For a review, see S. Washburn and R.A. Webb, Adv. Phys. **35**, 375 (1986), and references therein
2. B.L. Altshuler, Pis'ma Zh. Eksp. Theor. Fiz. **41**, 530 (1985) [JETP Lett. **41**, 648 (1985)].
3. P.A. Lee and A.D. Stone, Phys. Rev. Lett. **55**, 1622 (1985); P.A. Lee, A.D. Stone, and H. Fukuyama, Phys. Rev. B **35**, 1039 (1987)
4. H. Tamura and T. Ando, Phys. Rev. B **44**, 1792 (1991).
5. D.S. Fisher and P.A. Lee, Phys. Rev. B **23**, 6851 (1981).
6. M. Büttiker, Phys. Rev. Lett. **57**, 1761 (1986).
7. M. Büttiker, Phys. Rev. B **38**, 9375 (1988).
8. A.D. Stone and A. Szafer, IBM J. Res. Develop. **32**, 384 (1988).
9. H.U. Baranger and A.D. Stone, Phys. Rev. B **40**, 8169 (1989).
10. R. Landauer, IBM J. Res. Develop. **1**, 223 (1957); Phil. Mag. **21**, 863 (1970).
11. S. Datta, M. Cahay, and M. McLennan, Phys. Rev. B **36**, 5655 (1987).
12. M. Cahay, M. McLennan, and S. Datta, Phys. Rev. B **37**, 10125 (1988).
13. M. Cahay, S. Bandyopadhyay, M.A. Osman, and H.L. Grubin, Surf. Sci. **228**, 301 (1990).
14. S. Bandyopadhyay, M. Cahay, D. Berman, and B. Nayfeh, in *Proceedings of the 5th International Conference on Superlattices and Microstructures, Berlin, 1990.*
15. O.N. Dorokhov, Zh. Eksp. Theor. Fiz. **85**, 1040 (1983) [Sov. Phys. JETP. **58**, 606 (1983)].
16. P.A. Mello and J.-L. Pichard, J. Phys. France I **1**, 493 (1991).
17. M. Gaudin and P.A. Mello, J. Phys. G **7**, 1085 (1981).
18. P.A. Mello, J. Phys. A **23**, 4061 (1990).

19. P.A. Mello and B. Shapiro, Phys. Rev. B **37**, 5860 (1988).
20. P.A. Mello, Phys. Rev. Lett. **60**, 1089 (1988).
21. P.A. Mello, P. Pereyra, and N. Kumar, Annals of Phys. **181**, 290 (1988).
22. P.A. Mello and A.D. Stone, Phys. Rev. B **44**, 3559 (1991).
23. A.A. Abrikosov, Physica Scripta **T27**, 148 (1989).
24. K.B. Efetov and A.I. Larkin, Zh. Eksp. Theor. Fiz. **85**, 764 (1983)[Sov. Phys. JETP. **58**, 444 (1983)].
25. J.-L. Pichard, M. Sanquer, K. Slevin, and P. Debray, Phys. Rev. Lett. **65**, 1812 (1990).
26. H. Akera and T. Ando, Phys. Rev. B **41**, 11967 (1990).

Enhancement of Conductance Fluctuations in the Diffusive and Near the Ballastic Regimes

S. Iwabuchi[1], *H. Higurashi*[1], *and Y. Nagaoka*[2]

[1]ULSI Research Center, Toshiba Corporation,
1 Komukai Toshiba-cho, Saiwai-ku, Kawasaki 210, Japan
[2]Yukawa Institute for Theoretical Physics, Kyoto University,
Kitashirakawa, Sakyo-ku, Kyoto 606, Japan

Abstract Conductance fluctuations in the mesoscopic systems are studied from diffusive to ballistic regimes. It is shown that the conductance fluctuations are enhanced in the presence of the independent current-carrying states and/or independently fluctuating spatial domains not only in the diffusive but also near the ballistic regimes.

1 Introduction

Since the pioneering work on the theory of conductance fluctuations [1, 2, 3, 4], various modifications of conductance fluctuations has been studied. As far as two-terminal measurements are concerned, they are categorized into two groups. The first category can be expected when the time reversal symmetry of the system is broken. Systems with spin-orbit interaction or systems in the presence of the applied magnetic fields are the cases[1, 2]. In this case conductance fluctuations are reduced. The second category can be expected even for the systems with the time reversal symmetry, provided that the systems can have independently fluctuating 'states' such as current-carrying states (subbands or valleys) and spatial domains [6, 7]. In this case conductance fluctuations are enhanced.

In this article we discuss the latter category taking two dimensional electron systems with subband structures in the diffusive regime, and quantum wires near the ballistic regime as examples.

2 Enhancement of Conductance Fluctuations in Diffusive Regime

2.1 Model of the System

As an example, we consider a model for the quasi-two dimensional electron systems of size $L_x \times L_y \times L_z$ satisfying $L_z \sim \lambda_F \ll \ell \ll L_x,\ L_y$, being ℓ and λ_F elastic mean free path and Fermi wavelength, respectively [5]. Due to the geometrical restriction along z direction, the system has subband structures as shown in Fig.1. We only consider the situation that the average energy separation of subbands is comparable to the Fermi energy so that the electrons

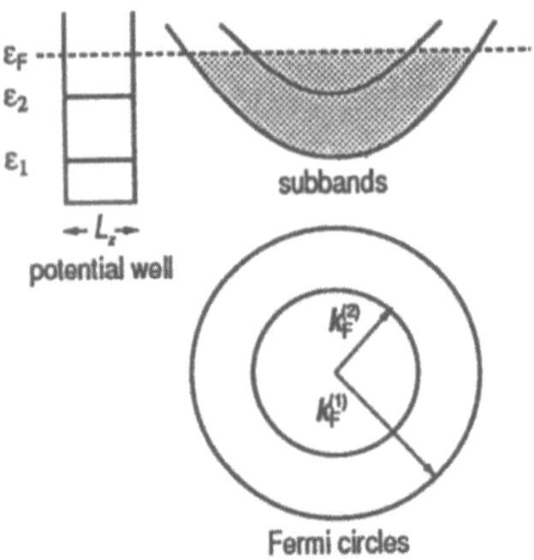

Figure 1: Model of two-subband electron system. Electrons form two kinds of Fermi circles. The shape of confinement potential is of no significance.

occupy two subbands. The Hamiltonian of the system is then given by

$$\mathcal{H} = \sum_{ks} (\epsilon_{ks} - \epsilon_F) C^\dagger_{k,s} C_{k,s} + \sum_{k,q} \sum_{s,s'} u_{s,s'}(q) C^\dagger_{k+q,s} C_{k,s'} \ , \tag{1}$$

where $C^\dagger_{k,s}$ ($C_{k,s}$) represents creation (annihilation) operator of an electron with wave vector $\boldsymbol{k}$ in subband s (1 or 2). The energy spectrum of an electron in subband s is given by $\epsilon_{k,s} = (\hbar k)^2/2m^* + \epsilon_s$, with $\epsilon_s = (\hbar s\pi/L_z)^2/2m^*$, and $u_{ss'}$ represents a short range impurity potential by which electrons undergo intra and intersubband scatterings. The total lifetime of the electrons in subband s is then given by $1/\tau_s = 1/\tau_{ss} + 1/\tau_{s\bar{s}} + 1/\tau_\phi$ ($(s,\bar{s}) = (1,2)$ or $(2,1)$), where $\hbar/2\tau_{ss'} \equiv \pi n_{\rm imp}|u_{ss'}|^2 N_o$, $N_o = m^*/2\pi\hbar^2$ and $n_{\rm imp}$ are the density of states of two dimensional electrons and impurity concentrations, respectively. Here the phase coherence time (inelastic scattering time), τ_ϕ is introduced phenomenologically. In this case one particle random-averaged Matsubara Green's functions are given by

$$G_s(\boldsymbol{k}, \epsilon_n) = \frac{1}{i\epsilon_n - \epsilon_{ks} + \epsilon_F + i\dfrac{\hbar}{2\tau_s}\mathrm{sgn}(\epsilon_n)}, \tag{2}$$

where $\epsilon_n = (2n+1)\pi/\beta$ ($\beta = 1/k_BT$, $n = 0, \pm1, \pm2, \cdots$).

2.2 Cooperons and Diffusons

Possible Cooperons and diffusons (randomons) in two-subband system are shown in Fig.2.[6] Matsubara frequencies are defined as $\omega_n \equiv \epsilon_n/\hbar$ and $\omega_\ell \equiv 2\pi\ell/\beta\hbar$ ($\ell = 0, \pm1, \pm2, \cdots$). As far as the average energy separation of subbands is comparable to the Fermi energy, randomons divergent for small $\boldsymbol{q}$ and ω_ℓ as $\tau_\phi \to \infty$ are the ones satisfying

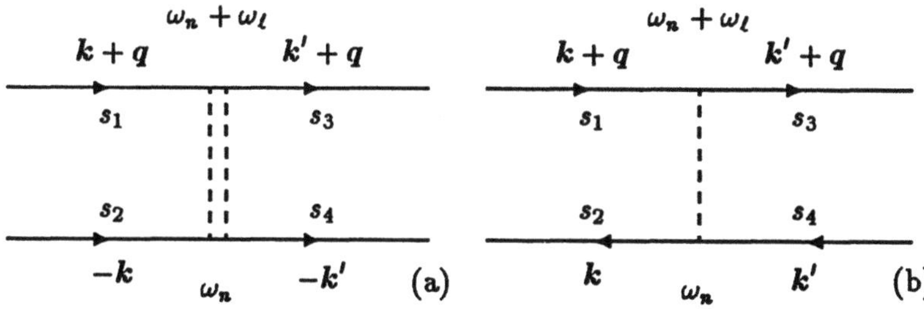

Figure 2: Possible randomons in two-subband system: (a)Cooperons (b)diffusons. Among them, only ones satisfying conditions, $s_1 = s_2$ and $s_3 = s_4$ are the most divergent as far as the average energy separation of subbands is comparable to the Fermi energy.

$$s_1 = s_2 = s \text{ and } s_3 = s_4 = s' . \tag{3}$$

Coupled Dyson equations for Cooperons ($C_{ss'}$) and diffusons ($D_{ss'}$) are analytically solved and are, in the small q and $\omega_\ell$limit, of form

$$C_{s,s'}(q,\omega_\ell) = \frac{\hbar}{2\pi N_o \tau_s \tau_{s'}} \left[\frac{\xi_{ss'+}(q)}{\eta_{s+}(q) + \omega_\ell} + \frac{\xi_{ss'-}(q)}{\eta_{s-}(q) + \omega_\ell} \right] , \tag{4}$$

where

$$\eta_{s\pm}(q) = \frac{1}{2} \left[X_s(q) + X_{\bar{s}}(q) \pm Y_s(q) \right] , \tag{5}$$

$$\xi_{ss'i}(q) = \alpha_s f_{si}(q) \delta_{ss'} - \frac{1}{tY_s(q)} (1 - \delta_{ss'}) \quad (i = \pm), \tag{6}$$

$$X_s(q) = D_s q^2 + t_s^{-1} , \tag{7}$$

$$Y_s(q) = \left[(X_s(q) - X_{\bar{s}}(q))^2 + 4t^{-2} \right]^{\frac{1}{2}} , \tag{8}$$

$$f_{s\pm}(q) = \frac{1}{2} \left[1 \pm \frac{X_s(q) - X_{\bar{s}}(q) + 2(\alpha_s^{-1} - 1)/t_s}{Y_s(q)} \right] , \tag{9}$$

$$\alpha_s = \frac{(1 - \frac{\tau_s}{\tau_{s\bar{s}}} - \frac{\tau_s}{\tau_\phi})(1 - \frac{\tau_{\bar{s}}}{\tau_\phi}) - \frac{\tau_{\bar{s}}}{\tau_{s\bar{s}}}(1 - \frac{\tau_s}{\tau_\phi})}{1 - (\tau_s + \tau_{\bar{s}})(\frac{1}{\tau_{s\bar{s}}} + \frac{1}{\tau_\phi}) + \frac{\tau_s \tau_{\bar{s}}}{\tau_\phi}(\frac{2}{\tau_{s\bar{s}}} + \frac{1}{\tau_\phi})} , \tag{10}$$

$$\frac{1}{t_s} = \frac{\frac{1}{\tau_{s\bar{s}}} + \frac{1}{\tau_\phi} - \frac{\tau_s}{\tau_\phi}(\frac{2}{\tau_{s\bar{s}}} + \frac{1}{\tau_\phi})}{1 - (\tau_s + \tau_{\bar{s}})(\frac{1}{\tau_{s\bar{s}}} + \frac{1}{\tau_\phi}) + \frac{\tau_s \tau_{\bar{s}}}{\tau_\phi}(\frac{2}{\tau_{s\bar{s}}} + \frac{1}{\tau_\phi})} , \tag{11}$$

$$\frac{1}{t} = \frac{\frac{1}{\tau_{s\bar{s}}}}{1 - (\tau_s + \tau_{\bar{s}})(\frac{1}{\tau_{s\bar{s}}} + \frac{1}{\tau_\phi}) + \frac{\tau_s \tau_{\bar{s}}}{\tau_\phi}(\frac{2}{\tau_{s\bar{s}}} + \frac{1}{\tau_\phi})} . \tag{12}$$

Here $D_s \equiv (\hbar k_F^s/m^*)^2 \tau_s/2$ is the diffusion constant of electrons in the subband s. The equation (4) is generalized expression of the previously reported one [6] in that they can be applied to the system with $D_1 \neq D_2$ in the case of arbitrarily strong intersubband impurity scattering as well as arbitrary value of τ_s/τ_ϕ. In fact, putting $D_s = D_{\bar{s}}$ and taking limit, $\tau_s/\tau_\phi \ll 1$, we obtain the previous results. Note that the time scales which characterize the effect of intersubband scattering are t and t_s rather than $\tau_{s\bar{s}}$.

It is instructive to see the following two limits under the condition $\tau_s/\tau_\phi \ll 1$ (s=1, 2):

(i) *weak scattering limit*: $t \to \infty$ ($\tau_{s\bar{s}} \to \infty$)

In this case, $\alpha_s \simeq 1$, $\eta_{s\pm} \simeq D_s q^2 + \tau_\phi^{-1}$, $f_{s\pm} \simeq \theta(\pm(\bar{s} - s))$, so that we have

$$C_{s,\bar{s}}(q, \omega_\ell) = \frac{\hbar}{2\pi N_o \tau_s^2} \frac{\delta_{s\bar{s}}}{D_s q^2 + |\omega_\ell| + \tau_\phi^{-1}} \tag{13}$$

This is the ordinary expression for Cooperons in the absence of intersubband scattering.

(ii) *strong scattering limit*: $t \to 0$ ($\tau_{s\bar{s}} \to \tau_s + \tau_{\bar{s}}$)

In this case, $\alpha_s \simeq 1$, $\eta_{s+} \simeq \infty$, $\eta_{s-} \simeq (D_1 + D_2)/2 \cdot q^2 + \tau_\phi^{-1}$, $f_{s\pm} \simeq 1/2$, so that we have

$$C_{s,s'}(q, \omega_\ell) = \frac{\hbar}{2\pi N_o \tau_s \tau_{s'}} \frac{1}{2} \frac{1}{Dq^2 + |\omega_\ell| + \tau_\phi^{-1}}, \tag{14}$$

where

$$D \equiv \frac{1}{2}(D_1 + D_2). \tag{15}$$

The intersubband impurity scattering reduces the interference effect to half, or 1/(the number of occupied subbands). Physical meaning of this is the following: In the case $\tau_{s\bar{s}} \to \infty$, the electrons stay in the same subband, so that the system behaves as if it consists of two independent electron systems. In the case, $\tau_{s\bar{s}} \to \tau_s + \tau_{\bar{s}}$, on the other hand, two kinds of subbands are coupled and mixed up completely due to the rather frequent intersubband scatterings. Consequently, the system behaves as one kind of electron system.

Concerning diffusons, their explicit form can be obtained from (4) by taking a limit $\tau_\phi \to \infty$. The effect of the intersubband scattering on diffusons is exactly the same for Cooperons.

Since, in mesoscopic systems, where $L_x < L_\phi$, wavevector q can be discretized as $q = (m_x\pi/L_x, m_y\pi/L_y)$ with $m_x = 1, 2, 3, \cdots$ and $m_y = 0, 1, 2, \cdots$, the phase coherence in each subband is cut by Min $\{\tau_c^s, \tau_T\}$ or Min $\{L_x, L_T^s\}$. Here the system traversal time, τ_c^s (energy correlation range: E_c^s) and the thermal diffusion time (length), τ_T (L_T^s) are defined by $\hbar/\tau_c^s \equiv E_c^s \equiv \hbar D_s(\pi/L_x)^2$, $\hbar/\tau_T \equiv \hbar D_s(\pi/L_T)^2 \equiv 2\pi k_B T$. This is in contrast to the macroscopic systems, where $\tau_c^s > \tau_\phi$; the phase coherence is cut by τ_ϕ since the localization effect will not be cut by τ_T. Therefore, in the presence of intersubband scattering, its effect reflects on the physical quantities depending on the ratio,

$$\frac{\text{Min}\,\{\tau_c^s, \tau_T\}}{t}. \tag{16}$$

2.3 Conductance Fluctuations

Let us calculate the conductance fluctuations in the mesoscopic electron system with intersubband scattering. Typical two-diffuson and two-Cooperon processes are shown in Figs.3(a) and 3(b), respectively. Conductance fluctuations due to these processes are calculated as

$$\mathrm{Var}(G) \equiv (\delta G)^2 = 2\left(\frac{e^2}{h}\right)^2 \left(\frac{8\pi}{L_x^2}\right)^2 \sum_{qs} \frac{1}{\beta\hbar} \sum_{\omega_n>0} \frac{1}{\beta\hbar} \sum_{\omega_{n'}>0}$$

$$\Bigg\{ \sum_{i=\pm} \left[3A_{si}(q) F_{si}(q,\omega_n+\omega_{n'})^4 + \frac{2B_s(q)}{\eta_{s+}(q)+\eta_{s-}(q)} F_{si}(q,\omega_n+\omega_{n'})^3 \right]$$

$$+2B_s(q)\,\big[\, F_{s+}(q,\omega_n+\omega_{n'})^3 F_{s-}(q,\omega_n+\omega_{n'})$$

$$+ F_{s+}(q,\omega_n+\omega_{n'})^2 F_{s-}(q,\omega_n+\omega_{n'})^2$$

$$+ F_{s+}(q,\omega_n+\omega_{n'}) F_{s-}(q,\omega_n+\omega_{n'})^3 \big] \Bigg\}\,, \qquad (17)$$

where

$$F_{si}(q,\omega_n) \equiv [\eta_{si}(q)+\omega_n]^{-1}\,, \qquad (18)$$

$$A_{si}(q) \equiv D_s^2 \alpha_s^2 f_{si}(q)^2 + \frac{D_s D_{\bar{s}}}{(tY_s(q))^2}\,, \qquad (19)$$

$$B_s(q) \equiv D_s^2 \alpha_s^2 f_{s+}(q) f_{s-}(q) + \frac{D_s D_{\bar{s}}}{(tY_s(q))^2} \qquad (20)$$

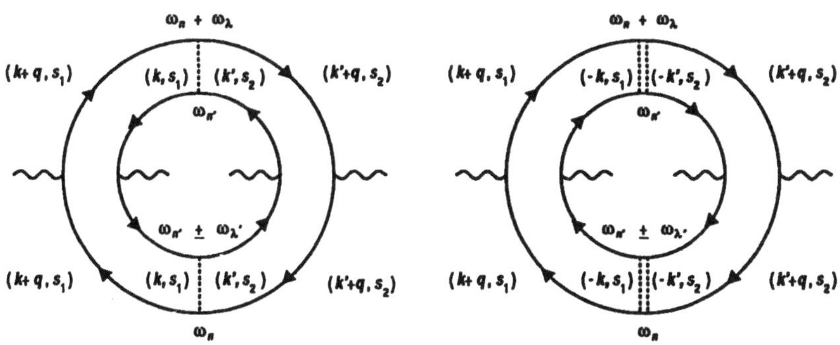

(a) (b)

Figure 3: Diagrams which contribute to the impurity averaged variance of conductance. Two-randomon processes: (a)diffusons (b)Cooperons.

2.3.1 zero temperature

Since $1/\tau_\phi \to 0$ at zero temperature, $\alpha_s \to 1$ and $t_s^{-1} \sim t^{-1} \to 1/[\tau_{s\bar{s}} - (\tau_s + \tau_{\bar{s}})]^{-1} \equiv t_o^{-1}$. We have then from (17)

$$\mathrm{Var}(G) = 2\left(\frac{e^2}{h}\right)^2\left(\frac{4}{\pi^2}\right)^2\left(\frac{\pi}{L_x^2}\right)^2\sum_s\sum_q D_s^2\left\{\left[\frac{f_{s+}(q)}{\eta_{s+}(q)}+\frac{f_{s-}(q)}{\eta_{s-}(q)}\right]^2\right.$$

$$\left.+\frac{D_{\bar{s}}}{D_s}\frac{1}{(t_0 Y_s(q))^2}\left[\frac{1}{\eta_{s+}(q)}-\frac{1}{\eta_{s-}(q)}\right]^2\right\} \qquad (21)$$

It is instructive to see the strong and weak scattering limits mentioned above:

(i) *strong scattering limit:* $\tau_c^s/t_0 \gg 1$

In this case, $\eta_{s+}(q) \sim Dq^2 + 2/t_0 \to \infty$, $\eta_{s-} \sim Dq^2$, $f_{s\pm}(q) \sim 1/2$ and $Y_s(q) \sim 2/t_0 \to \infty$, so that we have

$$\mathrm{Var}(G) = 2\left(\frac{e^2}{h}\right)^2\left(\frac{4}{\pi^2}\right)^2\sum_{m_x=1}^{\infty}\sum_{m_y=0}^{\infty}\frac{1}{\left[m_x^2+\left(\frac{L_x}{L_y}\right)^2 m_y^2\right]^2}. \qquad (22)$$

The situation remains unchanged for 3- and 4-randomon processes[1]. Therefore, the total conductance fluctuation amounts to $\delta G = 0.862e^2/h \equiv \delta G_2$[1]. This is the normal situation.

(ii) *weak scattering limit:* $\tau_c^s/t_0 \ll 1$

In this case, $\eta_{s+}(q) \sim D_s q^2$, $\eta_{s-}(q) \sim D_{\bar{s}} q^2$, $f_{s+}(q) \sim 1$, $f_{s-}(q) \sim 0$ and $Y_s(q) \sim (D_s - D_{\bar{s}})q^2$, so that we have

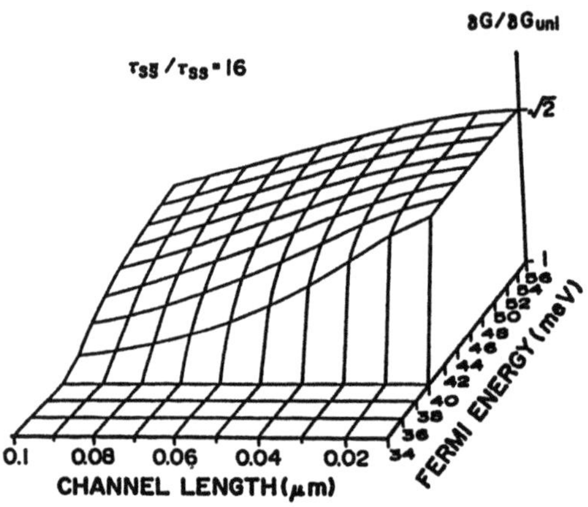

Figure 4: Conductance fluctuations in ultrasmall Si-MOSFET's at $T = 0$. Intrasubband impurity scattering time was tentatively chosen as $\tau_{11} = \tau_{22} = 8.1 \times 10^{-14}$sec, which is required for the substrate impurity concentration of Si-MOSFET's with around 0.1μm channel length. Channel shape is kept *isotropic*($L_x = L_y$).

$$\mathrm{Var}(G) = 2 \cdot 2 \left(\frac{e^2}{h}\right)^2 \left(\frac{4}{\pi^2}\right)^2 \sum_{m_x=1}^{\infty} \sum_{m_y=0}^{\infty} \frac{1}{\left[m_x^2 + \left(\frac{L_x}{L_y}\right)^2 m_y^2\right]^2}. \tag{23}$$

The situation also remains unchanged for 3- and 4-randomon processes. Therefore, we obtain $\delta G = \sqrt{2} \times 0.862 e^2/h$. This is the direct consequence of independent contribution of each subband to Var(G).

Conductance fluctuations are calculated for tentative parameters and are shown in Fig.4. Conductance fluctuations increase with decreasing channel length and increasing Fermi energy in spite of keeping conductor shape *isotropic* ($L_x = L_y$). It is because the amount of conductance fluctuations tends to increase with $t_0/\tau_c^s \propto \epsilon_F/L_x^2$ for a given τ_{ss}.

2.3.2 finite temperatures

At temperatures higher than E_c^s/k_B, zero temperature results are modified by energy averaging.[2] In the limit $E_c^s \ll k_B T$ ($\tau_c^s/\tau_T \gg 1$), we obtain the following results from (17):
(i)*strong scattering limit*: $\tau_T/t \gg 1$

$$\mathrm{Var}(G) = \left(\frac{e^2}{h}\right)^2 \cdot C \cdot \left[\frac{\frac{1}{2}\sum_s E_c^s}{k_B T}\right], \tag{24}$$

(ii)*weak scattering limit*: $\tau_T/t \ll 1$

$$\mathrm{Var}(G) = \left(\frac{e^2}{h}\right)^2 \cdot C \cdot \left[\frac{\sum_s E_c^s}{k_B T}\right], \tag{25}$$

where

$$C \equiv \frac{L_y}{L_x} \cdot \frac{4}{\pi^2} \sum_{n=0}^{\infty} \sum_{n'=0}^{\infty} \frac{1}{(n+n'+1)^3}. \tag{26}$$

As easily verified, again situation is the same for 3- and 4-randomon processes. Therefore, conductance fluctuations are enhanced by the intersubband scattering through the modification of the energy correlation range.

The similar results can be expected for the two-dimensional electron systems with degenerate valley structures. In this case, independently fluctuating 'states' is valleys. The fluctuation is enhanced up to $\sqrt{n_v} \cdot \delta G_2$ depending on Min$\{\tau_c, \tau_T\}/(\tau' - 2\tau)$, n_v, τ_c, τ' and τ being the number of valleys, system traversal time of electrons, intervalley scattering time and total lifetime of electrons.

3 Enhancement of Conductance Fluctuations near Ballistic Regime

3.1 Model of the System

As an example of the enhancement of conductance fluctuations near the ballistic regime, let us consider the mesoscopic quantum wire. We adopt the model introduced in Ref.[8], where the quantum wire is treated as a two dimensional square lattice of length L along x direction and width W along y direction as shown in Fig.5. The disordered region between the 1-st and the N-th columns is connected to the perfect leads on both sides, extended to $\pm\infty$ in the x direction. Impurities are put on the lattice points randomly at a certain concentration. Electronic states are described by the tight-binding method with the nearest neighbor transfers, t_x and t_y along x and y directions, respectively. Hamiltonian of the system is then of form

$$H = -\sum_{j,l}\sum_{\delta=\pm1}\{|j,l\rangle t_x\langle j+\delta,l| + |j,l\rangle t_y\langle j,l+\delta|\} + \sum_{j=1}^{N}\sum_{l=1}^{M}|j,l\rangle w_{j,l}\langle j,l|\,, \quad (27)$$

where $|j,l\rangle$ is the atomic orbital at site (j,l) and $w_{j,l}$ is the strength of the random potential at site (j,l) being assumed to be distributed uniformly between $-w/2$ and $w/2$. The wavefunction of an electron is then represented by $|\Psi\rangle = \sum_{j=1}^{N}\sum_{l=1}^{M} c_{j,l}|j,l\rangle$. The Schrödinger equation is then written as $\psi_j = T_j\psi_{j-1}$, giving

$$\psi_N = \Gamma_N\psi_0\,,\quad \Gamma_N \equiv \prod_{j=1}^{N} T_j\,. \quad (28)$$

Here we have used the same notations as those in Ref. [9]:

$$\psi_j^{\dagger} \equiv [\, c_{j+1,1}^{*}\,,\ \cdots,\ c_{j+1,M}^{*}\,,\ c_{j,1}^{*}\,,\ \cdots,\ c_{j,M}^{*}\,]\,, \quad (29)$$

$$T_j \equiv T + \Delta T_j\,, \quad (30)$$

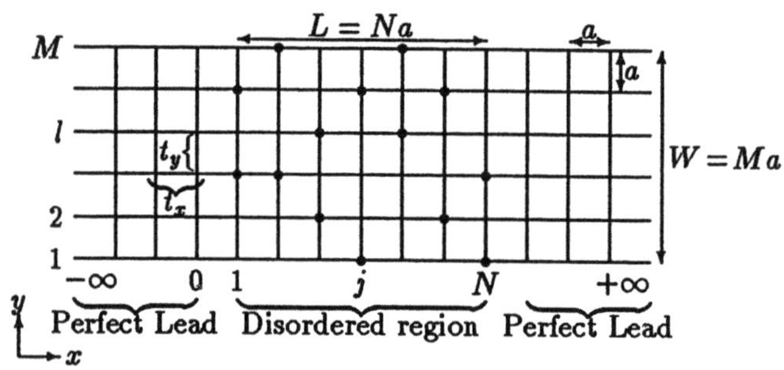

Figure 5: Schematic illustration of the system. System length and width are written as $L = Na$ and $W = Ma$ (a is lattice constant), respectively. Bullet sites represent impurities.

with

$$T = \begin{bmatrix} -\delta_{l,l'}\epsilon_F/t_x - (\delta_{l,l'-1} + \delta_{l-1,l'})t_y/t_x & -\delta_{l,l'} \\ \delta_{l,l'} & 0 \end{bmatrix}, \quad \Delta T_j = \begin{bmatrix} \delta_{l,l'} w_{j,l}/t_x & 0 \\ 0 & 0 \end{bmatrix},$$

where ϵ_F is the Fermi energy.

In the absence of impurities, the solution of the Schrödinger equation is

$$c_{j,l}^{(\pm,n)} = \frac{\exp(\pm ik_n aj)}{\sqrt{(M+1)\sin k_n a}} \sin\frac{\pi nl}{M+1}, \tag{31}$$

where k_n is Fermi wavevector of n-th subband (channel) and is determined by the energy dispersion relation,

$$\epsilon_F = -2t_x \cos k_n a - 2t_y \cos\frac{\pi n}{M+1}, \quad 0 \le k_n \le \pi/a. \tag{32}$$

The normalization factor in (31) is chosen arbitrarily since it has no influence on the final results.

3.2 Conductance Fluctuations

The conductance g of the sample with many channels is given by the Landauer formula [10, 11]

$$G \equiv \frac{2e^2}{h} g, \quad g = \sum_{\alpha,\beta}^{M} |t_{\alpha,\beta}|^2 = \mathrm{Tr}\, t^\dagger t, \tag{33}$$

where $t_{\alpha,\beta}$ is the amplitude of electron transmitting from channel α to β and M is the number of channels, corresponding to the total number of subbands occupied by electrons. Using (33), conductance fluctuations are given by

$$\delta G = \frac{2e^2}{h}\sqrt{\langle g^2\rangle - \langle g\rangle^2}, \tag{34}$$

where the angular bracket denotes the ensemble average over the impurity configurations. We therefore need to know the transmission matrix t of the sample.

The amplitudes of waves incoming and outgoing to each channel α ($A_\alpha^{(+)}$, $B_\alpha^{(-)}$ and $A_\alpha^{(-)}$, $B_\alpha^{(+)}$, respectively) are related with the $2M \times 2M$ transfer matrix $\mathcal{P}$ by

$$\begin{bmatrix} B^{(+)} \\ B^{(-)} \end{bmatrix} = \mathcal{P} \begin{bmatrix} A^{(+)} \\ A^{(-)} \end{bmatrix}, \quad \mathcal{P} \equiv \begin{bmatrix} t^{\dagger -1} & r' t'^{-1} \\ -t'^{-1} r & t'^{-1} \end{bmatrix}. \tag{35}$$

Here t, t' (r, r') are the $M \times M$ transmission (reflection) matrices and we have used for definiteness the vector notations for $A_\alpha^{(\pm)}$ and $B_\alpha^{(\pm)}$. Matrix $\mathcal{P}$ can be expressed in another way in terms of Γ_N [9]:

$$\mathcal{P} = \Lambda^{-N} \mathcal{C}^{-1} \Gamma_N \mathcal{C}. \tag{36}$$

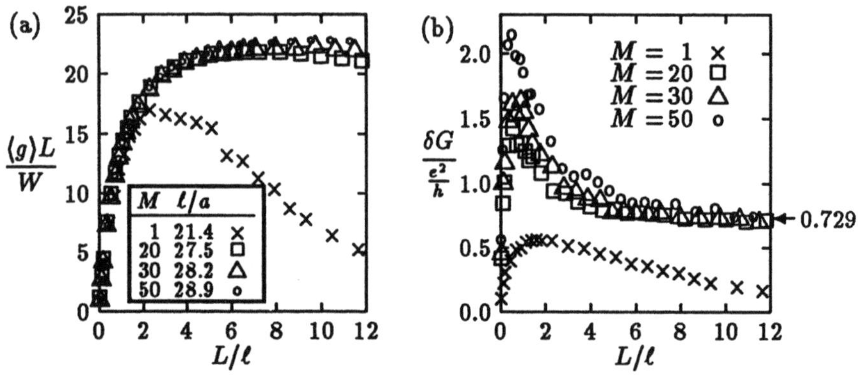

Figure 6: (a) $\langle g\rangle L/W$ as a function of L/ℓ for fixed $M(=1,\ 20,\ 30$ and $50)$. (b) Conductance fluctuations in unit of e^2/h as a function of L/ℓ for fixed M (= 1, 20, 30 and 50).

Here C and Λ are the $2M \times 2M$ matrices

$$C = \begin{bmatrix} [c_{1,l}^{(+,n)}] & [c_{1,l}^{(-,n)}] \\ [c_{0,l}^{(+,n)}] & [c_{0,l}^{(-,n)}] \end{bmatrix}, \quad \Lambda = C^{-1}TC = \begin{bmatrix} [\delta_{n,l}e^{ik_n a}] & 0 \\ 0 & [\delta_{n,l}e^{-ik_n a}] \end{bmatrix}. \tag{37}$$

Therefore, we can calculate δG using (33), (34), (35), and (36).

We calculated conductance fluctuations numerically for tentative parameters; $\epsilon_F = 0$, $t_y/t_x = 1/2$, $w/t_x = 1.84$ and $c_{imp} = 0.5$. Sample sizes treated are $M = 1$, 20, 30, and 50. In order to determine the mean free path ℓ, we utilize the Ohm's law in 2D system; $\langle g\rangle = k_F\ell W/2L$, which is valid for $L \gg \ell$. Estimated ℓ are listed in the inset of Fig.6(a) for several M. Using these, $\langle g\rangle L/W$ and δG are shown in Fig.6 as functions of L/ℓ.

In Fig.6(a), in the diffusive regime, Ohmic behaviors of conductances can be seen for $L \gtrsim 5\ell$ for all the samples except for $M = 1$. The Ohmic region is followed by the localization regime, at L larger than the localization length, ξ, although it is not shown in Fig.6(a). In strict one dimension, the situation is somewhat special. Since ξ is nearly equal to ℓ, conductance begins to decrease when $L \gtrsim \ell$. This is why there does not exist the Ohmic region for $M = 1$. In Fig.6(b), the fluctuations are almost independent of L in the Ohmic region of the diffusive regime, and the universal value is nearly equal to $\delta G_1 = 0.729e^2/h$ first found by Lee and Stone [1] for a quasi-one dimensional sample. As a consequence of the absence of Ohmic behavior, there is no universal region in the case $M = 1$.

As clearly seen in Fig.6(a), the Ohmic behavior is no longer kept as L approaches ℓ. Corresponding to this fact, the universality in δG is no longer kept: As seen in Fig.6(b) δG increases beyond the universal value and then decreases to zero towards the very ballistic regime, where the conductance is quantized as $\langle g\rangle = M$. Note that δG increases more for larger M.

First let us consider the initial increase in the fluctuation. The Lee-Stone theory [1] predicts $\delta G = \delta G_1$ for $L/W \gg 1$ and $\delta G \sim \sqrt{W/L}\cdot\delta G_{av}$ for $L/W \ll 1$ in two dimension as far as $L, W \gg \ell$. Therefore, δG increases in the Ohmic region as W/L increases. Physically, the factor W/L can be understood as the number of effective fluctuating domains each of which has the same amount of conductance fluctuation on the average ($\delta G_{av} \sim e^2/h$). Even near ballistic regime, where the theory based on randomons loses its validity, this view may be extended. In such a regime, however, the averaged conductance fluctuation of each domain cannot remain unchanged. In reality, it increases as a function of L/ℓ towards the ballistic regime. This is direct consequence of increase in the number of effective channels M_{eff} caused by the breakdown of the Ohm's law. The initial increase in δG is brought basically by these two mechanisms.

In order to understand peak structures at around ℓ, let us calculate the conductance fluctuations in very ballistic regime, $L/\ell \ll 1$, by treating the impurity potential perturbatively. We expand the transfer matrix Γ_N in (28) to the second order of ΔT_j. Inserting it into (36), we obtain

$$\delta G = \frac{2e^2}{h}\cdot\sqrt{M}\left[\sqrt{\frac{\ell}{2L}}(1-\frac{\ell}{L})e^{\ell/2L}\mathrm{Erfc}(\sqrt{\frac{\ell}{2L}})+\frac{\ell}{2L} - \left(2\sqrt{\frac{\ell}{2L}}e^{\ell/2L}\mathrm{Erfc}(\sqrt{\frac{\ell}{2L}})\right)^2\right]^{1/2} \tag{38}$$

$$\sim \frac{2e^2}{h}\cdot\sqrt{2M}\cdot\frac{L}{\ell} \qquad \text{for } \frac{L}{\ell} \ll 1, \tag{39}$$

with $\ell = 48t_x^2/(c_{\text{imp}}w^2)$, where $\mathrm{Erfc}(x)$ is the complementary error function. In deriving (38), we assumed that the interference between channels is sufficiently weak, and approximated Fermi wavevector k_n to an averaged Fermi wavevector, $k_F = \pi/(2a)$.

The result has a transparent meaning: In the very ballistic regime, where M_{eff} is almost maximum, M, $\delta(|t_{\alpha,\beta}|^2)$ decreases as L/ℓ, resulting in decrease in δG. In Fig.7, the conductance fluctuations scaled by $e^2/h\cdot\sqrt{M}$ are plotted. These results are consistent with numerical results in the very ballistic regime.

It should be noted that the peak structures in δG around $L \sim \ell$ do not depend on whether W is larger than ℓ or not, in contrast to the report by Tamura and Ando[12]. They reported peak structures in δG only when $W > \ell$.

Based on the above analysis, the peak structure in δG near $L \sim \ell$ can be thought as follows: As L decreases from the diffusive regime, *i.e.*, as the number of scattering event decreases, there occurs the competition between increase in M_{eff} and decrease in $\delta(|t_{\alpha,\beta}|^2)$. The former tends to increase δG and the latter tends to decrease δG. In the region, $L \gtrsim \ell$, the former effect dominates δG, resulting in the increase in δG. In the region, $L \lesssim \ell$, on the other hand, since the scattering event is extremely rare, the latter effect dominates δG. As a result, the conductance fluctuations decrease as $\delta G/\sqrt{M} \propto L/\ell$,

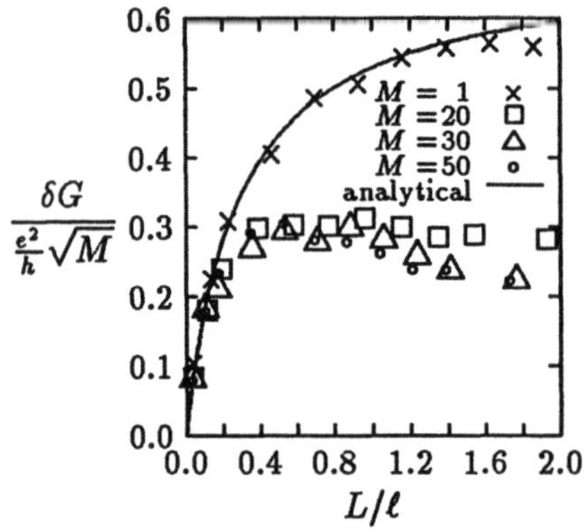

Figure 7: Conductance fluctuations scaled by $e^2/h\sqrt{M}$ as a function of L/ℓ. The solid line shows the analytical results by the perturbative calculation and the symbols the numerical results for fixed $M(= 1, 20, 30$ and $50)$.

although M_{eff} is almost maximum, *i.e.* total channel number M. Concerning this, a further study will be reported elsewhere.

4 Conclusions

We have shown that the conductance fluctuations are modified even in the system with the time reversal symmetry, provided that the system has independently fluctuating states (subbands or valleys) and/or independently fluctuating spatial domains. In such cases the modification is enhancement of the conductance fluctuation and are seen not only diffusive but also near ballistic regimes.

References

[1] P. A. Lee and A. D. Stone, Phys. Rev. Lett. **55**, 1622(1985).

[2] P. A. Lee, A. D. Stone and H. Fukuyama, Phys. Rev. **B35**, 1039(1987).

[3] B. L. Al'tshuler, JETP Lett. **41**, 648(1985).

[4] B. L. Al'tshuler and D. E. Khmel'nitskii, JETP Lett. **42**, 359(1986).

[5] S. Iwabuchi and Y. Nagaoka, J. Phys. Soc. Jpn. **58** (1989) 1325.

[6] S. Iwabuchi and Y. Nagaoka, *Proc. 3rd Int. Symp. Foundations of Quantum Mechanics* (1989) 288 ; *Proc. Symp. on New Phenomena in Mesoscopic Structures* p.169 (1989) ; *Proc. 8th Int. Workshop on Future Electron Devices* p.117 (1990).

[7] H. Higurashi, S. Iwabuchi and Y. Nagaoka, *Proc. 9th Int. Conf. on Electronic Properties of Two-Dimensional Systems*(Surface Science) to be published.

[8] K. Tankei, A. Sawada and Y. Nagaoka, J. Phys. Soc. Jpn. **58** (1989) 368.

[9] A. Sawada, K. Tankei and Y. Nagaoka, J. Phys. Soc. Jpn. **58** (1989) 639.

[10] R. Landauer, IBM J. Res. Dev. **1** (1957) 223.

[11] D. S. Fisher and P. A. Lee, Phys. Rev. **23** (1981) 6851.

[12] H. Tamura and T. Ando, Phys. Rev. **B** (submitted for publication).

Part V

Quantum Wires

Magnetotransport in Quantum Wires with Boundary Roughness

H. Akera and T. Ando*

Institute for Solid State Physics, University of Tokyo,
7-22-1 Roppongi, Minato-ku, Tokyo 106, Japan
*Present Address: Department of Engineering Science,
Faculty of Engineering, Hokkaido University, Sapporo 060, Japan

Abstract. A review is given on a quantum-mechanical calculation of the magnetoresistance and the Hall resistance of narrow wires with boundary roughness. Two different approaches, the Boltzmann transport equation and Landauer's formula, are used.

1. INTRODUCTION

Quantum wires are fabricated by introducing a confining potential in a two-dimensional electron system. A typical sample is made on a modulation-doped $GaAs/Al_xGa_{1-x}As$ heterostructure by the etching or split-gate techniques. Because of a long mean free path compared to the wire width, a number of new transport phenomena related to wire junctions have been observed, such as anomalies in the low-field Hall effect [1-3], a bend resistance [4], and a negative resistance [5]. Scattering from boundary roughness also plays an important role in long wires. The magnetotransport of such wires has been studied experimentally [6,7] and the importance of boundary roughness scattering has been pointed out based on the classical trajectory model [8]. If the wire width is comparable to the Fermi wavelength, strong quantum size effects are expected. The purpose of the present paper is to review effects of boundary roughness in such quantum wires from a theoretical point of view.

The classical theory [9] assumes that each electron follows a classical trajectory with the Fermi velocity and is reflected specularly with the probability p and otherwise scattered into a random direction. This electron-trajectory model has been introduced for the explanation of the magnetoresistance in metallic films [10,11]. In the absence of a magnetic field, the roughness itself cannot produce a nonzero resistivity because straight trajectories parallel to film surfaces have an infinite mean free path and dominate the current. In the presence of a magnetic field, the resistivity becomes nonzero because all electrons follow a curved trajectory and are scattered at a collision with the boundaries. When the cyclotron radius R_c becomes smaller than half of the film thickness W, the resistivity vanishes again because of the absence of back scattering. A detailed numerical calculation [12] performed for $p=0$ has shown that the resistivity increases with the magnetic field at low fields, exhibits a maximum at $W/R_c \approx 0.55$, and decreases down to the bulk resistivity at $W/R_c = 2$. The magnetoresistance observed in aluminum films [13] is well explained by this classical trajectory model.

In the language of quantum mechanics the boundary-roughness scattering leads to a singular current distribution among one-dimensional subbands. In fact, in the absence of a magnetic field most of the current is carried by a few low-lying subbands corresponding to the infinite mean free path of classical electrons following trajectories

Transport Phenomena in Mesoscopic Systems

along the wire direction. This peculiar distribution leads to a strong reduction of the Hall resistance measured in crossed wires as well as the positive magnetoresistance. Although some calculations of the Hall resistance were made quantum-mechanically [14] and classically [15], the importance of boundary roughness scattering has not been noted.

In this paper theories of the magnetotransport are reviewed in quantum wires in the presence of boundary roughness scattering [16,17]. The magnetoresistance is calculated using the Boltzmann transport equation in Sec. 2 and Landauer's conductance formula [18] in Sec. 3. The Hall resistance is calculated in crossed-wires geometry in Sec. 4. Summary and conclusion are given in Sec. 5.

2. MAGNETORESISTANCE

2.1 Model and Method

Consider a noninteracting two-dimensional system confined in a wire in the xy plane. In the presence of a magnetic field H perpendicular to the system, the Hamiltonian is given by

$$\mathcal{H} = \frac{1}{2m}\left(\mathbf{p} + \frac{e}{c}\mathbf{A}\right)^2 + V(y), \tag{2.1}$$

where $\mathbf{A} = (-Hy, 0)$. The confining potential is assumed to be as follows:

$$V(y) = \begin{cases} \frac{1}{2}\varepsilon_F(y+\frac{1}{2}W_0)^2/y_e^2, & \text{if } y < -\frac{1}{2}W_0; \\ 0, & \text{if } -\frac{1}{2}W_0 < y < \frac{1}{2}W_0; \\ \frac{1}{2}\varepsilon_F(y-\frac{1}{2}W_0)^2/y_e^2, & \text{if } y > \frac{1}{2}W_0. \end{cases} \tag{2.2}$$

We define an effective width W using the classical turning point of an electron traveling along the y direction with energy corresponding to half of the Fermi energy: $W = W_0+2y_e$. A hard-wall potential is obtained when $y_e \ll W$.

We shall use the Boltzmann transport equation to calculate the resistivity and therefore neglect any localization effects and fluctuations. The Boltzmann equation generalized for quantum wires gives the conductivity:

$$\sigma = \int d\varepsilon \left(-\frac{\partial f}{\partial \varepsilon}\right) \frac{4e^2}{h} \sum_{n=1}^{N} l_n(\varepsilon) \tag{2.3}$$

with l_n the mean free path of an electron in the nth subband and N the number of occupied subbands. Boundary roughness is described by deviations of boundaries: $\pm W_0/2+\Delta_\pm(x)$ and conventionally characterized by the root-mean-square deviation Δ and the correlation length Λ defined by

$$<\Delta_+(x)\Delta_+(x')>=<\Delta_-(x)\Delta_-(x')>= \Delta^2 \exp\left[-\frac{(x-x')^2}{\Lambda^2}\right] \tag{2.4}$$

In this paper Δ is assumed to be small compared with the Fermi wavelength λ_F and only the lowest order of Δ is taken into account in matrix elements of the perturbation:

$$(n'k'|\mathcal{H}'_\pm|nk) = \frac{\hbar^2}{2m} Y^\pm_{n'k'nk} \frac{1}{L} \int dx \Delta_\pm(x) \exp[i(k-k')x], \tag{2.5}$$

with

$$Y^{\pm}_{n'k'nk} = \mp\frac{2m}{\hbar^2}\int_{\pm W_0/2}^{\pm\infty} dy\ \eta_{n'k'}\frac{\partial V}{\partial y}\eta_{nk}, \tag{2.6}$$

where η_{nk} is the wave function for the motion along y of a state in the nth subband with wavenumber k and L is the wire length. In the case of the hard-wall potential ($\omega_e \to \infty$) the above integral leads to the well-known expression [19]:

$$Y^{\pm}_{n'k'nk} = \mp\frac{\partial\eta_{n'k'}}{\partial y}\frac{\partial\eta_{nk}}{\partial y}\bigg|_{\pm W/2}. \tag{2.7}$$

The Born approximation is employed in the calculation of transition rates.

2.2 Results

We first derive analytical expressions of the conductivity in some limiting cases. In the absence of a magnetic field and in the case of small correlation length $\Lambda \ll \lambda_F$, we can easily solve the Boltzmann equation and we have for the relaxation time $\tau_n = l_n/v_n$,

$$\tau_n^{-1} = \frac{\sqrt{\pi}}{2}\left(\frac{\hbar}{m}\right)^2 \Lambda\Delta^2 \sum_{m=1}^{N} \frac{(Y^{+}_{m-n+})^2 + (Y^{-}_{m-n+})^2}{v_m}. \tag{2.8}$$

In the case of the hard-wall potential, the relaxation time is proportional to $1/n^2$ because $|Y^{\pm}_{m-n+}| = 2\pi^2 mn/W^3$ where $n\pm$ means two states at the Fermi level in the nth subband. The roughness is not effective in the lower subbands and the conductivity is dominated by the current carried by the lowest subband. This is closely related to the vanishing of the resistivity at $H=0$ in the classical trajectory model. When the number of occupied subbands ($\approx 2W/\lambda_F$) is sufficiently large, we have

$$\frac{\sigma}{W} \simeq \frac{1}{3\pi^{3/2}}\frac{e^2}{h}\frac{\lambda_F^3}{\Lambda\Delta^2}\left(\frac{W}{\lambda_F}\right)^2. \tag{2.9}$$

In the presence of a magnetic field, electrons in the lower subbands start to suffer from roughness scattering due to the large Lorentz force. Therefore, the strong dependence of the relaxation time on the subband disappears and every subband is almost equally affected by scatterings, leading to a reduction in the conductivity, i.e., a positive magnetoresistance. A rough estimate of the conductivity near the resistance peak ($W \sim 0.5R_c$) is

$$\frac{\sigma}{W} \simeq \frac{3}{4\pi^{5/2}}\frac{e^2}{h}\frac{\lambda_F^3}{\Lambda\Delta^2}\frac{W}{\lambda_F}, \tag{2.10}$$

where the inverse relaxation time of each subband is assumed to be an average of the zero-field scattering rates over the occupied subbands: $\tau^{-1} = <\tau_n^{-1}>$. The classical calculation for $p=0$ [8] also predicts a linear dependence at the resistance peak, giving roughly $\sigma/W \sim (10e^2/h)W/\lambda_F$. The comparison with Eq. (2.10) suggests that $\Lambda\Delta^2/\lambda_F^3 \propto 1-p$. Equations (2.9) and (2.10) show in the limit $W/\lambda_F \to \infty$ that σ/W^2 at the vanishing field diverges while it remains finite at nonzero magnetic fields, in agreement with the classical result.

Approximate expressions can be derived also for parabolic confinement, i.e., for $y_e = W/2$. In this case, the matrix element of the roughness, Eq. (2.6), is appreciable

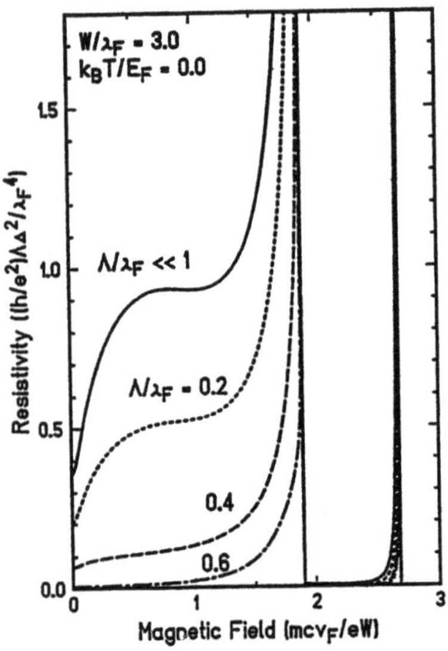

Fig. 1 Calculated magnetoresistance for wires with width $W/\lambda_F = 3$ at absolute zero for several correlation lengths. Along the horizontal axis the ratio of the wire width and the cyclotron radius is plotted

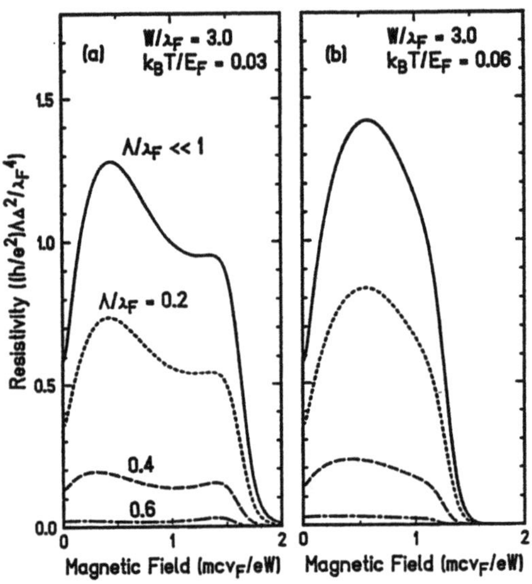

Fig. 2 Calculated magnetoresistance for wires with width $W/\lambda_F = 3$ at temperatures (a) $k_BT/E_F = 0.03$ and (b) 0.06

only among subbands which are close to each other in energy and $\tau_n \propto 2n - 1$ which leads to a weaker positive magneroresistance than in the hard-wall case.

Calculated results of the magnetic field dependence of the resistivity are presented in Fig. 1 when the temperature T is zero. The positive magnetoresistance is accompanied by quantum oscillations reflecting the density of states. At larger magnetic fields the resistivity decreases due to the formation of edge states. As the temperature increases (Fig. 2), the quantum oscillations are smeared out and the positive magnetoresistance is more clearly seen. For short correlation lengths and moderate temperatures, the resistivity exhibits a similar behavior as the classical calculations [12,8] and the experiments [7]. At correlation lengths larger than $0.6\lambda_F$ magnetoresistance disappears.

Figure 3(a) presents calculated results for a wider wire width $W/\lambda_F = 5.5$ at $k_BT/\epsilon_F = 0.02$. As the number of subbands increases, the resistivity dip at $H = 0$ becomes deeper and sharper. When the number of subbands decreases down to one, the resistivity becomes a decreasing function of the magnetic field even at the vanishing correlation length as shown in Fig. 3(b).

The conductivity divided by the wire width, or the averaged two-dimensional conductivity, is plotted as a function of the width for $\Lambda \ll \lambda_F$ in Fig. 4. The value at the vanishing magnetic field shows a parabolic dependence on the width with quantum oscillations, while that at $W/R_c = 0.5$ shows a linear dependence. A crossover occurs at $W/\lambda_F \sim 1$. The analytic expressions (2.9) and (2.10) are also given in the figure and explain the numerical results quite well.

Figure 5 shows how the shape of the confining potential affects the magnetoresistance. The effective width is fixed at $W/\lambda_F = 3$ and the percentage of the edge region,

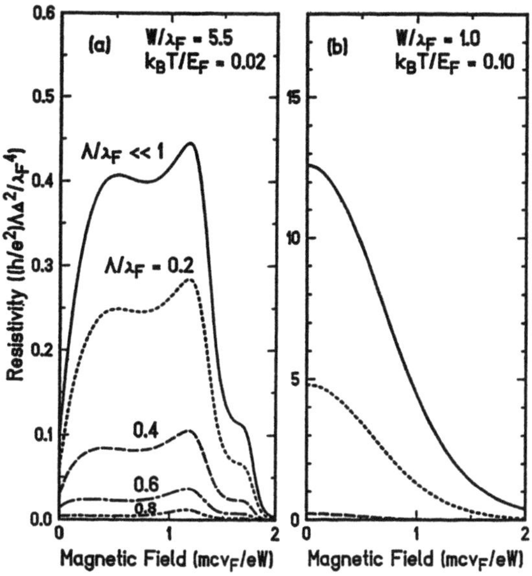

Fig. 3 Calculated magnetoresistance for wires with width (a) $W/\lambda_F = 5.5$ and (b) $W/\lambda_F = 1$.

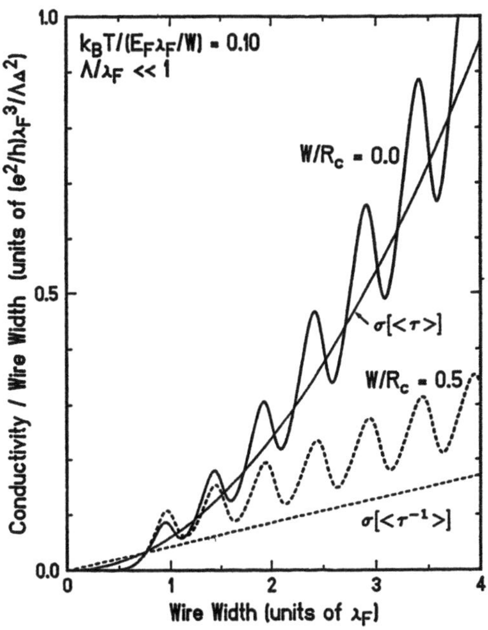

Fig. 4 The conductivity divided by the width is plotted at $H = 0$ (solid line) and at the magnetic field satisfying $W = 0.5R_c$ (dotted line) for $\Lambda \ll \lambda_F$. The asymptotic formula Eq. (2.9) (parabolic curve) and Eq. (2.10) (straight line) are also shown.

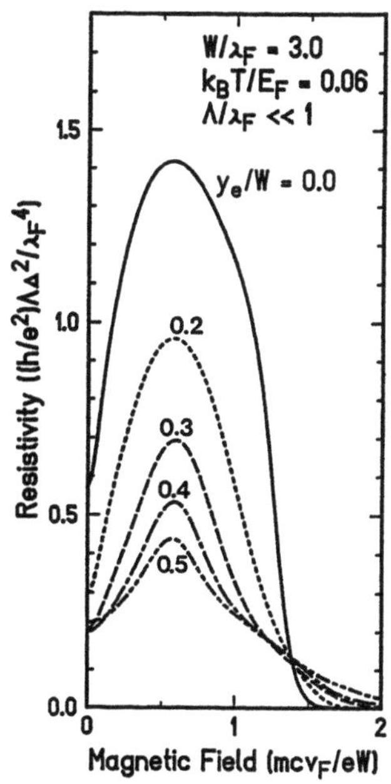

Fig. 5 Calculated magnetoresistance for wires with several confining potentials at $\Lambda \ll \lambda_F$. Wire width is fixed at $W = 3\lambda_F$ and an extent of soft walls y_e is varied. Calculated magnetoresistance for wires with width $W/\lambda_F = 3$ at temperatures (a) $k_BT/E_F = 0.03$ and (b) 0.06

y_e/W, is varied. As the confining potential becomes softer, the peak in the positive magnetoresistance becomes sharper and at the same time the resistance starts to have a longer tail at higher magnetic fields. The experimental results by Thornton *et al.* [7] can be best fitted by the result for $y_e \sim 0.3W$. However, some caution is necessary to compare the results with experiments due to the presence of localization effects in weak magnetic fields. See the next section for the importance of such phase coherence effects.

3. LANDAUER-FORMULA APPROACH

In the approach with use of the Landauer's formula, it is assumed that the wire is attached to two electron reservoirs, the current source and sink, via ideal leads. The conductance which is limited by scatterings within the wire is defined by the current divided by the voltage difference between the current source and sink. The phase coherence is assumed through the wire and the conductance is written in terms of transmission coefficients from one lead to the other through the wire. Effects of boundary roughness scatterings can be studied numerically within the lattice model [20,17] and the transmission coefficients are calculated using the technique of Green's function [21]. The conductance is calculated through the multi-channel version [22] of Landauer's formula [18] and its average is shown below over about 10000 different configurations of roughness.

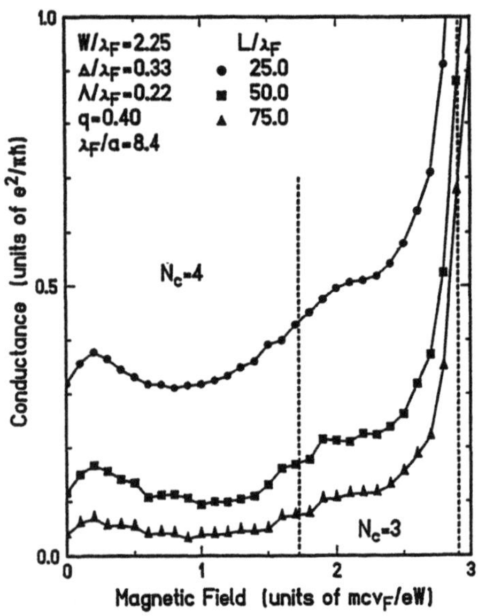

Fig. 6 Calculated conductance as a function of the magnetic field for wires with width $W/\lambda_F = 2.25$ at several lengths L. The magnetic depopulation occurs at the fields denoted by the vertical dotted lines and N_c is the number of occupied subbands.

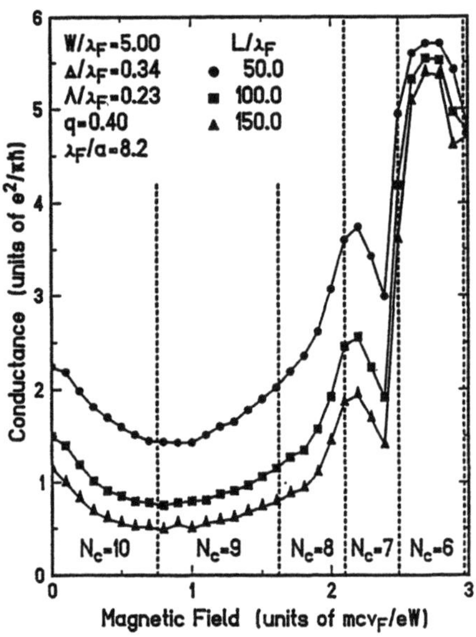

Fig. 7 Calculated conductance as a function of the magnetic field for wires with width $W/\lambda_F = 5$.

Figure 6 gives results of the numerical calculation for narrow wires with width $W/\lambda_F = 2.25$ at zero temperature. In weak magnetic fields, the conductance increases slightly with increasing magnetic field. This increase is due to the reduction of the localization effect in magnetic fields. As a matter of fact, it is more pronounced for stronger boundary roughness. With further increase of the field, the conductance starts to decrease, takes a minimum around $W/R_c \sim 1$, and then starts to increase again. This reduction in the conductance around $W/R_c \sim 1$ corresponds to the positive magnetoresistance obtained in the previous section with the use of the Boltzmann transport equation. Figure 7 shows that for wider wires with $W/\lambda_F = 5$ the negative magnetoresistance at low magnetic fields disappears. This reduction of the localization effect suggests a strong width dependence of the localization length for roughness scattering. The magnetic field which gives a minimum of the conductance is different between the present result and the result from the Boltzmann transport approach. This discrepancy is possibly another effect of the phase coherence included in the transmission approach.

4. HALL EFFECT

The Hall resistance is also strongly modified by the presence of the roughness scattering through the change of the current distribution among subbands. Here the crossed-wires geometry is chosen and right-angled sharp corners at the junction and the hard-wall confining potential are assumed. The transport within the junction is assumed to be ballistic and coherent, and transmission coefficients are calculated using the lattice model. Four arms of the crossed wires have boundary roughness and no impurities and their length is assumed to be much longer than the mean free path. The change of the current distribution due to scatterings along the wire, or the change of the chemical potentials of subbands is calculated using the Boltzmann equation.

Figure 8 shows an example of calculated Hall resistance [16]. At low magnetic fields the Hall resistance is reduced considerably from that in the absence of roughness. In this range, the current is mainly carried by the lowest subband and electrons in this subband have a large probability of going in the forward direction at the junction. The quenching of the Hall effect is due to this "collimation effect" by roughness. Experimentally the quenching of the Hall effect was observed together with the plateau at slightly higher magnetic fields (called the last plateau) [2]. Unfortunately, this last plateau can not be explained by the boundary-roughness scattering.

It is generally believed that both quenching and last plateau are a result of the presence of considerable rounding of corners at the junction of two wires [14,15,23]. In the absence of rounding of corners, there exists appreciable amount of probability of an electron being transmitted directly into voltage probes and the difference between the transmission probabilities into the left and right voltage probes increases in proportion to the magnetic-field strength due to the Lorentz force. The presence of gradual widening of the wire makes the electron wave collimated in the forward direction. Consequently, most electrons are scatterered from the junction boundaries several times before being transmitted into voltage probles. Effects of magnetic fields on electrons following such complicated trajectories are quite different from those in the absence of rounding and lead to the quenching of the weak-field Hall effect (sometimes a change in the sign). When the magnetic field becomes stronger and the cyclotron radius becomes comparable

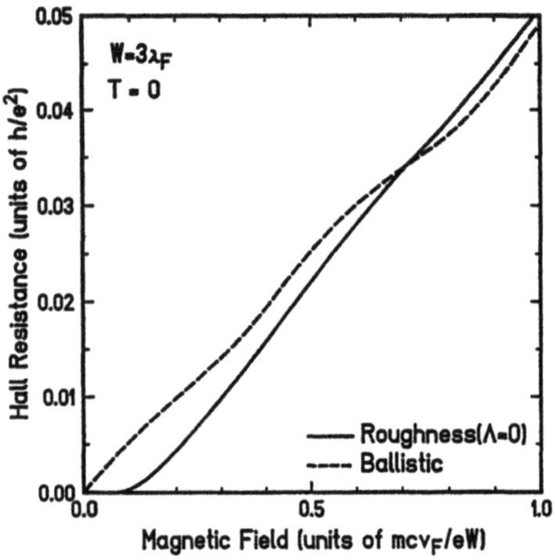

Fig. 8 Calculated Hall resistance in crossed-wires geometry in the presence of roughness and in its absence.

to the wire width, most electrons are transmitted into the left voltage probe. This leads to a saturation of the Hall effect or the last plateau.

5. SUMMARY AND CONCLUSION

Effects of boundary roughness scattering on the magnetoresistance and the Hall resistance of quantum wires are reviewed from a theoretical point of view. Roughness gives rise to a strong positive magnetoresistance, which is explained by the disappearance of the singularly long mean free paths of electrons in the lowest few subbands present at zero magnetic field. For narrow wires the effect is less prominent and vanishes when the width is less than the Fermi wavelength. At low temperatures where the phase coherence is important a sharp negative magnetoresistance appears at low magnetic fields due to localization effects for narrow wires, while this effect is too weak to be seperated from the large positive magnetoresistance for wide wires. The magnetoresistance is affected also by the shape of the confining potential and tends to be more sharply peaked at its maximum value when the confinement becomes softer. The Hall resistance is calculated in the crossed-wires geometry and shown to be suppressed at low magnetic fields because of the collimation caused by boundary roughness scattering.

ACKNOWLEDGMENTS

This work is supported in part by the Industry-University Joint Research Program "Mesoscopic Electronics" and by the Grant-in-Aid for Scientific Research on Priority Area "Electron Wave Interference Effects in Mesoscopic Structures" from the Ministry of Education, Science and Culture, Japan.

REFERENCES

1. G. Timp, A.M. Chang, P.M. Mankiewich, R.E. Behringer, J.E. Cunningham, T.Y. Chang, and R.E. Howard, Phys. Rev. Lett. **59** (1987) 732.
2. M.L. Roukes, A. Scherer, S.J. Allen, Jr., H.G. Craighead, R.M. Ruthen, E.D. Beebe, and J.P. Harbison, Phys. Rev. Lett. **59** (1987) 3011.
3. C.J.B. Ford, S. Washburn, M. Büttiker, C.M. Knoedler, and J.M. Hong, Phys. Rev. Lett. **62** (1989) 2724.
4. G. Timp, H.U. Baranger, P. de Vegvar, J.E. Cunningham, R.E. Howard, R.E. Behringer, and P.M. Mankiewich, Phys. Rev. Lett. **60** (1988) 2081.
5. Y. Takagaki, K. Gamo, S. Namba, S. Takaoka, K. Murase, S. Ishida, K. Ishibashi, and Y. Aoyagi, Solid State Commun. **68** (1988) 1051; **69** (1989) 811.
6. T.L. Cheeks, M.L. Roukes, A. Scherer, and H.G. Craighead, Appl. Phys. Lett. **53** (1988) 1964.
7. T.J. Thornton, M.L. Roukes, A. Scherer, and B.P. Van de Gaag, Phys. Rev. Lett. **63** (1989) 2128.
8. A.B. Pippard, *Magnetoresistance in Metals* (Cambridge Univ. Press, Cambridge, 1989) pp. 196–236.
9. K. Fuchs, Proc. Cambridge Philos. Soc. **34** (1938) 100.
10. D.K.C. MacDonald and K. Sarginson, Proc. Roy. Soc. London **A203** (1950) 223.
11. M.Ya. Azbel, Zh. Eksp. Teor. Fiz. **44** (1963) 1262 (Sov. Phys. JETP **17** (1963) 851).
12. E. Ditlefsen and J. Lothe, Philos. Mag. **14** (1966) 759.
13. K. Forsvoll and I. Holwech, Philos. Mag. **9** (1964) 435.
14. H.U. Baranger and A.D. Stone, Phys. Rev. Lett. **63** (1989) 414.
15. C.W.J. Beenakker and H. van Houten, Phys. Rev. Lett. **63** (1989) 1857.
16. H. Akera and T. Ando, Phys. Rev. B **41** (1990) 11967.
17. H. Akera and T. Ando, Phys. Rev. B **43** (1991) 11676.
18. R. Landauer, IBM J. Res. Dev. **1** (1957) 223; Philos. Mag. **21** (1970) 863.
19. For effects of boundary roughness on the electron transport and their theoretical treatment, see, for example, T. Ando, A.B. Fowler, and F. Stern, Rev. Mod. Phys. **54** (1982) 437.
20. T. Ando, Phys. Rev. B **42** (1990) 5626.
21. T. Ando, Phys. Rev. B **44** (1991) 8017.
22. D.S. Fisher and P.A. Lee, Phys. Rev. B **23** (1981) 6851.
23. H.U. Baranger, D.P. DiVincenzo, R.A. Jalabert, and A.D. Stone, Phys. Rev. B **44** (1991) 10637.

Shape Dependence of Ballistic Transport in Mesoscopic Systems

A. Okiji and H. Kasai

Department of Applied Physics, Osaka University, Suita, Osaka 565, Japan

Shape dependence of ballistic transport is investigated in mesoscopic systems within the framework of single electron approximation. In the quasi-one-dimensional quantum wire with a wide or a bent part, there appears inter-channel coupling which produces various pattern behaviours in the conductance. In the quantum wire with a small ring geometry, the inter-channel coupling leads to the periodic behaviour of the conductance as a function of the magnetic field, in addition to the oscillation originating from the intra-channel coupling. In the system of semi-infinite two-dimensional region with two quantum wires as an injector and a collector of electrons, the pattern behaviour of the observed electron focusing spectra can be explained qualitatively by the wavelike nature of the electron.

1 Introduction

The quantization of conductance in the ballistic transport phenomena is one of the remarkable findings, which has been discovered recently in the two-dimensional electron system of GaAs-AlGaAs heterostructures [1,2]. An explanation of the observed quantization of conductance has been given on the basis of the quantization of electronic motion perpendicular to the current direction. For simplicity, let us consider a quasi-one-dimensional straight wire. Although the electronic states along the wire are described in terms of plane wave functions, in the direction perpendicular to the current the electronic states are described by localized orbitals, and the electronic states have subband structures. When the number of subbands intersecting the Fermi level is N_c (the number of open channels), the conductance of the system can be given by $G = (2e^2/h)N_c$ on the basis of the Landauer formula [3] for the conductance. The conductance takes a constant value as long as the number of open channels is unchanged, even if the Fermi energy is increased. On the other hand, as soon as the Fermi level intersects the next subband ($(N_c + 1)$-th subband), the conductance increases by the amount of $2e^2/h$. As a result, the conductance shows step-wise variations as a function of the Fermi energy, *i.e.*, the quantization of conductance.

The observed quantization of conductance, however, turns out to be not so accurate as expected from the above simple explanation. The accuracy of the quantization may depend on various factors such as the geometry of constriction, the finite temperature and so on. Therefore, it is interesting and necessary to investigate effects of these things on the ballistic transport in addition to the effect of the coulomb interaction between electrons [4]. In the present paper, shape dependences of the ballistic transport are considered in quasi-one dimensional quantum wires within the framework of the single electron approximation [5-10]. If the width of the wire varies along the direction parallel to the current, there appears the coupling between different subbands (inter-channel coupling), which may easily disturb the above-mentioned quantization and produce various pattern behaviours in the conductance.

With regard to the interference effect, the coherent electron-focusing spectra have been measured experimentally [11,12] in the system of semi-infinte two-dimensional region with two quantum wires as an injector and a collector of electrons. In the experiment, the external magnetic field is applied perpendicularly to the two-dimensional system, and the magnetic field dependence is measured of transmission probability for electrons from the injector to the collector. The transmission probability takes maxima and minima as a function of applied magnetic fields. It is interesting to note that the values of magnetic field, at which the transmission probability takes maxima, deviate from those expected from a simple classical model for cyclotron motion of electrons in the two-dimensional system. Classically expected maximum seems to be divided into two maxima in the spectra. The pattern behaviour of the observed spectra can be explained qualitatively by the results of quantum calculations of electronic wave functions in the stationary state and of the transmission probability.

In order to describe the ballistic transport phenomena mentioned above, we introduce a Hamiltonian with two variables x and y,

$$H = -\frac{1}{2m^*}\left(\left(-i\hbar\frac{\partial}{\partial x} + eBy\right)^2 + \left(-i\hbar\frac{\partial}{\partial y}\right)^2\right) + V(x,y), \tag{1.1}$$

where m^* is the effective electron mass, $-e$ is the electron charge, B is the magnitude of magnetic field and $V(x,y)$ is a confining potential.

2 Ballistic Transport in Quantum Wires

2.1 Collimation Effect

The Fermi energy dependence of conductance is calculated numerically for the case where the effective confined length L in the right end of the wire is different from the effective confined length L_0 in the left end of the wire (see Fig.2-1). It is noted that the energy value of the bottom of n-th subband is given by $\hbar\omega(y)(n+1/2) = \hbar^2(n+1/2)/(m^*L^2(y))$ in the region where the effective confined length is $L(y)$, if the confining potential is of parabolic type. Then, it is confirmed from the numerical results [5] that the subband structure

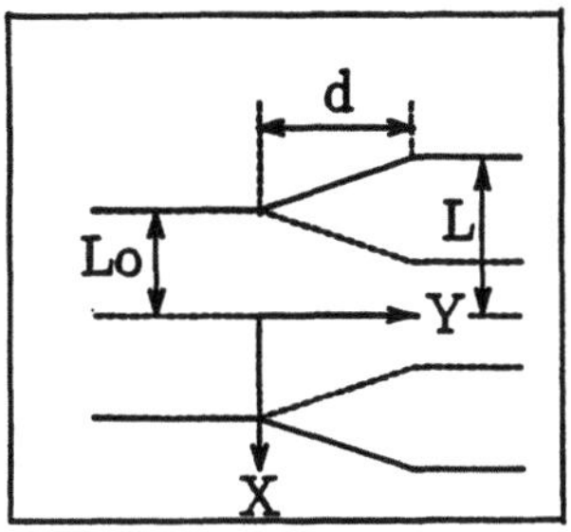

Fig.2-1

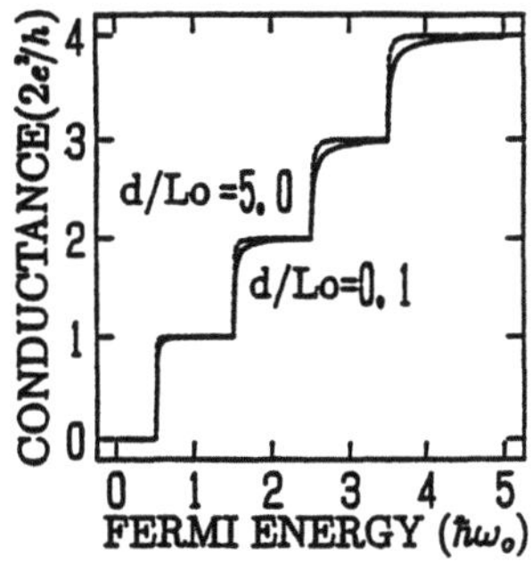

Fig. 2-2

Fig.2-1: A quasi-one-dimensional quantum wire of which the effective confined length in the right hand side (L) is different from that in the left hand side (L_0). d is the length of the region where the effective confined length depends linearly on y. From ref.[5].

Fig. 2-2: The conductance as a function of the Fermi energy for two different values of d, $d = 5.0L_0$ (dashed line) and $d = 0.1L_0$ (solid line) in the case of $L = 1.2L_0$. From ref.[5].

in the region having narrower effective confined length determines the pattern behaviour of conductance as a function of the Fermi energy. Namely, the step-wise increase of the conductance occurs as soon as the Fermi level makes contact with the next subband defined in the region having narrower effective confined length. The width of the plateau is given by the energy difference between the two adjacent subbands defined in the region having narrower effective confined length. The energy difference between the two adjacent subbands becomes larger and hence the width of the plateau becomes larger as the effective confined length gets narrower.

In Fig.2-2, the pattern behaviour of conductance is shown for two quantum wires with different length d of the region where the effective confined length depends linearly on y. The collimation effect can be seen in the figure. Namely, the step-wise increase of the conductance becomes sharper as d increases. As d becomes larger and the change of the effective confined length along the wire becomes slower, an electron can transfer more smoothly from the left end to the right end of the wire. This means that for larger value of d, the excitation of the transverse mode is suppressed at the cost of the component of kinetic energy parallel to the current and that the reflection decreases at the region where the effective confined length depends on y.

2.2 A Straight Wire with a Wide Part

The conductance of a quantum wire with an extended part (see Fig.2-3) is shown in Fig.2-4 as a function of the Fermi energy. In the figure one can see sharp dip structures near $E_F = 2.1\hbar\omega_0$ and $2.86\hbar\omega_0$ in the case of $B = 0$.

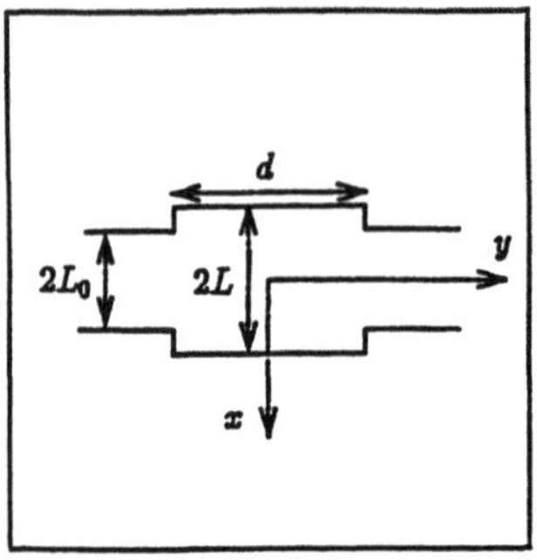

Fig.2-3

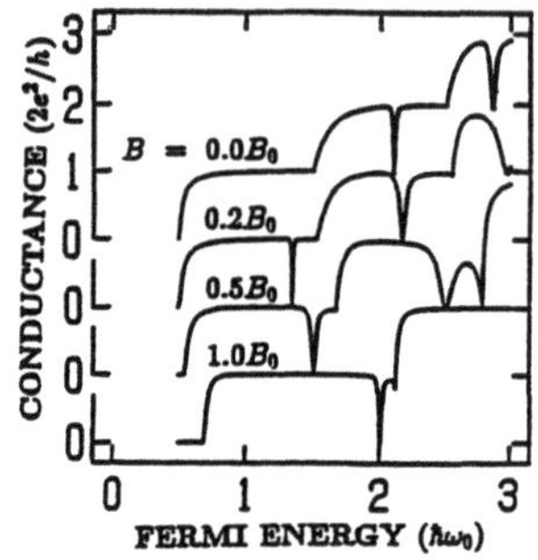

Fig.2-4

Fig.2-3: Schematic view of the model for a quasi-one dimensional quantum wire with wide part of step wise variation. $2L$ ($2L_0$) is the effective width of the wire in the region of $|y| < d/2$ ($|y| > d/2$). From ref.[6].

Fig.2-4: The Fermi energy dependence of the conductance of the quantum wire with extended (wide) part of step-wise variation in the various magnetic field, $B = 0, 0.2B_0, 0.5B_0$ and $1.0B_0$, where $B_0 = \hbar/(eL_0^2)$. Used parameters are $d = 2L_0$ and $L = 1.2L_0$. From ref.[6].

Here, ω_0 is defined by $\omega_0 = \hbar/(m^* L_0^2)$ and L_0 is the effective confined length in asymptotic regions ($y \to \pm\infty$). The origin of these dip structures can be considered as follows. In the region of $1.74\hbar\omega_0 < E_F < 2.5\hbar\omega_0$, the propagating modes (open channels) are n=0 and n=1 in the asymptotic regions ($y \to \pm\infty$). Since there exists no coupling between even and odd channels in the absence of the external magnetic field, the total conductance G can be decomposed into contributions from an even channel n=0 and an odd channel n=1, *i.e.*, $G = G_{even} + G_{odd}$. Then the dip structure near $E_F = 2.1\hbar\omega_0$ of Fig.2-4 can be considered to originate from the reduction of G_{even} due to the enhancement of the reflection probability for the electron initially in n=0 channel, which occurs if the energy of the incident electron takes the value corresponding to the energy of the bound state existing in the wide part. As a result, the number of channels contributing to the conductance is reduced by one. If the Fermi energy differs from that in the case of Fig.2-4, the localizing character of the electron wave function vanishes and the electron in n=0 channel can transfer from the left hand side of the wide part to the right hand side of it without reflection. As a result, the conductance recovers its magnitude. In the case of $E_F = 2.86\hbar\omega_0$, the electron in n=1 channel is perfectly reflected by the wide part. Thus the number of channels contributing to the conductance is reduced by one. Due to the coupling between n=0 and n=1 channels induced by the external magnetic field, additional dip structures appear in the region of $E_F < 1.5\hbar\sqrt{\omega_0^2 + \omega_c^2}$ in the presence of the external magnetic field. Here, ω_c is a cyclotron frequency given by $\omega_c = eB/m^*$. Sharp dip structures similar to those mentioned above appear as well in the conductance of a quantum wire with a wide part of smooth variation [6].

2.3 A Bent Wire

The conductance of a bent quantum wire (see Fig.2-5) is shown in Fig.2-6 as a function of the Fermi energy. For simplicity, the confining potential of hard wall type is assumed in the present calculations [7,13,14]. There are extremely narrow dips in the conductance just below the threshold energies for the next channel. The Fermi energy of the center of the first dip is $E_F = 3.9965 E_1$ and its width is approximately $0.0021 E_1$. Here, $E_1 = \hbar^2 \pi^2 / (2m^* w^2)$ is the minimum energy required for propagation in the first subband in asymptotic regions I and III. The decomposition of conductance G into the contributions from the single channels reveals the following features: In the first dip, transmission of the channel n=1 vanishes ($T_{11} \simeq 0, R_{11} \simeq 1$). On the other hand, for $E_F \simeq 8.9975 E_1$ (close to the center of the second dip), transmission from the channel n=2 vanishes almost completely (reflection is $R_{21} + R_{22} \simeq 0.93$), while transmission from the channel n=1 vanishes to a lesser extent (reflection is $R_{11} + R_{12} \simeq 0.07$). Here, $T_{mn}(R_{mn})$ is the transmission probability (reflection probability) from m-th to n-th channel. The presence of a magnetic field changes the coupling between channels in straight region I, III and in bent region II, and the composition of the vanished channel is therefore different. Although the inter-channel coupling originating from the finite curvature of the wire leads to variety of structures in the transmission and the reflection probabilities, it has little influence on the total conductance except for the dip structures (for details, see ref.[7]).

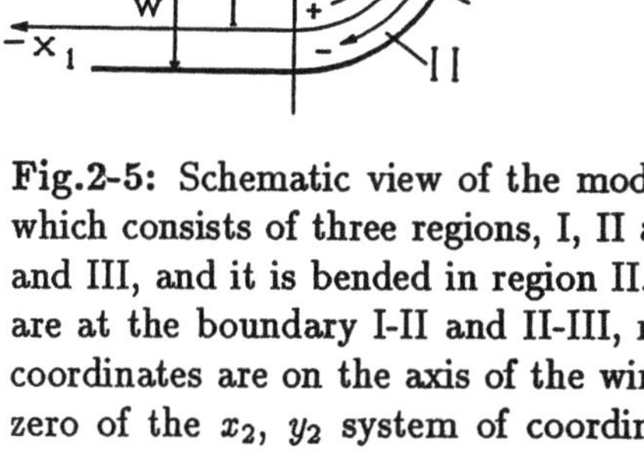

Fig.2-5

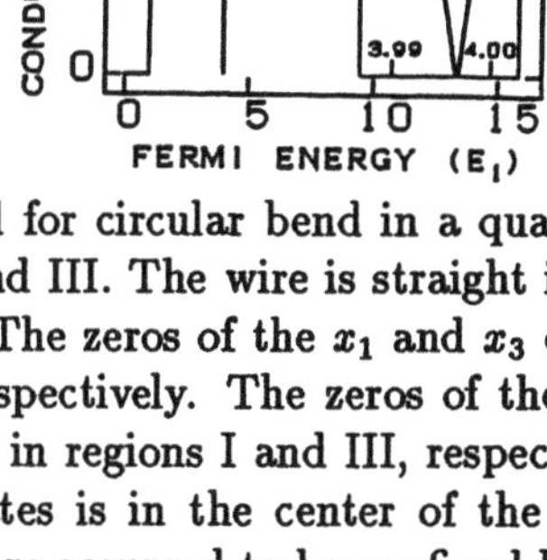

Fig. 2-6

Fig.2-5: Schematic view of the model for circular bend in a quantum wire, which consists of three regions, I, II and III. The wire is straight in regions I and III, and it is bended in region II. The zeros of the x_1 and x_3 coordinates are at the boundary I-II and II-III, respectively. The zeros of the y_1 and y_3 coordinates are on the axis of the wire in regions I and III, respectively. The zero of the x_2, y_2 system of coordinates is in the center of the cylindrical symmetry for region II. The electrons are assumed to be confined by the hard wall potential in the present calculations for simplicity. From ref.[7].

Fig. 2-6: Conductance of the bent quantum wire $R_0 = 0.5w, \Theta = \pi/2$ plotted as a function of Fermi energy for zero magnetic field. Insets: The detailed structure of the first step. From ref.[7].

2.4 Quantum Wire with a Small Ring Geometry

An explanation of Aharonov-Bohm (AB) oscillations [15] is usually given on the basis of one-dimensional ring model [16], where transverse motion of electron perpendicular to the current direction is neglected completely. On the basis of the model, one can strictly define an area enclosed by the ring in which magnetic flux passes through. As a result, a period of the AB oscillations of conductance can be obtained as a correct value of h/e. In reality, however, a ring has a finite width in this system and electronic wave functions in the ring part spread out in the transverse direction beyond the geometrical width of the ring. Then, it is not so easy to define the area enclosed by the ring exactly in which the magnetic flux passes through. A period of conductance oscillation seems not to be well-defined in this case. Furthermore, because of finite width of the wire, electronic states in the direction perpendicular to the current direction have subband structures as mention above. The coupling between electrons in different subbands is induced at the ring part, which produces various pattern behaviours in the conductance.

The magnetic field dependence of conductance of a quantum wire with a small ring part (see Fig. 2-7) is shown in Fig.2-8 in the case where the lowest

Fig. 2-7

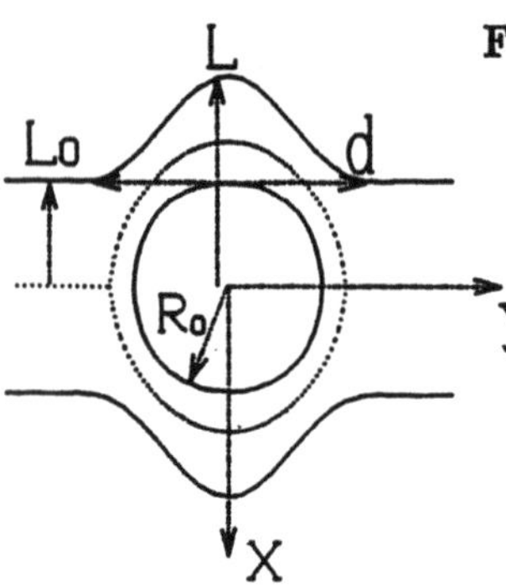

Fig. 2-8

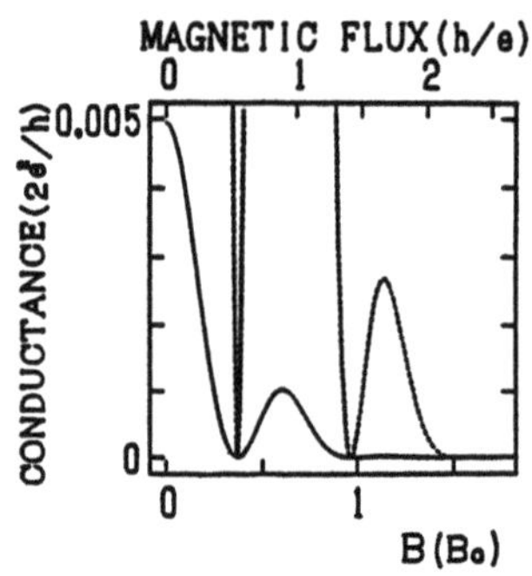

Fig. 2-7:A quasi-one-dimensional quantum wire with a small ring part. The effective confined length in the asymptotic region is given by L_0 and that in the region of ($|y| < d/2$) is given by $L(y) = (L - L_0)\cos^2(\pi y/d) + L_0$. d is the length of the region where the effective confined length depends on y. The potential barrier of Gaussian type, $V_s(x, y) = V_0 \exp[-(x^2 + y^2)/R_0^2]$, is introduced in the system in addition to the confining potential of parabolic type. The center of the potential barrier (a circle enclosed by a solid line) is located at $(x, y))$=(0,0). The bottom of the potential energy valley is denoted by a dashed line. From ref.[8].

Fig. 2-8: Conductance of the quasi-one-dimensional system of Fig. 2-7 in the magnetic field for $E_F = \hbar\omega_0$. The dashed line is enlarged one of the solid line by 100 times. Used parameters are $d = 4.0L_0, L = 1.4L_0, V_0 = 5\hbar\omega_0$ and $R_0 = L_0$. From ref.[8].

subband is an only propagating mode (open channel). Although the absolute value of conductance is small, the oscillating behaviour can be seen clearly. If we define tentatively the effective area S enclosed by the ring part as an area enclosed by the potential energy minimum curve, then, the magnetic flux passing through the area enclosed by the ring part is given by $\Phi = BS$ approximately. In the figure the dependence of conductance on Φ is also indicated. A period of the oscillation turns out to be given by h/e approximately. The conductance takes a minimum at $B = 0.37B_0$ and a maximum at $B = 0.61B_0$. In order to clarify the origin of such structures, wave functions are calculated at $B = 0.37B_0$ and $B = 0.61B_0$. In both cases, electrons traveling from the left hand side of the ring part are strongly reflected by the potential barrier. Bound states appear near the ring part on its left hand side. As a result, the transmission probability of electrons through the ring part to its right hand side becomes small. Even in the ring part of the wire, only lowest subband is an open channel in the present case. Therefore, there is little influence of the inter-channel coupling on the electronic transport. Thus, the situation seems to have a resemblance to the one-dimensional ring model mentioned above. That is, the periodic oscillations in the present case originate from a coupling within a single channel (intra-channel coupling) between electronic wave function propagating through the upper part of the ring and that propagating through the lower part of the ring. It is noted that the difference between δ_{up}/π and δ_{down}/π becomes odd integer for $B = 0.37B_0$ and even integer for $B = 0.61B_0$, respectively, which leads to the minimum and maximum values of the conductance (for details, see ref.[8]). Here, δ_{up} (δ_{down}) is the phase of the wave function propagating through the upper (lower) part of the ring.

As a reference, the magnetic field dependence of conductance is shown in Fig.2-9 together with the differential conductance $G_\ell = (2e^2/h)\Sigma_m T_{m\ell}$ in the

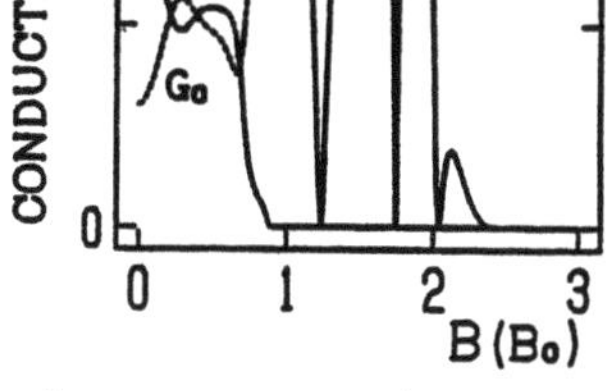

Fig. 2-9: Conductance in the magnetic field for $E_F = 2.0\hbar\omega_0$. The total conductance is denoted by G and the differential conductance by $G_\ell = (2e^2/h)\sum_m T_{m\ell}$, where the sum runs over propagating modes (open channels). Used parameters are $d = 4.0L_0$, $L = 1.4L_0$, $V_0 = 5\hbar\omega_0$ and $R_0 = L_0$. From ref.[8].

case of $E_F = 2.0\hbar\omega_0$. It is clear that the oscillations of the total conductance G originate from the oscillation of G_0. It is important to point out that the dip structures of the conductance show a periodic behaviour possibly originating from the inter-channel coupling in the region of $0.6B_0 < B < 2.0B_0$, and that the value of the period turns out to be of the order of h/e. Since the bottom of the lowest subband comes nearer to E_F for $B > 2.0B_0$, the conductance oscillation as a function of magnetic flux in this region can be considered to originate from the intra-channel coupling, which is similar to the case of Fig.2-8.

3 Electron Focusing Spectra

In order to give an account of the experimental findings of the coherent electron focusing spectra[11,12], we have done numerical calculations of the wave functions in stationary states for the geometry of semi-infinite two-dimensional region with two narrow channels as an injector and a collector in the external magnetic field (see Fig. 3-1)[10]. For simplicity, we take into account a single channel for the propagating mode (open channel). Even in this case, up to 20 modes are necessary to attain stability in the numerical results. In Fig.3-2, we plot the transmission probability as a function of magnetic field. One can see the oscillating behaviour in the transmission probabilities as a function of

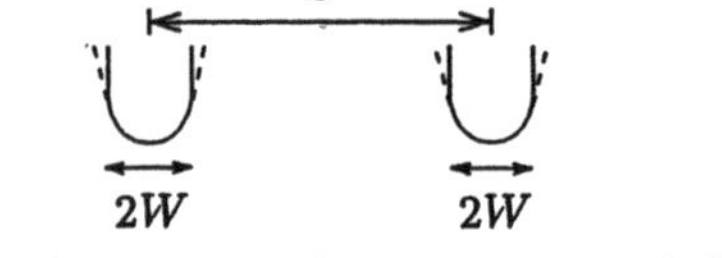

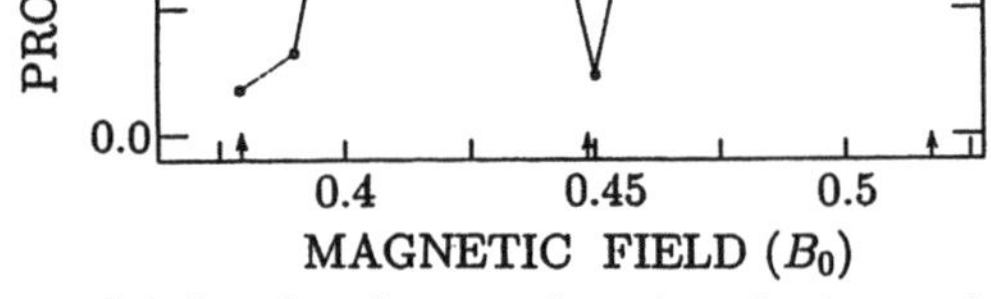

Fig. 3-1: Schematic view of the model for the electron-focusing device and the confinement potential $V(x,y)$ of the quantum wires. In the calculations we take $V(x,y) = 0$ in the semi-infinte two-dimensional region ($y > 0$), $V(x,y) = \frac{1}{2}m^*\omega_A^2 x^2$ in the injector ($y < 0$, $|x| < W$) and $V(x,y) = \frac{1}{2}m^*\omega_A^2(x-S)^2$ in the collector ($y < 0$, $|x - S| < W$). From ref.[10].

Fig. 3-2: Plotts of the transmission probability for several values of magnetic field. Maximum and minimum positions expected by a simple classical trajectory of an electron are indicated by arrows ↓ and ↑, respectively. Preliminary results for wider collector case are represented by dashed line. Used parameters are $\omega_A = 2\omega_0$, $S = 58L_0$ and $W = 4.2L_0$ for $E_F = 2\hbar\omega_0$. From ref.[10].

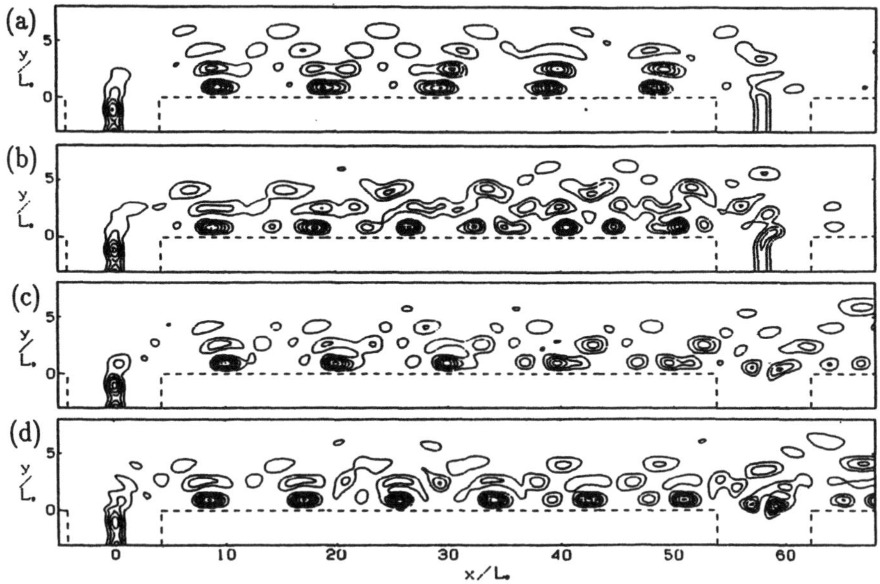

Fig. 3-3: The modulus squared of the wave function in the case of $B = 0.400B_0$ (a), $B = 0.43B_0$ (b), $B = 0.379B_0$ (c) and $B = 0.45B_0$ (d). From ref.[10].

magnetic field. As for the periodic pattern behaviour of the value of magnetic field at which the transmission probability takes minima, it seems to correspond to the classical cyclotron motion (skipping motion) of electrons ejected from an injector of infinitesimally small width. However, as for the pattern behaviour of the value of magnetic field at which the transmission probability takes maxima, the characteristic behaviour appears as the wavelike nature of electron. In order to see these facts, we draw in Fig.3-3 the modulus squared of the wave function for $B = 0.40B_0, 0.43B_0, 0.379B_0$ and $0.45B_0$. In the case where the transmission probability is large, the electronic density distribution shows two kinds of pattern behaviours. The first one, as shown in Fig.3-3 (a), is formed by periodic peaks along the boundary between the injector and the collector, and this remainds us the focused classical skipping orbit of the electron. The second one, as shown in Fig.3-3 (b), implies that the transmission probability can be enlarged by redistribution of the electron near the collector. Thus, the peak is considered to be produced by the wavelike nature of the electron. Note that the densities of probability in the case of $B = 0.475B_0$ and $0.500B_0$ are similar to the pattern behaviours of Fig.3-3 (a) and Fig.3-3 (b), respectively. In the case of Fig.3-3 (c) and (d), on the other hand, the electronic density distribution around the collector looks like just that of the mode n=1, which is the evanescent wave (closed channel) at the collector and is not able to transmit into it. In order to confirm this fact, we investigate the wide collector system where parameters $\omega_A = \omega_0$ and $W = 6.0L_0$ are taken only for the collector permitting the propagating modes n=0 and 1. The preliminary results are shown in Fig.3-2(a), where the transmission probability at $B = 0.379B_0$ or $B = 0.450B_0$ is increased remarkably. Then, the difference is decreased between the maximum and the minimum values of the transmission probability. This tendency has been observed in the experiment [12].

Acknowledgments

The present authors are grateful to A. Nakamura, S. Suga, H. Nakanishi, K. Vacek, K. Mitsutake, Y. Maki and N. Negishi for discussions. This work was supported in part by the Grant-in-aid for Scientific Research on Priority Area, "Electron Wave Interference Effects in Mesoscopic Structures" from the Ministry of Education, Science and Culture.

References

[1] B. J. van Wees, H. van Houten, C. W. J. Beenakker, J. G. Williamson, L. P. Kouwenhoven, D. van der Marel and C. T. Foxon, Phys. Rev. Lett. **60** (1988) 848.

[2] D. A. Wharam, T. J. Thornton, R. Newburg, M. Pepper, H. Ahmed, J. E. F. Frost, D. G. Hasko, D. C. Peacock, D. A. Ritchie and G. A. C. Jones, J.Phys.C **21** (1988) L209.

[3] R. Landauer, Philos. Mag. **21** (1970) 863.

[4] A. Nakamura and A. Okiji, J. Phys. Soc. Jpn. **60** (1991) 1873.

[5] A. Okiji, H. Kasai and A. Nakamura, Prog. Theor. Phys. Suppl. (1992) in press

[6] H. Kasai, K. Mitsutake and A. Okiji, J. Phys. Soc. Jpn. **60** (1991) 1679.

[7] K. Vacek, H. Kasai and A. Okiji, J. Phys. Soc. Jpn. **61** (1992) No.1

[8] A. Okiji, H. Kasai and K. Mitsutake, submitted to J. Phys. Soc. Jpn.

[9] A. Nakamura, Y. Maki and A. Okiji, J. Phys. Soc. Jpn. **60** (1991) 749.

[10] A. Okiji, N. Negishi and A. Nakamura, submitted to J. Phys. Soc. Jpn.

[11] H. van Houten , B. J. van Wees, J. E. Mooij, C. W. Beenakker, J. G. Williamson and C. T. Foxon, Europhys. Lett. **5** (1988) 721.

[12] H. van Houten, C. W. Beenakker, J. G. Williamson, M. E. I. Broekaart, P. H. M. van Loosdrecht, B. J. van Wees, J. E. Mooij, C. T. Foxon and J. J. Harris, Phys. Rev. **B39** (1989) 8556.

[13] F. Soles and M. Macucci, Phys. Rev. **B41** (1990) 10354.

[14] R. L. Schult, D. G. Ravenhall and H. W. Wyld, Phys. Rev. **B39** (1989) 5476.

[15] Y. Aharonov and D. Bohm, Phys. Rev. **115** (1959) 485.

[16] Y. Gefen, Y. Imry and M. Ya. Azbel, Phys. Rev. Lett. **52** (1984) 129.

Application of the Kirchhoff-Helmholtz Formula to the Evaluation of Wave Functions in Mesoscopic Systems

K. Kawamura, T. Ueta, and H. Sawano

Department of Physics, Faculty of Science and Technology,
Keio University, 3-14-1 Hiyoshi, Kohoku-ku, Yokohama 223, Japan

Abstract. The Kirchhoff-Helmholtz (KH) integral Formula is useful to solve the Schrödinger equation for electrons with arbitrary boundary conditions. The Green function which is involved in KH formula in the presence of a magnetic field is expressed in a closed form in term of the Whittaker function. As examples, the wave functions are computed in the devices of the Hall resistance and the coherent magnetic focusing.

1. Introduction

The Kirchhoff-Helmholtz (KH) formula, which is called the Kirchhoff-Huygens formula in optics, can be applied to study electron waves in topologically disordered systems. The topological disorder mentioned here is a disorder which is not described by perturbation potential in the Schrödinger equation. As an example, one of the present authors (KK) studied the electron scattering off a screw dislocation by using KH formula [1].

The KH formula transforms the differential equations in a space of the d-dimension to an integral equation in a space of the $(d-1)$-dimension. This feature is practically important, since it saves the region of memories on a computer. In fact, we can study the Aharonov-Bohm problem by this method to get practically meaningful informations [1], although the vector potential field is extended to infinity.

In the following sections, we will briefly describe the application of KH formula and BEM to study the electrons in mesoscopic systems in the ballistic regime.

2. Kirchhoff-Helmholtz Formula

The time-independent Schrödinger equation

$$\frac{1}{2m}\left[-i\hbar\nabla - q\mathbf{A}(\mathbf{r})\right]^2\psi(\mathbf{r}) = E\psi(\mathbf{r}) \tag{1}$$

has the following solution for $\mathbf{r}$ within a closed region Ω bounded by the surface Σ;

$$\psi(\mathbf{r}) = \int_\Sigma \left[G(\mathbf{r},\mathbf{r}')\mathbf{n}\cdot\mathbf{v}'\psi(\mathbf{r}') - (\mathbf{n}\cdot\mathbf{v}'^*G(\mathbf{r},\mathbf{r}'))\psi(\mathbf{r}')\right]dS' \tag{2}$$

where $\mathbf{v} = \frac{1}{m}(-i\hbar\nabla - q\mathbf{A}(\mathbf{r}))$ is the velocity operator and $\mathbf{n}$ represents the unit vector normal to Σ. The Green function $G(\mathbf{r},\mathbf{r}')$ is an outgoing solution to the equation

$$\frac{1}{2m}\left[-i\hbar\nabla - q\mathbf{A}(\mathbf{r})\right]^2 G(\mathbf{r},\mathbf{r}') = EG(\mathbf{r},\mathbf{r}') - \delta(\mathbf{r}-\mathbf{r}'). \tag{3}$$

The expression (2) is the KH formula relevant to (1). Notice that the right-hand side involves the boundary values of $\psi(\mathbf{r})$ and $\mathbf{n}\cdot\nabla\psi(\mathbf{r})$, whose number is twice of the necessary conditions to get a unique solution to the Schrödinger equation. When $\mathbf{r}$ stands for the coordinates on the surface Ω, we can derive the following integral equations

$$\left(1-\frac{\theta}{2\pi}\right)\psi(\mathbf{r})=\int_{\Sigma}\left[G(\mathbf{r},\mathbf{r}')v_n'\psi(\mathbf{r}')-(v_n'G(\mathbf{r},\mathbf{r}'))\psi(\mathbf{r}')\right]dS'$$

in which $\mathbf{r}$ are coordinates of points on the boundary and θ is the angle of the edge formed by the boundary line at the point $\mathbf{r}$.

The boundary values which are not included in the boundary conditions for the Schrödinger equation are given by solving this equation.

3. Green Function in a Magnetic Field

In order to use KH formula, the Green function in a free space should be generated on a computer. Closed form expressions for the Green functions in the absence of a magnetic fields, however, had not been known, to the knowledge of the present authors, until one of them (TU) [2] derived an expression in terms of the Whittaker function, one of the confluent hypergeometric functions.

Since the translational symmetry is broken by the vector potential, the Green function $G(\mathbf{r}_1,\mathbf{r}_2)$ depends on both the coordinates $\mathbf{r}_1$ and $\mathbf{r}_2$ in the presence of a magnetic field. If we had to compute the Green function for every sets of $(\mathbf{r}_1,\mathbf{r}_2)$, we should spend long computation time and a large memory region on a computer.

Ueta, however, found [2] that a Green function $G(\mathbf{r}_1,\mathbf{r}_2)$ can be factorized as follows;

$$G(\mathbf{r}_1,\mathbf{r}_2)=\exp\left(i\frac{q}{\hbar}\int_{\mathbf{r}_2}^{\mathbf{r}_1}\mathbf{A}\cdot d\mathbf{r}\right)G_B(\mathbf{r}_1-\mathbf{r}_2) \tag{4}$$

where the integral is performed along a straight line which connects the points represented by $\mathbf{r}_1$ and $\mathbf{r}_2$.

To derive this formula, we introduce the translation operator [3]

$$T(\mathbf{R})=\exp\left(\frac{i}{\hbar}\mathbf{R}\cdot(\mathbf{p}+q\mathbf{A})\right)$$

in which $\mathbf{p}$ represents a momentum operator and $\mathbf{R}$ are lattice vectors.

It is shown by using this operator that the Green function G_s for the symmetric gauge $A(\mathbf{r})=\frac{1}{2}\mathbf{B}\times\mathbf{r}$ obeys the following relation;

$$G_s(\mathbf{r}_1,\mathbf{r}_2)=\exp\left(i\frac{q}{2\hbar}\mathbf{B}\cdot(\mathbf{r}_2\times\mathbf{r}_1)\right)G_B(\mathbf{r}_1-\mathbf{r}_2). \tag{5}$$

The function $G_B(\mathbf{r}_1-\mathbf{r}_2)$ which is $G_s(\mathbf{r}_1-\mathbf{r}_2,\mathbf{0})$ depends on the two coordinates through their differences and it is independent of the gauge.

The Green function G for an arbitrary gauge $\tilde{\mathbf{A}}$, which is combined with the symmetric gauge $\mathbf{A}$ by $\tilde{\mathbf{A}}=\mathbf{A}+\nabla\Lambda$ is related to the Green function G_s by

$$G(\mathbf{r}_1,\mathbf{r}_2)=\exp\left\{i\frac{q}{\hbar}\left(\Lambda(\mathbf{r}_1)-\Lambda(\mathbf{r}_2)\right)\right\}G_s(\mathbf{r}_1,\mathbf{r}_2).$$

Insertion of (5) in the right-hand side of this equation proves (4).

Another factor which depends on $\mathbf{r}_1$ and $\mathbf{r}_2$ separately is gauge-dependent but it is easily computable exponential function.

It follows from (3),

$$\left[\frac{\partial}{\partial z}\left(z\frac{\partial}{\partial z}\right)-\frac{1}{4}z+\varepsilon\right]G_B(z)=\frac{m}{2\pi\hbar^2}\delta(z). \tag{6}$$

To derive this equation, new variables have been defined by

$$z=\frac{1}{2}\left(\frac{\mathbf{r}_1}{l}-\frac{\mathbf{r}_2}{l}\right)^2, \varepsilon=\frac{E+i\delta}{\hbar\omega}$$

where l is the magnetic length and ω is the cyclotron frequency. The equation (6) has the solution

$$G_B(z;\varepsilon)=-\frac{m}{2\pi\hbar^2}\frac{\pi}{\cos\pi\varepsilon}\frac{1}{\Gamma(\varepsilon+\frac{1}{2})}\frac{W_{\varepsilon,0}}{\sqrt{z}}$$

where $W_{\varepsilon,0}$ is the Whittaker function.

4. Electron Waves in Some Devices

When the area of boundary Σ is finite and small, KH formula can be easily applied to compute the wave function. A typical example is the cross geometry of devices in the experiments of the Hall resistance anomalies. In this case, the boundary Σ in Eq.(2) is formed by the walls of the conducting paths and the orifices of the wave guides [See Figure 1a]. When the boundary values at the orifices are determined, the transmission amplitudes to various wave guides and then, by the Landauer-Bütticker formula, the Hall resistance is computed. As an example, the present authors examined the picture in the billiard model [4] which successfully explains the Hall resistance anomalies. It has been found that the billiard model provides good picture only when the incident wave belongs to the lower transverse modes.

Although we should employ infinitely large area of boundary to study the diffraction of electron waves injected into a semi-infinite plane in the absence of magnetic field, the integral over the semicircle in Figure 1b vanishes as R^{-1} when its radius R tends to infinity. This is because the outgoing waves in the absence of a magnetic field decays as R^{-1} even if they are well collimated [6].

Analysis of the electron diffraction from a narrow wave guide has been published in a separate form [1].

In the presence of a magnetic field, electron waves which are injected in a semi-infinite region propagates along the classical path shown in Figure 1c. Since electron waves do not decay because of the focusing action of a magnetic field we should employ

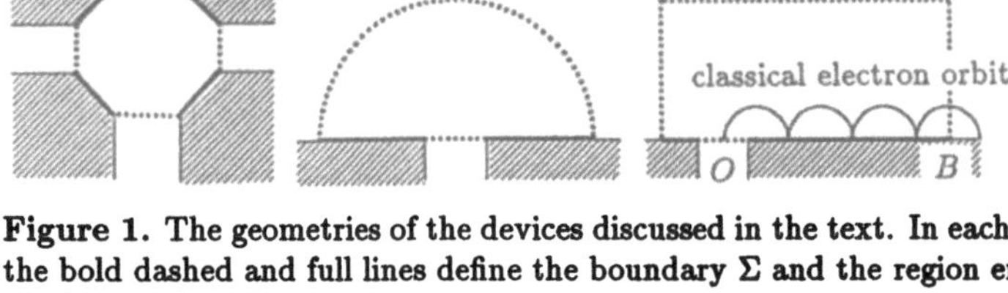

Figure 1. The geometries of the devices discussed in the text. In each figure, the bold dashed and full lines define the boundary Σ and the region enclosed by them defines Ω.

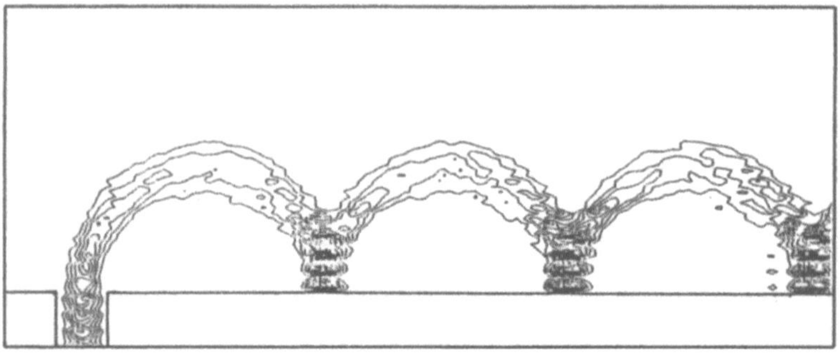

Figure 2. Cycrotron orbit of electron wave injected from the wave guide attached on the left-hand side of the bottom. The incident wave belongs to the fundamental mode.

some procedures as the impedance matching at the boundary AB in Figure 1c with a linear combination of edge states at the boundary.

The example of contours of electron intensity shown in Figure 2 has been obtained putting $OB = 15 \times r_c$, where r_c represents the the cyclotron radius.

Since the complete sets of wave functions within the semi-infinite region in the presence of the magnetic field is not known yet, the procedure of the impedance matching is not trivial. Some technical problem about this procedure will be discussed in the near future.

Acknowledgments

The present work is supported partially by a Grant-in-Aid for Scientific Research on Priority Area, "Electron Wave Interference Effects in Mesoscopic Structures " from the Ministry of Education, Science and Culture.

References

[1] K, Kawamura, Z. Physik B **30** 1 (1978).

[2] T. Ueta, *Proc. Int. Sympo. on Science and Technology of Mesoscopic Structures*, Nara Nov. 6-8, 1991 (Springer Verlag, Tokyo, 1992) to be published.

[3] R. E. Peierls, Z. Phys. **80**, 763 (1933).

[4] C. J. B. Ford, S. Washburn, M. Büttiker, C. M. Knoedler, J. M. Hong, Phys. Rev. Lett. **62**, 2724 (1989).

[5] K. Kawamura, T. Ueta, and H. Sawano, Jpn. J. Appl. Phys. (in press).

[6] A. Sommerfeld, *Partial Differential Equations in Physics* (Academic Press Inc., New York, 1964) P 189.

Quantum Mechanics of an Electron on a Curved Surface

Y. Nagaoka[1] *and M. Ikegami*[2]

[1]Yukawa Institute for Theoretical Physics, Kyoto University, Kyoto 606-01, Japan

[2]Department of Physics, Nagoya University, Nagoya 464-01, Japan

Abstract. The effect of the bend of a surface to which electrons are confined is studied. It is shown that there appears an attractive potential due to the curvature in the two-dimensional effective Hamiltonian of an electron. Some examples of the curved surface are examined. An interesting behavior of electrons is found in the case of a circular cone, that is "fall" of electrons to the vertex.

1. Introduction

Recent technical development made it possible to realize a clean surface where electrons trapped on it behave as free particles in two dimensions. In such a system various interesting quantum mechanical behaviors of electrons are expected. In this paper, we considered a clean curved surface, and examine how the motion of electrons is affected by the shape of its bend.

Several authors have worked on this problem from different points of view, which happens to give different answers. Among them, Jensen and Koppe [1] considered a very narrow region along a curved surface in which a particle is confined by a potential. Ogawa, Fujii and Kobushukin [2] introduced a curved surface in a higher-dimensional space and considered a particle constrained on it. In the former theory a small but finite width of the surface is taken into account, whereas in the latter the width is put zero from the beginning. The former model is more realistic to treat actual cases.

In this paper we give a review of the former theory from a different point of view with some extension. We also give some applications, and show some interesting phenomena due to the effect of surface bending [3].

2. Simple Cases

As a model of an electron on a curved surface, we consider an electron confined between two parallel hard walls with a constant width d. In actual cases, the width may not be constant, and there appears an effective potential depending on the width. However, we take the width constant to avoid undesirable complexity. The motion of the electron perpendicular to the surface is quantized by the hard wall potential, which results in the subband structure. We consider an electron in the lowest subband, and derive the effective Hamiltonian for its motion parallel to the surface in the limit d tends to zero.

Before deriving the effective Hamiltonian for general cases of the curved surface, it seems instructive to consider some simple cases: a cylindrical surface and a spherical surface.

Springer Series in Solid-State Sciences, Vol. 109 Eds.: H. Fukuyama and T. Ando
Transport Phenomena in Mesoscopic Systems

Cylindrical Surface

We consider an electron confined between two cylindrical surfaces with a common axis and radii $R \pm d/2$. By using the cylindrical coordinates (r, θ, z), the Hamiltonian is written as

$$H = -\frac{\hbar^2}{2m}\left(\frac{\partial^2}{\partial r^2} + \frac{1}{r}\frac{\partial}{\partial r} + \frac{1}{r^2}\frac{\partial^2}{\partial \theta^2} + \frac{\partial^2}{\partial z^2}\right) , \tag{1}$$

where m is the mass of the electron. The wave function $\Psi(r, \theta, z)$ obeys the Schrödinger equation

$$H\Psi = E\Psi. \tag{2}$$

Due to the hard walls located at $r = R \pm d/2$, the wave function should satisfy the boundary condition

$$\Psi(R \pm d/2, \theta, z) = 0 . \tag{3}$$

If we take the wave function as

$$\Psi(r, \theta, z) = \frac{1}{\sqrt{r}}\Phi(r, \theta, z), \tag{4}$$

the Schrödinger equation (2) is rewritten as

$$\hat{H}\Phi = E\Phi, \tag{5}$$

where

$$\hat{H} = -\frac{\hbar^2}{2m}\left(\frac{\partial^2}{\partial r^2} + \frac{1}{4r^2} + \frac{1}{r^2}\frac{\partial^2}{\partial \theta^2} + \frac{\partial^2}{\partial z^2}\right) \tag{6}$$

In the case of small d, we can take the adiabatic approximation to solve the problem. In this approximation, we put the wave function as

$$\Phi(r, \theta, z) = \chi(r;\ \theta, z)\psi(\theta, z) \tag{7}$$

and take χ as the solution of

$$\hat{H}\ \chi(r;\ \theta, z) = H_{\text{eff}}(\theta, z)\chi(r; \theta, z), \tag{8}$$

where the coordinate θ is treated as a parameter. Here $\chi(r;\ \theta, z)$ should satisfy the boundary condition

$$\chi(R \pm d/2;\ \theta, z) = 0. \tag{9}$$

For small d, the wave function χ and the effective Hamiltonian for the lowest subband are given by

$$\chi(r) = \sqrt{\frac{2}{d}}\cos\left[\frac{\pi}{d}(r - R)\right] \tag{10}$$

$$H_{\text{eff}} = \frac{\pi^2\hbar^2}{2md^2} - \frac{\hbar^2}{2m}\left(\frac{1}{R^2}\frac{\partial^2}{\partial \theta^2} + \frac{\partial^2}{\partial z^2}\right) - \frac{\hbar^2}{8mR^2} \tag{11}$$

The result is exact in the limit d tends to zero, and holds as far as d is much smaller than the radius R and the wavelength λ of the electron motion along the θ-direction:

$$d \ll R, \lambda. \tag{12}$$

The first term of the effective Hamiltonian (11) is very large, but as far as the width is constant, it is constant and can be ignored. The second term is the kinetic energy of the electron motion along the surface. Attention should be paid to the third term. This term is also constant on the cylindrical surface where the radius R is constant. However, the result may be generalized to the case where the radius changes spacially. It should be noticed that this term arises from the local property of the surface. Therefore, it is expected that this term remains even when the curvature changes spacially, and that there appears an effective attractive potential which depends on the radius of curvature R:

$$V_{\text{eff}} = -\frac{\hbar^2}{8mR^2} \, . \tag{13}$$

Spherical Surface

In order to see what happens in general cases where the surface is curved in two directions, we next study a simple case of a spherical surface; i.e., we consider an electron confined between two spheres with a common center and radii $R \pm d/2$.

By taking the spherical coordinates (r, θ, ϕ), the Hamiltonian is given by

$$H = -\frac{\hbar^2}{2m}\left\{\frac{\partial^2}{\partial r^2} + \frac{2}{r}\frac{\partial}{\partial r} + \frac{1}{r^2}\left[\frac{1}{\sin\theta}\frac{\partial}{\partial\theta}\left(\sin\theta\frac{\partial}{\partial\theta}\right) + \frac{1}{\sin^2\theta}\frac{\partial^2}{\partial\phi^2}\right]\right\} \tag{14}$$

In this case, we put the wave function as

$$\Psi(r,\theta,\phi) = \frac{1}{r}\,\Phi\,(r,\theta,\phi). \tag{15}$$

Then the Schrödinger equation reduces to

$$\hat{H}\Phi = E\Phi, \tag{16}$$

where

$$\hat{H} = -\frac{\hbar^2}{2m}\left\{\frac{\partial^2}{\partial r^2} + \frac{1}{r^2}\left[\frac{1}{\sin\theta}\frac{\partial}{\partial\theta}(\sin\theta\frac{\partial}{\partial\theta}) + \frac{1}{\sin^2\theta}\frac{\partial^2}{\partial\phi^2}\right]\right\} \tag{17}$$

An important difference between the two Hamiltonians (6) and (17) should be noticed; i.e., the term proportional to r^{-2} in the former is missing in the latter. Thus, by taking the same procedure as before, we can easily see that there appears no effective potential in this case.

This result shows that the effective potential disappears if the radii of curvature are the same in two directions. It should also reduce to Eq.(13) when one of the radii of curvature tends to infinity. Then it is expected that in general cases the effective potential is given by

$$V_{\text{eff}} = -\frac{\hbar^2}{8m}\left(\frac{1}{R_1} - \frac{1}{R_2}\right)^2 \, , \tag{18}$$

where R_1 and R_2 are the two principal radii of curvature, which may change spacially. In the next section, we will prove Eq.(18) in a more general way.

3. General Cases

To treat general cases, we introduce the Gaussian normal coordinates (τ, q^1, q^2) defined in the neighborhood of the surface, where τ and $q \equiv (q^1, q^2)$ are the coordinates normal to and along the surface. We define the metric as

$$ds^2 = d\tau^2 + h_{ij}(\tau, q)dq^i dq^j \quad (i, j = 1, 2). \tag{19}$$

Then the Hamiltonian is expressed as

$$H = -\frac{\hbar^2}{2m} h^{-1/2} \left(\frac{\partial}{\partial \tau} h^{1/2} \frac{\partial}{\partial \tau} + \frac{\partial}{\partial q^i} h^{ij} h^{1/2} \frac{\partial}{\partial q^j} \right) ,$$

where h and h^{ij} are respectively the determinant and the inverse of h_{ij}.

If we transform the wave function $\Psi(\tau, q)$ into $\Phi(\tau, q)$ by

$$\Psi(\tau, q) = f^{-1} \Phi(\tau, q), \tag{20}$$

where

$$f = h^{1/4}, \tag{21}$$

then the Hamiltonian for Φ is given by

$$\hat{H} = -\frac{\hbar^2}{2m} \left(\frac{\partial^2}{\partial \tau^2} - \frac{1}{f} \frac{\partial^2 f}{\partial \tau^2} + h^{-1/4} \frac{\partial}{\partial q^i} h^{ij} h^{1/2} \frac{\partial}{\partial q^j} h^{-1/4} \right) \tag{22}$$

Due to the hard walls, the wave function Φ should satisfy the boundary condition

$$\Phi(\pm d/2, q) = 0. \tag{23}$$

Next we introduce the extrinsic curvature tensor $\boldsymbol{K}$ whose elements are given by

$$\kappa_{ij} \equiv -\frac{1}{2} \frac{\partial h_{ij}}{\partial \tau} .$$

Then, after some manipulations, the second term in the brackets of Eq.(22) is calculated as

$$\frac{1}{f} \frac{\partial^2 f}{\partial \tau^2} = -\frac{1}{2} (Tr\boldsymbol{K})^2 + K_G , \tag{24}$$

where K_G is the Gaussian curvature. By using the principal curvatures κ_1 and κ_2, each term of Eq.(23) is expressed as

$$\left. \begin{aligned} \frac{1}{2} Tr\boldsymbol{K} &= \frac{1}{2}(\kappa_1 + \kappa_2) \\ K_G &= \kappa_1 + \kappa_2 \, . \end{aligned} \right\} \tag{25}$$

Thus the Hamiltonian (22) is rewritten as

$$\hat{H} = -\frac{\hbar^2}{2m} \left(\frac{\partial^2}{\partial \tau^2} + h^{-1/4} \frac{\partial}{\partial q^i} h^{ij} h^{1/2} \frac{\partial}{\partial q^j} h^{-1/4} \right) - \frac{\hbar^2}{8m} (\kappa_1 - \kappa_2)^2 \, . \tag{26}$$

As before, we here introduce the adiabatic approximation which is exact in the limit d tends to zero. Then the effective Hamiltonian for the motion of electrons along the surface is given simply by replacing the curvatures κ_1 and κ_2 by those of the surface itself. They are expressed as $\kappa_i = 1/R_i$, where R_i is the principal radius of curvature of the surface. In this way we finally obtain Eq.(18) as the effective potential for general cases.

It is worthwhile to note here that this term can not be obtained in strictly two-dimensional cases. In the traditional method of quantization of the systems with constraints, i.e., the Dirac method, the coordinate τ and the momentum p conjugate to τ are first put zero. Evidently this is incompatible with the uncertainty principle for the motion perpendicular to the surface. This is the reason why the present result is different from the result of Ogawa et al. [2] based on the Dirac method.

The result is readily generalized to the case in the presence of an external magnetic field, if the field is of ordinary strength. We can show that the effective Hamiltonian in a magnetic field is obtained by replacing the momentum p_i along the surface by

$$p_i + \frac{e}{c}A_i \, , \tag{27}$$

where A_i is the component of the vector potential in the q^i direction. In order that this result holds, the magnetic field should satisfy the following condition:

$$d \ll \ell_B \equiv \sqrt{\frac{c\hbar}{eB_{||}}} \, , \tag{28}$$

where $B_{||}$ is the component of the magnetic flux density parallel to the surface.

4. Applications

One-Dimensional Bend

When the surface is bended only in one direction, we can take the Cartesian coordinates (x, y) along the surface, where the x- and y-axes are respectively parallel and perpendicular to the bending direction. Then the effective Hamiltonian becomes separable, and for the x direction it is given by

$$H_{\text{eff}} = -\frac{\hbar^2}{2m}\frac{d^2}{dx^2} - \frac{\hbar^2}{8mR(x)^2} \, . \tag{29}$$

The problem is nothing but an elementary exercise of quantum mechanics of one particle in an attractive potential.

In a simple case of a round corner as shown in Fig.1 [4], the effective potential becomes a square-well one given by

$$V_{\text{eff}}(\lambda) = \begin{cases} -\dfrac{\hbar^2}{8mR^2} & |x| < \dfrac{1}{2}R\alpha \\ 0 & |x| > \dfrac{1}{2}R\alpha \, , \end{cases} \tag{30}$$

where the origin is taken at the middle point of the corner. In this case, it is found that one bound state is formed at the corner.

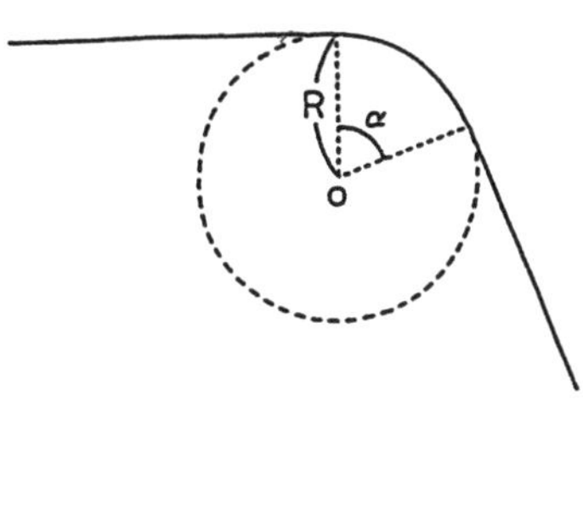

Fig.1 A cross section of the corner; two planes are connected with a portion of a cylinder whose radius is R.

Bend with Rotational Symmetry

When the surface has the rotational symmetry, we take the coordinates (r, ϕ), where r is the distance from the center of symmetry along the ridge, and ϕ is the angle about the symmetry axis (Fig.2). Then the effective Hamiltonian is given by

$$\begin{aligned} H_{\text{eff}} = & -\frac{\hbar}{2m}\left[\frac{\partial^2}{\partial r^2} + \frac{1}{\rho(r)^2}\frac{\partial^2}{\partial \phi^2}\right] + \frac{\hbar^2}{8m}\left\{2\frac{\rho''(r)}{\rho(r)} - \left[\frac{\rho'(r)}{\rho(r)}\right]^2\right\} \\ & - \frac{\hbar^2}{8m}\left[\frac{1}{R(r)} - \frac{\sin\theta(r)}{\rho(r)}\right]^2 \end{aligned} \tag{31}$$

Here $\theta(r)$ is the angle between the symmetry axis and the direction normal to the surface and $R(r)$ and $\rho(r)$ are the radii of curvature along the ridge direction and the direction normal to it. $\rho'(r)$ and $\rho''(r)$ stand for the first and second derivatives of $\rho(r)$.

In the case of a circular cone, we have

$$\theta(r) = \frac{\pi}{2} - \alpha, \qquad R(r) = \infty, \qquad \rho(r) = r\tan\alpha. \tag{32}$$

Then the Hamiltonian becomes

$$H_{\text{eff}} = -\frac{\hbar^2}{2m}\left(\frac{\partial^2}{\partial r^2} + \frac{1}{r^2\sin^2\alpha}\frac{\partial^2}{\partial\phi^2}\right) - \frac{\hbar^2}{8mr^2\sin^2\alpha}\,. \tag{33}$$

There appears an attractive potential proportional to the inverse square of r. For the eigenstate with the angular momentum $\hbar\ell$, the Hamiltonian of the radial motion is given by

$$H_{\text{eff}}^{(\ell)} = -\frac{\hbar^2}{2m}\frac{d^2}{dr^2} + \frac{\hbar^2}{8m}\frac{4\ell^2 - 1}{r^2\sin^2\alpha}\,. \tag{34}$$

Thus for $\ell = 0$ electrons may be fallen to the vertex; i.e., the extent of the ground state wave function is zero and the ground state energy $-\infty$ [5].

In actual cases, this can never happen, since the top of the cone is rounded with a finite radius. In the round region, there is no attractive potential. Therefore, the minimum value of the potential will be of the order $-\hbar^2/ma^2$, where a denotes the extent of the round region. The extent of the ground state wave function will be of the order a, and the ground state energy of the order $-\hbar^2/ma^2$.

It is interesting to see how many electrons are trapped at the top of the cone. If more than two electrons are trapped, the Coulomb repulsion will act between them. However, the inverse square potential overwhelms the Coulomb interaction in the region

$$r < r_c \simeq a_B\ , \tag{35}$$

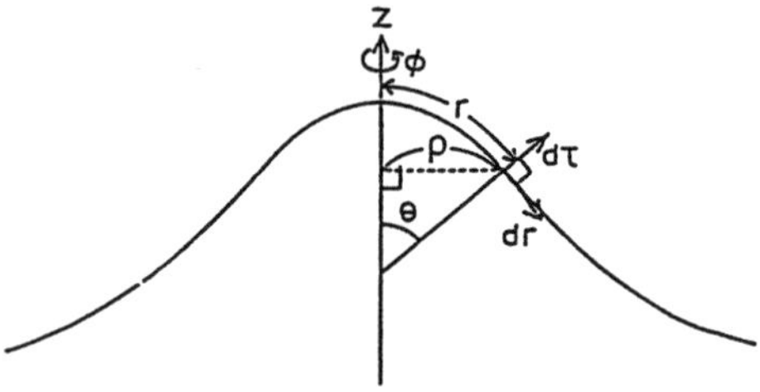

Fig.2 A cross section of the interface with rotational symmetry.

where a_B denotes the Bohr radius. Therefore, if the condition

$$a < a_B \tag{36}$$

is satisfied, the accumulation of electrons at the top of the cone will be expected. This may be called as "quantum lightning rod".

Acknowledgement
The authors would like to thank Professor S. Takagi for valuable discussions. This work was supported in part by the Grant-in-Aid for Scientific Research on Priority Area "Electron Wave Interference Effects in Mesoscopic Systems" from the Ministry of Education, Science and Culture.

References

[1] H. Jensen and H. Koppe, Ann. of Phys. **63** (1971), 586.

[2] N. Ogawa, K. Fujii and A. Kobushukin, Prog. Theor. Phys. **83** (1990), 894.

[3] M. Ikegami and Y. Nagaoka, Prog. Theor. Phys. Suppl. No.106 (1991), 235.

[4] R.C.T. da Costa, Phys. Rev. **A23** (1981), 1982.

[5] L.D. Landau and E. M. Lifshitz, Quantum Mechanics: non-relativistic theory, 2nd ed. (Oxford Pergammon, 1965), §35.

Nonlocal Coupling of Magnetoplasma Modes in Lateral Quantum-Wire Superlattices

S. Katayama

Institute of Physics, College of General Education, Niigata University, Niigata 950-21, Japan

Nonlocality in dispersion relations of magnetoplasmons and higher harmonics of cyclotron resonance in a lateral superlattice of quantum wires is studied theoretically. It is shown that the nonlocal coupling of the magnetoplasmons and the higher harmonics of cyclotron resonance gives rise to significant effects on far-infrared transmission spectra for relatively wide GaAs quantum wires.

1. Introduction

The recent experimental studies of far-infrared [FIR] radiation spectroscopy [1-3] as well as of Raman scattering [4-6] have directly revealed the electronic excitations of a laterally confined two-dimensional (2D) i.e. quasi-1D electron system realized in semiconductor quantum wires. Theoretical works [7-11] have also explored the dispersion relations of the collective electronic excitations in the quantum wires. In our recent studies [12,13], we have adopted the theory of the FIR absorption in Si-MOS systems [14] to examine the FIR transmission spectra of the deep-mesa-etched quantum-wire superlattices [15] without a metal grating. Our calculation has given a good description for the transmission spectra in the absence and presence of a magnetic field in which we have considered only the plasma waves oscillating perpendicular to the 1D wires. The present work develops a nonlocal theory of the FIR transmission resonances due to the magnetoplasma absorption in which the plasma charge density oscillates parallel and perpendicular to the 1D wires. The nonlocal i.e. finite wavenumber q effects express the influence of the wave field on the dynamics of electrons during the traverse of characteristic length L. Therefore the effects can be classified by a parameter $X = qL$ with $q = 2\pi/\lambda$, λ being the wavelength of the associated waves. It is known in bulk materials that these determine the characteristic feature of lineshape of the Azbel-Kaner cyclotron resonance as well as of the geometric resonances in the magnetoplasma and magnetoacoustic absorption [16]. From these studies, we could probe the dynamics of electrons from both the collective plasma description and the single-particle picture. The recent advances of the nano-meter fabrication techniques make it possible to observe these effects in the quasi-1D and quasi-2D electron gas (EG) in the lateral quantum-wire superlattice by means of a metallic grating coupler technique [17].

In this case L is referred to a radius of cyclotron orbit (R) or a width of wire (t). We demonstrate these effects by calculating the transmission spectra with use of the nonlocal conductivities which are evaluated by solving the semiclassical Boltzmann equation, taking into account the lateral confinement of the 2DEG in the quantum wires.

2. FIR Transmission

Here we calculate the FIR transmission spectra of the periodic arrays of semiconductor quantum wires i.e. the lateral quantum-wire superlattice as shown in Fig. 1. The system consists of the laterally confined quasi-2DEG at z= d bounded by the two insulating layers (1, 2) and the substrate in which multiple-quantum-wire system is fabricated with width t and period a by the deep-mesa-etched technique. The hatched area indicates metallic strips evapolated onto the horizontal 1D wires with period b along the y-axis. The incident unpolarized FIR radiation with the electric field vector $\mathbf{E}^{0i}$ at $z < 0$ is in nearly normal to the layers, and a magnetic field **B** is applied along the z-axis. We suppose that the insulating layers and substrate are transparent in the FIR frequency regime. The energy loss of the FIR radiation occurs at electron layer due to the coupling between the induced currents of the 2D free carriers and the electric fields $E_i(z)$. Then the FIR transmission intensity T(B) is evaluated by an energy consideration of the incident radiation including such energy dissipations. It should be noted that $E_i(z)$ at $z > 0$ contains the non-radiative components with wavenumbers $q_{nx}=(2\pi n/a)$ and $q_{my}=(2\pi m/b)$ [n,m = ±1, ±2, ±3, . .] produced by the grating lattice, in addition

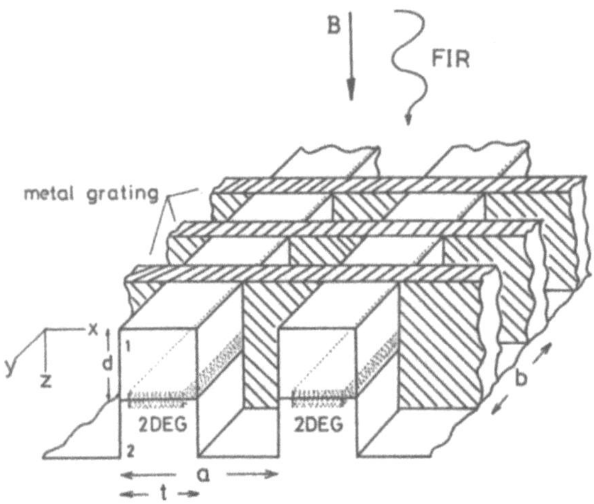

Fig. 1 Geometry of FIR transmission. a, b, and t denote the period of surface grating, metallic grating, and the width of quantum-wires. The magnetic field B is oriented normal to the surface.

to the inherent radiative component $q_{nx} = q_{my} = 0$ [12]. The explicit form of T(B) is given by

$$T(B) = 1 - \mathrm{Re}[\sum_{hh'} \frac{4\pi}{c} \sigma_{ii}^{eff}(q_h, q_{h'})$$

$$\times \frac{E_i^*(q_h)\, E_i(q_{h'})}{|E_i^{0i}(0, z<0)|^2}], \tag{1}$$

where q_h denotes the wave-vector (q_{nx}, q_{my}). Since the electric field component polarized parallel to the 1D wires i.e. i = y in eq.(1) is only transmitted because of a metallic grating coupler, our required effective conductivity component becomes

$$\sigma_{yy}^{eff}(q_h, q_{h'}) = \sum_{h''} \varepsilon_{yy}^{-1}(q_h, q_{h''}:\omega)\sigma_{yy}(q_{h''}, q_{h'})$$

$$+ \sum_{h''} \varepsilon_{yx}^{-1}(q_h, q_{h''}:\omega)\sigma_{xy}(q_{h''}, q_{h'}), \tag{2}$$

ε^{-1} being the inverse of the dielectric function [12]. The second term on the right hand side of eq.(2) comes from the magnetic field induced polarization current.

3. Numerical Results and Discussions

To evaluate approximately ε^{-1} in eq.(2), the components of ε with $\pm q_h$ $(= \pm q_{h'}) = (0,0)$, $(q_{1x}, 0)$, $(0, q_{1y})$, (q_{1x}, q_{1y}) are retained. Further, we approximate σ_{ij} the nonlocal 2D conductivities which are calculated by using the semiclassical Boltzmann equation, taking into account the lateral confinement of the 2DEG to the simple parabolic potential [18]. Following Cohen et al.[19] and Chaplik and Heitmann [20], we can find

$$\sigma_{ij} = \frac{2\sigma_0}{v_F^2} \sum_{n=-\infty}^{\infty} [\begin{smallmatrix} a_{nx}^* \\ b_{ny}^* \end{smallmatrix}] J_n(\bar{x}) [\begin{smallmatrix} a_{nx} \\ b_{ny} \end{smallmatrix}] J_n(\bar{X}) \frac{1}{1 + i(n\Omega - \omega)\tau}, \tag{3}$$

where J_n and v_F denote the Bessel function and the Fermi velocity, $\sigma_0 = N_s e^2 \tau / m^*$, N_s, m^*, and τ being the quasi-2D electron density, the electron effective mass, and the phenomenological relaxation time, respectively, $\Omega = (\omega_c^2 + \Omega_0^2)^{1/2}$ with the cyclotron frequency $\omega_c = eB/m^*c$ and the effective harmonic frequency Ω_0 for the parabolic potential, and

$$a_{nx} = \gamma[\frac{n\bar{q}_x \bar{v}_F}{\bar{X}\bar{Q}} - i \frac{\bar{q}_y \bar{v}_F}{\bar{Q}} \frac{\partial}{\partial \bar{X}}], \tag{4a}$$

$$b_{ny} = \frac{1}{\gamma}[\frac{n\bar{q}_y \bar{v}_F}{\bar{X}\bar{Q}} + i \frac{\bar{q}_x \bar{v}_F}{\bar{Q}} \frac{\partial}{\partial \bar{X}}]. \tag{4b}$$

Using $\gamma = \sqrt{\Omega/\omega_c}$, $\bar{q}_x = \gamma q_x$, $\bar{q}_y = q_y/\gamma$, $\bar{v}_F = v_F/\gamma$, and $\bar{Q}^2 = \bar{q}_x^2 + \bar{q}_y^2$. Further, $\bar{X} \cong \bar{Q}\bar{v}_F/\Omega$. As $\gamma \to 1$ when B increases, the various

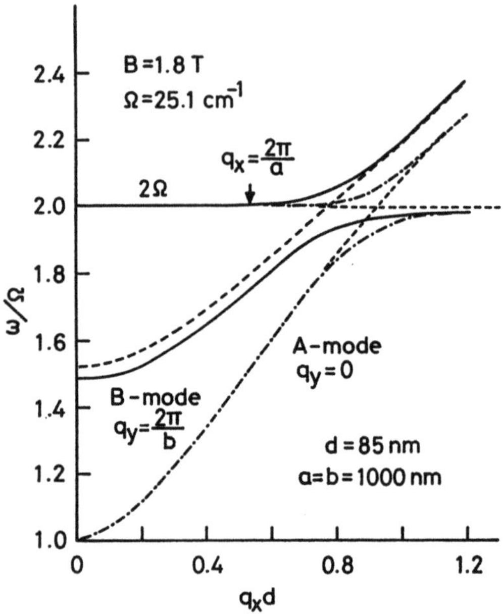

Fig. 2 The anticrossing of the dispersions of the magnetoplasma modes and 2Ω at B=1.8 T. The dot-dashed and solid lines are the dispersions for q_y =0 and 2π/b, respectively. The dashed lines denote the dispersion of the non-interacting magnetoplasmons and 2Ω.

quantities behave as ones in the isotropic 2D case. From eq.(2), we expect that the transmission resonances occur at energies of the collective magnetoplasma modes i.e. poles of ε^{-1} as well as at nΩ. The used parameters for numerical computations are ε_0= 13.8, N_s = 4.4×10^{11} cm^{-2}, m^* = 0.0707 m_0, τ = 1.48 psec and $\hbar\Omega_0$ = 1.0 meV. Further, a= b= 1.0μm, t = 0.54 μm, d = 85nm.

To show anticrossing of the dispersion relations for the magnetoplasmons and 2Ω as was observed in experiments [17], we plot the dispersion relation as a function of $q_x d$ for B= 1.8 T in Fig. 2. Since $\bar{X}$ ≃ 0.5, the energies of each mode are estimated from $\varepsilon^{-1}\to\infty$ by expanding σ_{ij} in ε^{-1} to the lowest order of $\bar{X}^2$. The dot-dashed and solid lines indicate the dispersion relations for q_y= 0 and 2π/b which are referred to the A- and B-mode, respectively. The frequencies of the non-interacting magnetoplasmons and 2Ω are also depicted by the dashed lines. In the A-mode, the charge density oscillates perpendicular to the 1D wire, while the B-mode has the finite oscillating amplitude of charge density parallel to the 1D-wire direction together with the perpendicular component. It is seen that there exists a coupling of 2Ω and both the A- and B-mode. The coupling of the B-mode and 2Ω is quite significant in the near vicinity of q_x= 2π/a (see arrow in Fig. 2).

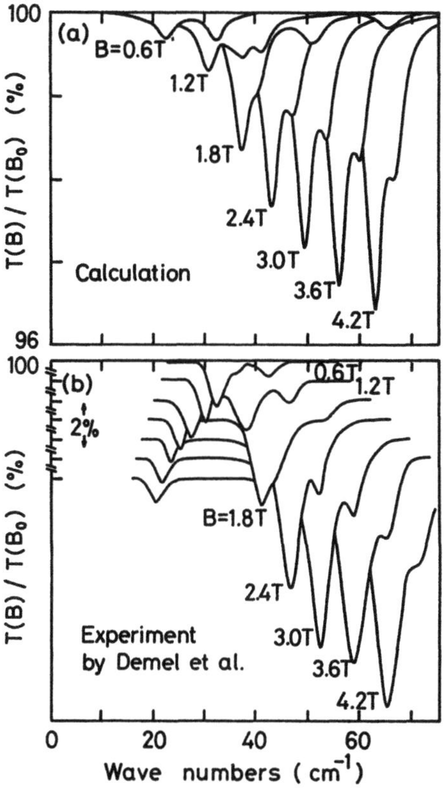

Fig.3(a) Calculated FIR transmission spectra at B= 0.6 - 4.2 T by including the nonlocal effects. The other parameters are the same as in Fig. 2.
3(b) Transmission spectra observed by Demel et al.[17] at 2.2 K.

In Figs. 3(a) and (b), we plot the calculated and observed $T(B)/T(B_0)$ for seven different magnetic fields where $T(B_0)$ is the reference of intensity at a strong magnetic field B_0. In the present calculation, we ignore the contributions from $\sigma_{yy}^{eff}(0,0)$. It is evident that there appear clearly two resonances in high frequency regime above B= 2.4 T in both calculated and experimental spectra. The variation of the resonance frequencies as well as of the separation of the two resonances as B increases are reproduced well by the present calculaltions. The change of resonance energies as B increases may come from the increment of the magnetoplasma excitation energy through Ω. However there exists a systematic deviation from the observed resonance frequencies to low-frequency side. In the present analysis, we regard these resonances as ones due to collective modes in lateral superlattice of quantum wires, while Demel et al.[17] have pointed out that the resonances originate from the localized plasmon in a well with the effective width w. From the present Fourier analysis, one can not

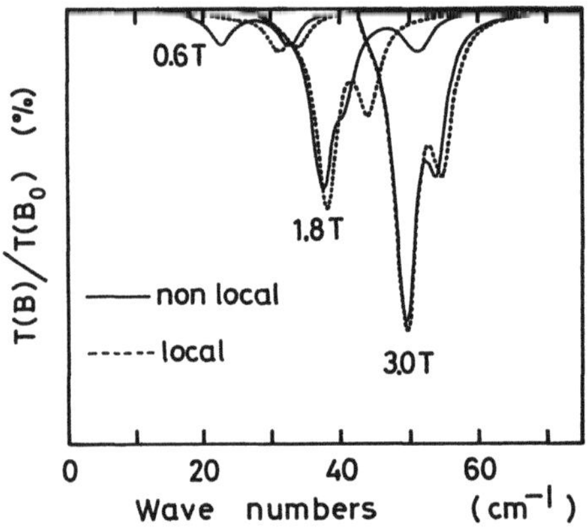

Fig. 4 Nonlocal effects on the transmission spectra in comparison with the results in the local limit at B= 0.6, 1.8, and 3.0 T. The other parameters are the same as in Fig. 3.

estimate w. Although there remain such unsolved points, the two resonances in spectra above B= 2.4 T may be ascribed to the A- and B-mode at $q_x = 2\pi/a$ on the basis of this direct comparison. Note that the difference of the calculated transmission intensity in the two resonances is due to the difference in the strength of the electric fields.

Figure 4 shows the nonlocal effects on the lineshape of transmission spectra in explicit comparison with the lineshape in the local limit for B = 0.6, 1.8, and 3.0 T. The dashed lines indicate the results for the local limit, whereas the solid lines displays the nonlocal one. By the nonlocal effects, the B-mode resonance is shifted to lower energy side as a shoulder of the A-mode resonance and a new weak resonance at 2Ω appears when B = 1.8 T. Though it seems for the nonlocal effects at low field B= 0.6 T to be quite large in our calculation, this is due to the cut in the series expansion of σ_{ij} to the order of $\bar{X}^2$. To make more quantitative comparison for the lower fields, we have to consider the terms in σ_{ij} with more higher order than $\bar{X}^2$ [16].

The puzzle in Fig. 3(b) is that there occurs a quenching of cyclotron resonance in the laterally bounded 2DEG as was observed in the absence of metal grating [3]. If $\sigma^{eff}_{yy}(0,0)$ is included in the present approximation, the cyclotron resonance appears in our spectra as one in the pure 2DEG [14,21]. To interpret occurrence of the quenching, we should consider the off-diagonal contributions of $\varepsilon^{-1}(q_h,q_{h'}:\omega)$ with $q_h \neq q_{h'}$, which are ignored in the present approximation. In the observed spectra below 35 cm^{-1}, we also see a very intriguing resonance whose energy decreases as B increases. Demel et al.[17] have found that the resonance is attributable to the edge-like mode propagating along the 1D wires [22].

In conclusion, our calculations suggest that the observed FIR transmission spectra in GaAs quantum wires [17] exhibit the significant nonlocal effects. In particular, the nonlocal coupling of the magnetoplasmons and the higher harmonics of cyclotron resonance can be observed by a combination of an appropriate magnitude of B and accessible grating spacings. In this way, we will utilize the nonlocal effects in order to probe the electron dynamics of a laterally bounded 2DEG i.e. quasi-1DEG in a quantum-wire superlattice.

Acknowledgement

This work was supported in part by a Grant-in-Aid for Scientific Research on Priority Area, "Electron Wave Interference Effects in Mesoscopic Structures" from the Ministry of Education, Science and Culture.

References

1. See the review article: D. Heitmann, in Physics and Applications of Quantum Wells and Superlattices, edited by E.E.Mendez and K.von Klitzing (Plenum, 1987) p. 317, and also see T. Demel, D. Heitmann, and P. Grambow, in Spectroscopy of Semiconductor Microstructures, edited by G. Fasol, A. Fasolino, and P. Lugli (Plenum, 1989) p.75.
2. U. Merkt, Superlattices and Microstructures **6**, 341 (1989).
3. T. Demel, D. Heitmann, P. Grambow, and K. Ploog, Phys. Rev. **B38**, 12732 (1988).
4. J.S.Weiner, G.Danan, A. Pinczuk, J. Valladares, L.N. Pfeiffer, abd K. West, Phys. Rev. Lett. **63**, 1641 (1989).
5. T. Egeler, G. Abstreiter, G. Weinmann, T. Demel, D. Heitmann, P. Grambow, and W. Schlapp, Phys. Rev. Lett. **65**, 1804 (1990).
6. A.R. Goni, A.Pinczuk,J.S.Weiner,J.M.Calleja, B.S. Dennis, L.N. Pfeiffer, and K.W.West, Phys. Rev. Lett. **67**, 3298 (1991).
7. A.V. Chaplik, Superlattices and Microstructures, **6**, 329 (1989).
8. Q. Li and S. Das Sarma, Phys. Rev. **B40**, 5860 (1989), Phys. Rev. **B43**, 11768 (1991).
9. Wei-ming Que and G. Kirczenow, Phys. Rev. **B40**, 5998 (1989).
10. U.Wulf, E. Zeeb, P. Gies, R.R. Gerhardts, and W. Hanke, Phys. Rev. **B41**, 3113 (1990).
11. V.A. Shchukin and K.B. Efetov, Phys. Rev. **B43**, 14164 (1991).
12. S. Katayama, J. Phys. Soc. Jpn. **60**, 1123 (1991).
13. S. Katayama, Proc. 12th In'l Conf. of Electronic Propertis in Two-Dimensional Electron System (1991, Nara) to be published in Surface Science.
14. See the review: T. Ando, A.B. Fowler, and F. Stern, Rev. Mod. Phys. **54**, 437 (1982), and references therein.
15. T. Demel, D. Heitmann, P. Grambow, and K. Ploog, Appl. Phys. Lett. **53**, 2176 (1988).

16. See, for example, P.M. Platzman and P.A.Wolf, Waves and Interactions in Solid State Plasmas, (Academic Press, 1973), pp.151-203.
17. T. Demel, D. Heitmann, P.Grambow, and K.Ploog, Phys. Rev. Lett. **66**, 2657 (1991).
18. K.-F. Berggren, T.J.Thornton, D.J. Newson, and M. Pepper, Phys. Rev. Lett. **57**, 1769 (1986).
19. M.H.Cohen, M.J. Harrison and W.A. Harrison, Phys. Rev. **117**, 937 (1960).
20. A.V. Chaplik and D.Heitmann, J.Phys.C: Solid State Phys. **18**, 3357 (1985).
21. D.Liu and S. Das Sarma, Phys. Rev. **B44**, 9122 (1991).
22. A.L.Fetter, Phys.Rev. **B33**, 3717 (1986), G. Eliasson, J.-W.Wu, P. Hawrylak, and J.J. Quinn, Solid State Commun. **60**, 41 (1986).

Part VI

Quantum Hall Effects

Roles of Bulk Extended States in Quantum Wires in High Magnetic Fields

T. Ando

Institute for Solid State Physics, University of Tokyo,
7-22-1 Roppongi, Minato-ku, Tokyo 106, Japan

Two examples of recent theoretical works are reviewed which have been done for the purpose of understanding the relationship between the bulk Hall current picture and edge current picture of the quantum Hall effect. The first example is on the localization in quantum wires, which gives information on mixing of edge states with bulk states and on the edge-state wavefunction inside the wire. As the second example, the transmission probability of edge states is calculated and its energy and size dependence is shown to be determined by localization of bulk Landau states.

1. INTRODUCTION

A key ingredient of the quantum Hall effect is the localization of Landau states in the presence of random potential fluctuations. It has been shown in various different ways that the Hall conductivity is quantized into integer multiples of $-e^2/h$ when states around the Fermi level are all localized. Several reviews have been published already [1-3]. In quantum wires, on the other hand, the Hall effect is understood in terms of the transmission probability of edge channels [4]. As a matter of fact, the Hall resistance is easily shown to be quantized when backscattering of edge states is negligible. The relationship between these two pictures based on bulk Landau states and edge states still remains unclear. The purpose of the present paper is to give a brief review of some of theoretical works performed for the purpose of shedding some light on this problem.

This paper is organized as follows. In Sec. 2 a short review is given of explanations of the quantum Hall effect. In Sec. 3 localization of edge states in quantum wires is discussed. It is shown that the width dependence of the localization length gives how edge-state wavefunctions decay inside the wire. In Sec. 4 the backscattering probability of edge states is calculated numerically. It will be shown that the dependence of the Hall-step width on the wire width is determined by the energy dependence of the localization length of bulk Landau states.

2. QUANTUM HALL EFFECT

In the picture based on the bulk Hall current, we start with the Hall conductivity. The Kubo formula for the Hall conductivity is written as

$$\sigma_{xy} = -\frac{\hbar}{iL^2} \sum_{\alpha,\beta\neq\alpha} \frac{f(E_\alpha)}{(E_\alpha - E_\beta)^2} \left[(\alpha|j_x|\beta)(\beta|j_y|\alpha) - (x \leftrightarrow y)\right], \tag{2.1}$$

where L is the linear dimension of the system, α and β denote eigenstates, E_α and E_β their energy, and $f(E)$ is the Fermi distribution function. Consider a two-dimensional (2D) system in magnetic field H on torus by imposing periodic boundary conditions in

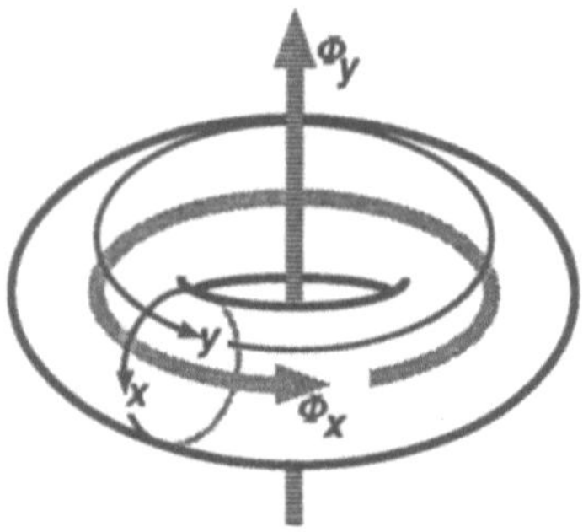

Fig. 1 A two-dimensional system on a torus. The magnetic field is everywhere perpendicular to the surface and the Aharonov-Bohm magnetic flux Φ_x and Φ_y is introduced through the openings.

both x and y directions and introduce magnetic flux Φ_x and Φ_y through the openings of the torus as shown in Fig. 1. This Aharonov-Bohm flux gives rise to vector potential $A_x = \Phi_x/L$ and $A_y = \Phi_y/L$. Because such vector potential independent of position can be eliminated through a change in the phase of the wavefunction (a gauge transformation) whenever the flux is integer multiples of the flux quantum $\Phi_0 = ch/e$, all the physical quantities including σ_{xy} are a periodic function of Φ_x and Φ_y with period Φ_0 or of A_x and A_y with period Φ_0/L.

In terms of this vector potential, the current operator is written as

$$j_x = c\frac{\partial \mathcal{H}}{\partial A_x}, \qquad j_y = c\frac{\partial \mathcal{H}}{\partial A_y}, \tag{2.2}$$

where $\mathcal{H}$ is the Hamiltonian of the system. Therefore, the Hall conductivity becomes

$$\sigma_{xy} = -\frac{\hbar c^2}{iL^2}\sum_\alpha f(E_\alpha)\left[\left(\frac{\partial\alpha}{\partial A_x}\middle|\frac{\partial\alpha}{\partial A_y}\right) - (x \leftrightarrow y)\right] \tag{2.3}$$

We take its average over the magnetic flux Φ_x and Φ_y [5,6].

$$\sigma_{xy} = -\frac{e^2}{\hbar}\sum_\alpha f(E_\alpha)\frac{1}{2\pi i}\iint dA_x dA_y \left[\left(\frac{\partial\alpha}{\partial A_x}\middle|\frac{\partial\alpha}{\partial A_y}\right) - (x \leftrightarrow y)\right] \tag{2.4}$$

Now, the quantity

$$N_\alpha = \frac{1}{2\pi i}\iint dA_x dA_y \left[\left(\frac{\partial\alpha}{\partial A_x}\middle|\frac{\partial\alpha}{\partial A_y}\right) - (x \leftrightarrow y)\right] \tag{2.5}$$

is a topological quantity called "winding number" and is always an integer [7-9].

Note that this does not necessarily mean that the Hall conductivity is always quantized into integer multiples of $-e^2/h$. We have to take a sort of average over states whose energy is within a few times of temperature around the Fermi level. Because each state can give an integer different from others, the resulting average is usually not an integer. The situation is quite different when states around the Fermi level are all localized. For localized states, the integer N_α vanishes and the Hall conductivity becomes quantized at sufficiently low temperatures. Thus, the localization or the presence of a spectral gap is essential in the quantization of the Hall conductivity.

Another explanation of the quantum Hall effect is based on edge current. Figure 2 gives schematically the energy levels when an electron is confined within a wire with width W. When the orbit center is far away from boundaries, the energy is independent

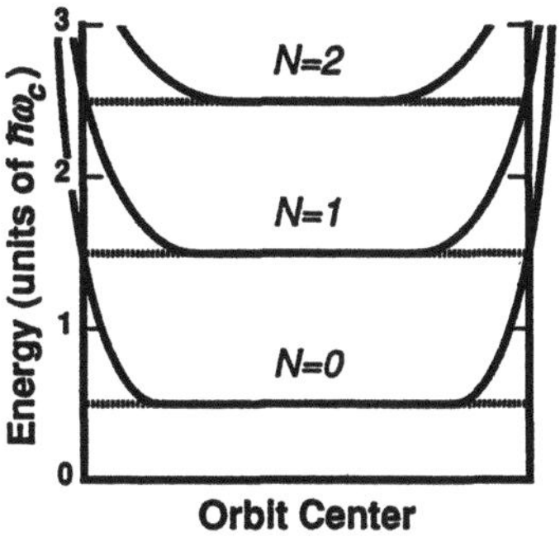

Fig. 2 A schematic illustration of energy levels in a wire with width W infinitely long in the x direction.

of the orbit position and given by those of bulk Landau levels. When it approaches a boundary, the energy becomes higher due to the formation of edge states. The states localized in the vicinity of each boundary correspond to those for an electron moving in the same direction and have the velocity $v_j(k) = \partial E_j(k)/\hbar\partial k$, where k is the wave vector along the wire related to the orbit coordinate Y through $Y = l^2 k$ and $E_j(k)$ is the energy of the jth level with l being the magnetic length defined by $l^2 = c\hbar/eH$. The total current is given by

$$I = \frac{(-e)}{h}\sum_{j=1}^{q}\int_{k_j^{\min}}^{k_j^{\max}} dk\,\frac{\partial E_j(k)}{\partial k} = \frac{(-e)}{h}\sum_{j=1}^{q}[E_j(k_j^{\max}) - E_j(k_j^{\min})], \tag{2.6}$$

where $k^{\max}$ and $k_j^{\min}$ are the maximum and minimum wave vectors, respectively, occupied by electrons. In the presence of current the chemical potential at the left and right boundaries are different, giving rise to a Hall voltage. Usually, the left Hall probe measures the average of the chemical potential of left edge states and the right probe that of the chemical potentials of right edge states. The Hall voltage is given by

$$(-e)V_H = q^{-1}\sum_{j=1}^{q}[E_j(k_j^{\max}) - E_j(k_j^{\min})], \tag{2.7}$$

where q is the number of edge states below the Fermi level. This immediately leads to the quantization of the Hall resistance

$$R_H = \frac{V_H}{I} = \frac{1}{q}\frac{h}{e^2}. \tag{2.8}$$

In 2D the Fermi level always lies within the broadened bulk Landau levels, where edge states are mixed with bulk states. Therefore, the above is certainly not a correct explanation but is shown with the use of Landauer's formula [4,10] to be essentially valid as long as the mixing does not cause backscattering between edge states moving in opposite directions.

The essential point of this edge-current explanation is that the Hall field is not required at all and that the current is carried only by edge states. In the bulk Hall current picture discussed above, on the other hand, the Hall field is essential and the current is carried by bulk extended states in the vicinity of the center of each Landau level [11]. The relationship between these two pictures remains still unclear. The following sections are devoted to a brief review of two examples which have been done for the purpose of shedding some light on this interesting and important question.

3. MIXING OF EDGE STATES WITH BULK STATES

We consider a 2D system on a square lattice (site number M in the y direction) and introduce effects of scattering through randomness of site energy distributed uniformly with a certain width. A magnetic field H is included in the form of a Peierls' phase factor in the nearest-neighbor transfer integral. For sufficiently large M, the lattice system corresponds to a continuum system with width $W=(M+1)a$, where a is the lattice constant. Note that we should make the magnetic flux passing through a unit cell as small as possible in order to avoid peculiarities characteristic of lattice systems [12]. The inverse localization length $\alpha(E)$ and the density of states can be calculated using the well-known technique of Green's function. Some representative results [13] are given in the following.

Figure 3 shows the subband-energy dispersions calculated in the lattice model ($M=16$) for $\hbar\omega_c/E_W=10$ ($W/l=7.02$), where $\omega_c=eH/mc$ and E_W is the energy of the lowest one-dimensional subband given by $E_W=(\hbar^2/2m)(\pi/W)^2$. The energy of the bottom of the low-lying subbands calculated in the lattice model is slightly smaller than $(N+1/2)\hbar\omega_c$ ($N=0,1,\ldots$). This slight difference is due to the nonparabolicity present in the lattice model and becomes smaller with increasing M for a given W. In the lattice model the Landau levels have an oscillatory dependence on the wave vector k especially for higher-lying levels [14-16]. This Harper broadening [17] is not important for the low-lying levels as long as the quantity $\hbar\omega_c/E_W M$ remains small. In actual calculations, we shall choose M in such a way that the magnetic flux contained in the unit cell does not exceed 6% of the flux quantum Φ_0.

Corresponding results for the density of states and the inverse localization length are given in Fig. 4. The density of states becomes closer to that in 2D at the ener-

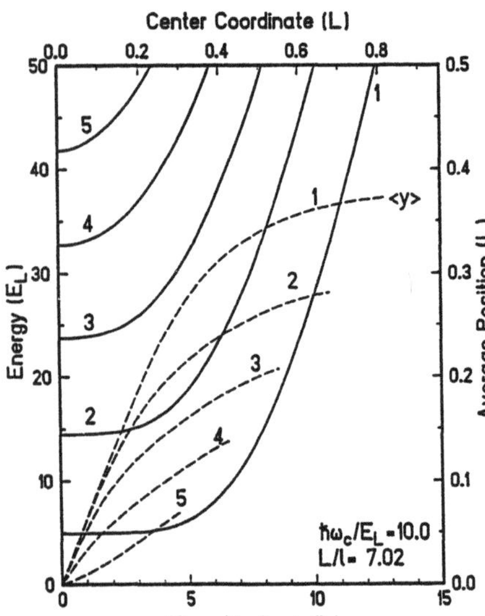

Fig. 3 Examples of calculated energy dispersion in a quantum wire in the magnetic field $\hbar\omega_c/E_W=10$. The solid lines represent the subband energy and the dashed lines the average value of y.

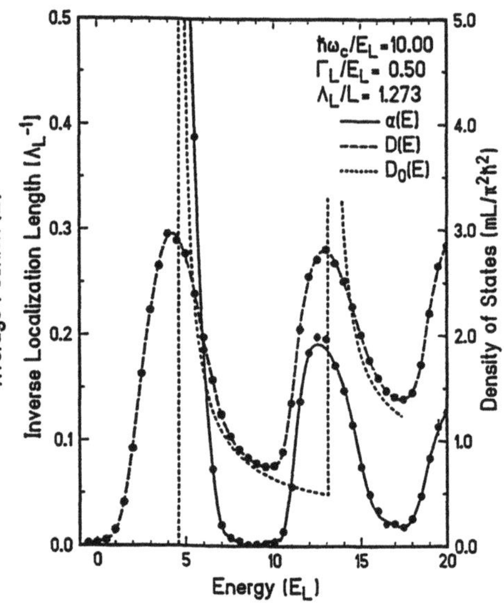

Fig. 4 Calculated inverse localization length (solid line) in units of Λ_W^{-1} and density of states (dashed line) of a quantum wire in magnetic field $\hbar\omega_c/E_W=10$ and $\Gamma_W/E_W=0.5$.

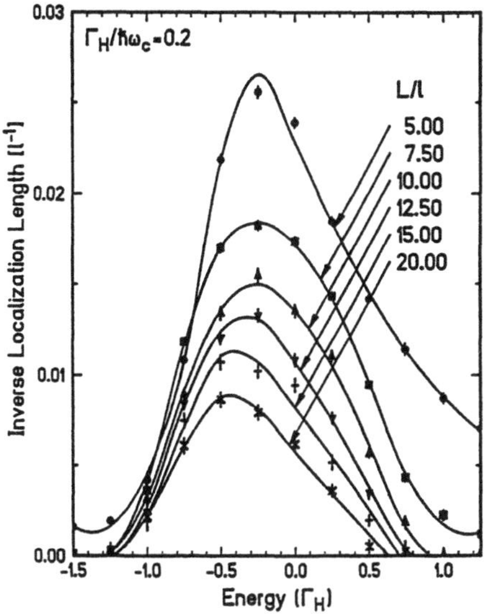

Fig. 5 Calculated width dependence of the inverse localization length for $\Gamma/\hbar\omega_c = 0.2$. The energy is measured from that of the first excited Landau level in two dimension and normalized by the broadening Γ.

gies of the bulk Landau levels, while it is given well by that of the unperturbed edge states for energies between the Landau levels. The scattering strength is characterized by broadening Γ_W/E_W, where $\Gamma_W/\hbar$ represents the probability of backscatterings for electrons with wave vector $k=\pi/W$ in the lowest subband or the corresponding mean free path Λ_W/W. The broadening of the bulk Landau level Γ calculated in the self-consistent Born approximation [18,19] is given by $\Gamma=[(4/3)\hbar\omega_c\Gamma_W]^{1/2}$. In the presence of randomness the peak in the density of states is shifted to the lower energy side due to quantum mechanical level repulsions between different Landau levels.

We should note that the energy dependence of $\alpha(E)$ is completely different from that in the cylinder geometry. In the cylinder geometry, the inverse localization length takes a local minimum near the center of each Landau level and becomes large at the spectral gaps between adjacent Landau levels ($\alpha \sim l^{-1}$) [20,21]. In the quantum wire, on the other hand, the inverse localization length becomes extremely small when the energy is away from the density of states of the bulk Landau levels. This is a direct consequence of the fact that the edge states are most extended in the wire. There is a remarkable difference in the behavior between the lowest and first excited Landau levels, reflecting the obvious fact that the edge states are present only above the lowest Landau level. In the energy region corresponding to the first excited Landau level, $\alpha(E)$ becomes large due to strong mixings [22, 23] with bulk states and its energy dependence reflects the form of the density of states of the bulk Landau level.

One interesting question is which of the bulk states at the Landau-level center or the edge states are more extended in the limit of 2D. In all the examples given above, the calculated inverse localization length is not that of the bulk states but that of the edge states. Figure 5 gives inverse localization length $\alpha(E)$ in the energy range corresponding to the first excited Landau level for varying width W/l for a fixed value of Γ ($\Gamma/\hbar\omega_c = 0.2$). The energy is measured from the bottom of the first excited subband in the absence of randomness and normalized by Γ. The actual energy of the density-of-states peak, shifted to the lower energy side due to the quantum-mechanical level-repulsion effect, is close to -0.25Γ. It is clear that $\alpha(E)$ decreases with increasing width, showing that the edge states become more and more extended for wider wires.

In the cylinder geometry for which there are no edge states, the inverse localization length of the bulk extended states near the center of the first excited Landau level has been shown to be given by $\alpha l \sim 0.4(W/l)^{-1}$ [21]. This value is still much larger than $\alpha(E)$ of the edge states for $W/l \lesssim 20$. Therefore, the edge states are more extended than the bulk extended states at the Landau-level center.

The dependence of $\alpha(E)$ on W is expected to reflect how the amplitude of the edge-state wavefunction decays inside the wire away from the boundary. We should note that the decay rate of the wavefunction does not exhibit strong dependence on energy. This fact suggests that it is not determined by the localization length of the Landau levels in 2D. Further, the wavefunction inside the wire is not symmetric about the center of the bulk Landau level, i.e., the edge states below the Landau level decays more slowly inside the wire away from the boundary. This asymmetry probably reflects the spatial extent of the edge states in the absence of randomness. For the edge state with lower energy, the confinement is weaker and mixings with the bulk states can become appreciable. With increasing energy the edge state becomes more and more confined at the boundary and less mixed with the bulk states. The asymmetry may also be a reflection of the fact that there exist additional edge states associated with the first excited Landau level for $E > (3/2)\hbar\omega_c$.

4. BACKSCATTERING OF EDGE STATES

We consider a rectangular system with length L confined within width W [24]. At $x = \pm L/2$, the system is connected to an infinitely long ideal wire which is eventually connected to a reservoir. In order to calculate transmission probability for the whole energy range of the broadened 2D Landau levels, we shall lower the bottom energy in the ideal wires by ΔE ($\Gamma < \Delta E < \hbar\omega_c - \Gamma$ with $\omega_c = eH/mc$). The transmission probability is calculated using the Green function technique already described [25].

Numerical calculations are performed for rectangular systems $L/W = 1/2$, 1, and 2. The system size is $W/l = 5$, 10, 15, 20, 25, 30, 35, 40, 50. The transmission probability for a given energy is calculated by averaging over those of typically 10^4 different samples.

Figure 6 gives some examples of calculated energy dependence of the transmission probability T for the lowest Landau level. The broadening of the Landau level is $\Gamma/\hbar\omega_c = 0.2$ and the system size is $W/l = 20$. The results show clearly that a change in the choice of the shift ΔE causes negligible effects.

In order to deduce the effective energy-width over which the transmission probability increases from zero to unity, we fit the results to the function $T(E) = 1/\{\exp[\beta(E - E_{th})] + 1\}$. The parameter β or $\Gamma(dT/dE)_{max}$ represents the inverse of the effective energy width and E_{th} the effective threshold energy. The fitted results are shown by solid lines in Fig. 6. The resulting $\Gamma(dT/dE)_{max}$'s are listed in each figure.

Figure 7 gives the steepness parameter as a function of the width in logarithmic scales. It is clear that the results can be fitted to a straight line except those for smallest $W/l = 5$, showing that $\Gamma(dT/dE)_{max} \propto W^{1/s}$. The exponent is almost independent of the form of the system and obtained as $s = 2.2 \pm 0.1$.

The threshold energy becomes larger with increasing L for a given W but the threshold for systems with different L/W's becomes closer with increasing system size.

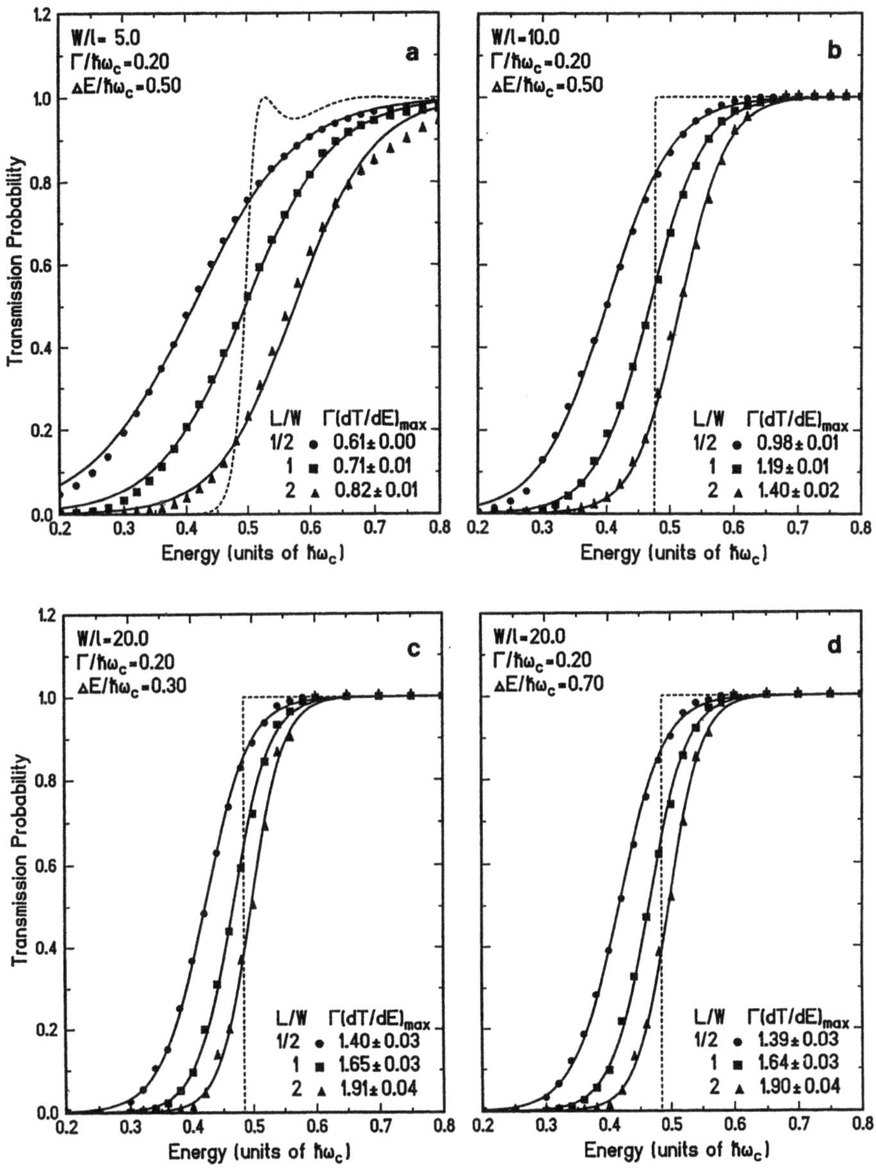

Fig. 6 Calculated energy dependence of the transmission probability for the lowest Landau level. The dotted lines represent the probability in the absence of randomness and the solid lines the fitted curves. The maximum energy derivatives are listed in the figure. (a) $W/l=5$ and $\Delta E/\hbar\omega_c=0.5$. (b) $W/l=10$ and $\Delta E/\hbar\omega_c=0.5$. (c) $W/l=20$ and $\Delta E/\hbar\omega_c=0.3$. (d) $W/l=20$ and $\Delta E/\hbar\omega_c=0.7$.

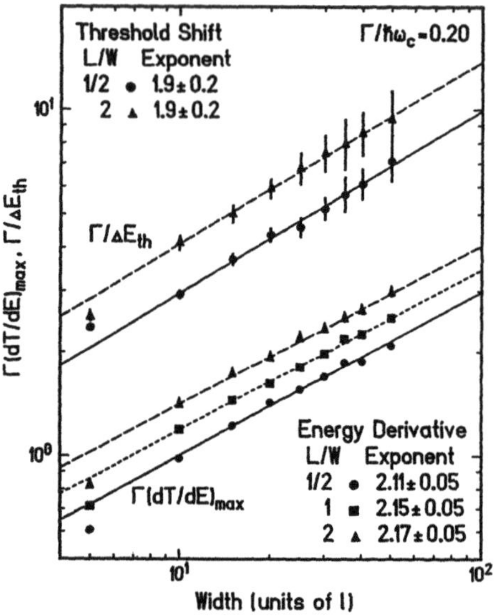

Fig. 7 Log-log plot of the maximum energy derivative of the transmission probability as a function of the system width. The straight lines represent $\Gamma(dT/dE)_{\max} \propto W^{1/s}$ with exponent s listed at the bottom-right corner of the figure. The dots denoted as $\Gamma/\Delta E_{\rm th}$ represent the threshold energy difference between $L/W = 1$ and $1/2$, and the squares the difference between $L/W = 2$ and 1. The exponents s' defined by $\Delta E_{\rm th} \propto W^{-1/s'}$ are listed at the top-left corner.

Figure 7 contains also the difference in the threshold energy $\Delta E_{\rm th}^{(2)} = E_{\rm th}^{L/W=2} - E_{\rm th}^{L/W=1}$ and $\Delta E_{\rm th}^{(1/2)} = E_{\rm th}^{L/W=1} - E_{\rm th}^{L/W=1/2}$. Although statistical errors are much larger, $\Delta E_{\rm th}$ is also fitted to a power law dependence, $\Delta E_{\rm th} \propto W^{-1/s'}$. The resulting $s' = 1.9 \pm 0.2$ agrees with s within statistical errors.

Numerical Thouless-number study of localization for Landau levels has shown that states are all localized except those lying in the vicinity of the center in 2D [26]. The critical exponent s, defined by $\alpha \propto |E|^s$ with α the inverse localization length and the energy origin at the center, was obtained as $s \sim 2$ for the lowest Landau level. This result was confirmed later by finite-size scaling calculations [20,21]. More recent finite-size scaling calculations within a random-matrix model which is not the actual 2D system in magnetic fields gave the exponent $s \approx 2.34$ [27]. There have been some experimental attempts of determining the exponent based on the temperature dependence of ρ_{xy} [28] and σ_{xy} [29].

Let E_- (<0) represent the energy below the Landau level where the transmission becomes appreciable and denote E_+ (>0) the energy above which the backscattering becomes negligible. For sufficiently large systems where localization length α^{-1} is the only relevant length scale, these energies are determined by the condition such that $h_\pm[\alpha(E_\pm)L, \alpha(E_\pm)W] = 0$ with $h_\pm$ a certain function. This leads to $E_\pm/\Gamma = \pm C_\pm (W/l)^{-1/s}$ for $\alpha(E) \propto |E|^s$ with $C_\pm$ an appropriate constant depending only on the ratio L/W. Therefore, the width of the energy region over which the transmission probability varies from zero to unity is given by $E_+ - E_- \propto W^{-1/s}$, i.e., $\Gamma(dT/dE)_{\max} \propto (E_+ - E_-)^{-1} \propto W^{1/s}$. The same is applicable to the threshold shift and $\Delta E_{\rm th} \propto W^{-1/s}$. This means that the exponent s and s' obtained above are both equal to that for the inverse localization length of bulk 2D Landau states.

The critical exponent obtained in the present study is ~ 2.2. This is consistent with that of almost all the existing theory either numerical or analytic because actual errors are likely to be larger than $\sim \pm 0.1$ estimated statistically. Quite recently, Koch *et al.*

investigated the system-size dependence of the width of the transition region between quantized Hall plateaus and deduced a critical exponent ($s \sim 2.3$) for the localization length of bulk Landau states [30].

5. SUMMARY AND CONCLUSION

Two examples of recent theoretical works which have been done for the purpose of understanding the relationship between the bulk Hall current picture and the edge current picture were presented. The first example dealt with localization in quantum wires. Unfortunately it failed to give any information on relative importance of edge and bulk Landau states in the conduction along the wire since the obtained localization length turned out to be that of edge states. It has been shown that mixing of edge states associated with the lowest Landau level with bulk first excited states strongly modifies both bulk and edge states in the case of abrupt barriers. The degree of the mixing may depend on the detailed form of the confining potential. The second example dealt with backscattering of edge states and in particular with the width of the transition region where the Hall resistance changes from one quantized value to another. It has been demonstrated through the comparison of the critical exponent that both bulk current and edge-state pictures give the same answer within the numerical accuracy.

ACKNOWLEDGMENTS

This work was supported in part by the Industry-University Joint Research Program "Mesoscopic Electronics" and by the Grants-in-Aid for Scientific Research on Priority Areas, "Electron Wave Interference Effects in Mesoscopic Structures" and "Computational Physics as a New Frontier in Condensed Matter Research," from the Ministry of Education, Science and Culture, Japan.

REFERENCES

1. T. Ando: Prog. Theor. Phys. Suppl. **84** (1985) 69.
2. *The Quantum Hall Effect*, edited by R.E. Prange and S.M. Girvin (Springer, New York, 1987).
3. H. Aoki: Rep. Prog. Phys. **50** (1987) 655.
4. M. Büttiker: Phys. Rev. Lett. **57** (1986) 1761.
5. Q. Niu, D.J. Thouless, and Y.-S. Wu, Phys. Rev. B **31** (1985) 3372.
6. H. Aoki and T. Ando, Phys. Rev. Lett. **57** (1986) 3093.
7. D.J. Thouless, M. Kohmoto, M.P. Nightingale, and M. den Nijs, Phys. Rev. Lett. **49** (1982) 405.
8. J. Avron, R. Seiler, and B. Simon, Phys. Rev. Lett. **51** (1983) 51.
9. B. Simon, Phys. Rev. Lett. **51** (1983) 2167.
10. R. Landauer, IBM J. Res. Dev. **1** (1957) 223; Philos. Mag. **21** (1970) 863.
11. H. Aoki and T. Ando, Solid State Commun. **38** (1981) 1079.
12. D.R. Hofstadter: Phys. Rev. B **14** (1976) 2239.
13. T. Ando, Phys. Rev. B **42** (1990) 5626.
14. L. Schweitzer, B. Kramer and A. MacKinnon, J. Phys. C **17** (1984) 4111.
15. A. MacKinnon, L. Schweitzer and B. Kramer, Surf. Sci. **142** (1984) 189.
16. V. Gudmundsson, R.R. Gerhardts, R. Johnston, and L. Schweitzer, Z. Phys. B **70** (1988) 453.

17. P.G. Harper, Proc. Phys. Soc. A **68** (1955) 874.
18. T. Ando and Y. Uemura: J. Phys. Soc. Jpn. **36** (1974) 959.
19. T. Ando: J. Phys. Soc. Jpn. **36** (1974) 1521; **37** (1974) 622; **37** (1974) 1233.
20. H. Aoki and T. Ando: Phys. Rev. Lett. **54** (1985) 831.
21. T. Ando and H. Aoki: J. Phys. Soc. Jpn. **54** (1985) 2238.
22. T. Ohtsuki and Y. Ono: J. Phys. Soc. Jpn. **58** (1989) 1705.
23. Y. Ono, T. Ohtsuki and B. Kramer: J. Phys. Soc. Jpn. **58** (1989) 956.
24. T. Ando, J. Phys. Soc. Jpn. (February 1992).
25. See, for example, T. Ando: Phys. Rev. B **44** (1991) 8017 and references cited therein.
26. T. Ando: J. Phys. Soc. Jpn. **52** (1983) 1893; **53** (1983) 3101; **53** (1983) 3126; Phys. Rev. B **40** (1989) 9965.
27. B. Huckestein and B. Kramer: Phys. Rev. Lett. **64** (1990) 1437.
28. H.P. Wei, D.C. Tsui, M.A. Paalanen and A.M.M. Pruisken: Phys. Rev. Lett. **61** (1988) 1294.
29. J. Wakabayashi, M. Yamane and S. Kawaji: J. Phys. Soc. Jpn. **58** (1989) 1903.
30. S. Koch, R.J. Haug, K. von Klitzing and K. Ploog: Phys. Rev. Lett. **67** (1991) 883.

Probes as Part of the System in the Quantum Hall Regime

Y. Ono

Department of Physics, Toho University,
Miyama 2-2-1, Funabashi, Chiba 274, Japan

Abstract. Electronic states in 2D systems with finite width subject to a quantizing magnetic field are studied numerically in the presence of probes attached to the system as its parts. We find not only the bending of the edge state path along the probe parts but also the change of amplitude of wave functions accompanied by a current vortex in those states located between bulk and edge states.

1. Introduction

According to the Kubo formula the Hall conductivity σ_{yx} is expressed in terms of a sum of contributions of all the states below the Fermi level [1-3]. This expression for the Hall conductivity was utilized by Aoki and Ando [4] to explain the quantum Hall effect (QHE) from the view point of localization and delocalization of electronic states in 2D bulk disordered systems subject to strong magnetic fields. It has been shown by several authors [5-8] that, in the case of confined systems, σ_{yx} can be described solely by the informations at the Fermi level. This means that σ_{yx} is expressed in terms of edge states, when the magnetic field is strong enough to make gaps between bulk Landau subbands and when the Fermi level is in one of these gaps. The two equivalent expressions for σ_{yx}, i.e. the bulk and edge expressions, are understood to be related through a kind of "Stokes' theorem " [9]. What we learn from the above mentioned equivalence between the two expressions for σ_{yx} is that the bulk delocalized and edge states are related to each other in some sense. In fact Hajdu et al [10] have shown that the bulk delocalized states are connected to edge states through a certain parameter. This has been confirmed numerically in [11].

In normal measurements of σ_{yx}, samples of Hall bar geometry with voltage or current probes are used. As the probes modify the sample shape at the edges, the edge states are seriously affected by the presence of probes. From the above argument, it will be clear that the bulk delocalized states are also affected by probes. In this sense, it seems to be very important to understand the effects of probes on electronic states in the quantum Hall regime. This will be the case even when the system size is much larger than the mesoscopic one. In the present work we study numerically the electronic states in 2D systems with finite width subject to a strong magnetic field by regarding the probes as part of the system.

2. Model System

In this paper we consider only the effect of the sample shape change due to the attachment of probes. The role of probes as particle source and sink is outside the scope of the present work.

The probes are treated as small extended regions with dimensions $x_p \times y_p$ attached on both sides of a Teller system [12], having a width L_x and a length L_y, around $y = \frac{1}{2}L_y$. The Teller system satisfies the periodic boundary condition in the y-direction and the Dirichlet boundary condition in the x-direction. In order to realize this situation in the numerical calculations, we prepare first a Teller system with a width $L_x + 2x_p$ and a length L_y and put a high uniform potential in the regions specified by $(x - x_p)(x - L_x - x_p) > 0$ and $(y - \frac{1}{2}L_y + \frac{1}{2}y_p)(y - \frac{1}{2}L_y - \frac{1}{2}y_p) < 0$. If the potential height is much larger than $\hbar\omega_c$ (ω_c the cyclotron frequency), then it mimics a hard wall. As a result, we have a Hall bar system with two probes. In the practical numerical calculation we use the following Hamiltonian,

$$H = \frac{1}{2m}(\boldsymbol{p} + \frac{e}{c}\boldsymbol{A})^2 + V_w(\boldsymbol{r}), \qquad (1)$$

where the first term is the usual free electron Hamiltonian in a magnetic field and the wall potential $V_w(\boldsymbol{r})$ has a constant value ($= 10\hbar\omega_c$) in the above mentioned regions and vanishes otherwise. Matrix elements of the Hamiltonian are calculated by using basis functions satisfying Dirichlet conditions at $x = 0$ and $x = L_x + 2x_p$ and the periodic boundary condition in the y-direction (period L_y); see for example [13]. The electronic states are obtained by diagonalizing this Hamiltonian matrix [14]. If we want to study disordered systems, a random potential is added to the above Hamiltonian.

The dimensions of the system used in the following calculations are specified by $L_x = 33$, $L_y = 49$, $x_p = 8$ and $y_p = 16$ in unit of $\Delta X \equiv 2\pi\ell^2/L_y$. In unit of the magnetic length ℓ, L_y is equal to 17.5ℓ.

3. Electronic States

By analyzing the eigenfunctions obtained numerically, we find that the states are extended straight in the y-direction when the eigenenergy is quite near to $0.5\hbar\omega_c$ and that the electron path is bent along distorted edges when the eigenenergy is around $0.6\hbar\omega_c$ or more. In both cases the amplitude of the eigenfunctions are almost constant along the "path". These behaviors are reasonable and understandable from the classical point of view. Namely, when an electron goes through the region far from the probes, its path is not affected. On the other hand, the skipping orbit along edges is bent by the presence of probes. If the speed of the electron is constant along its path, the corresponding wave function will have a constant amplitude.

However, when the eigenenergy is slightly away from $0.5\hbar\omega_c$, we have eigenstates as depicted in Fig. 1, for which the region of finite amplitude is stretching almost straight in the y-direction but the amplitude is not uniform along that region. The example shown in Fig. 1 has an eigenenergy $0.50179\hbar\omega_c$ and its amplitude is enhanced near the probe. We find also those states having reduced amplitudes near the probe, whose eigenenergies are slightly higher than those which have enhanced amplitudes near the probe, as long as the distance from the nearest edge is almost the same.

In order to understand what is happening in the region of the enhanced amplitude in the example shown in Fig. 1, we have calculated the current distribution of this state, which is shown in Fig. 2. In Fig. 2 the local

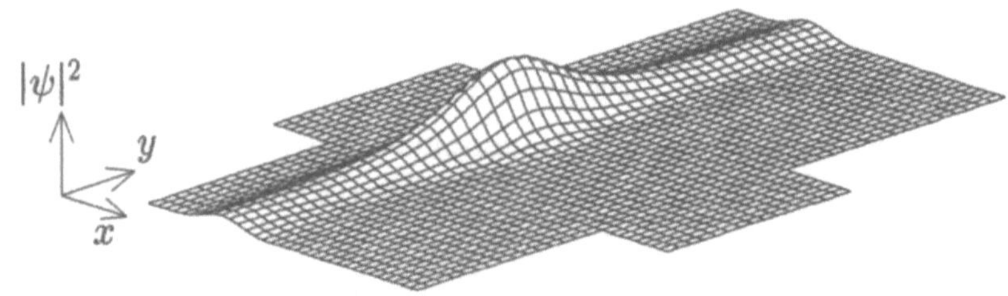

Fig. 1: Square amplitude of an eigenfunction in a two dimensional electron system with probes subject to a strong magnetic field. The system is confined in the x-direction and periodic in the y-direction. Its size is specified by the width $L_x = 33$, the length $L_y = 49$, and the probe size is given by $x_p = 8$ and $y_p = 16$, in unit of $\Delta X \equiv 2\pi\ell^2/L_y$. The eigenenergy of this state is equal to $0.50179\hbar\omega_c$.

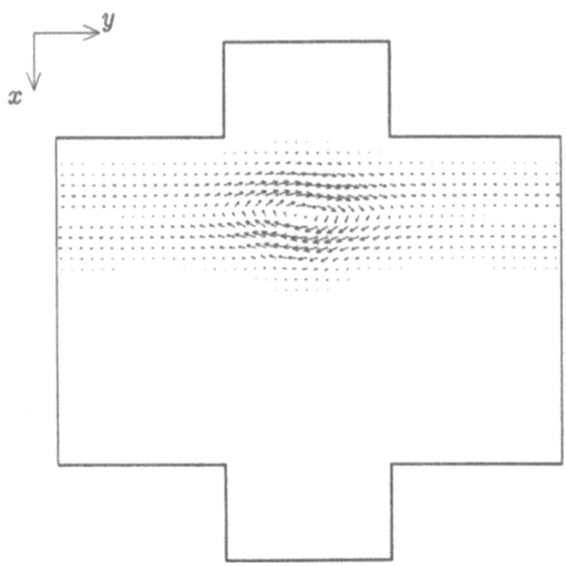

Fig. 2: The diamagnetic current distribution in the state depicted in Fig. 1. The length of the arrow represents the strength of the current. Thick edge lines mean the Dirichlet type boundary condition and thin ones the periodic boundary condition.

current direction and strength are expressed by the direction and length of an arrow, respectively. The vortex structure in the region of the enhanced amplitude is seen clearly. From Figs. 1 and 2, it is plausible to say that an electron in the intermediate region between the bulk and the edge feels a distortion of the edges due to the presence of probes as an effective potential.

4. Analogy with Gate Dip

In order to clarify the idea of the effective potential, we have studied electronic states in a Hall bar geometry without probes but with a gate dip potential specified by $V_{\text{Gdip}}(x, y) = -V_0$ for $\frac{1}{2}(L_y - y_p) < y < \frac{1}{2}(L_y + y_p)$ and 0 otherwise. We omit the details of the calculation, since it is analogous to that in the case with a gate barrier [15].

What we have found can be summarized as follows; (i) the edge states are bent nearer to the edge in the dip region, (ii) the edge state whose energy degenerates with one of localized states in the dip region is hybridized with the localized states, and (iii) the aforementioned localized states show vortex structures in their current distribution. Hybridized states have a similar property as that shown in Figs. 1 and 2 in the sense that they look like normal edge states outside the dip region while showing vortex structures in the dip region.

An example of such a hybridization is shown in Fig. 3, where the system dimensions L_x and L_y are the same as in Fig. 1. The gate dip length y_p is also the same as the probe length y_p in Fig. 1. The absolute value of the gate dip potential V_0 is chosen to be $0.6\hbar\omega_c$. The eigenenergy of the state in Fig. 3 is equal to $0.90301\hbar\omega_c$. It should be noted that the vortex structures are extended over whole range of the dip region in contrast to the vortex seen in the system with probes; the latter is restricted to the region near the probe and not extended in the x-direction. Although the behaviors of electronic states in two cases, i.e. the cases of the gate dip and of the probes, may not be exactly the same, it can be said that the distortion of the edge shape due to the presence of probes yields a similar effect to electrons as an effective attractive potential.

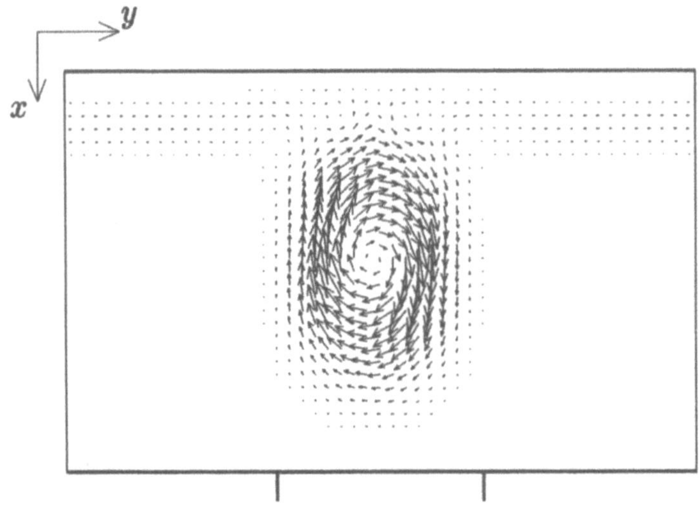

Fig. 3: An example of the diamagnetic current distribution in the presence of a gate dip. The different boundary conditions are represented by different thickness of the edge lines as in Fig. 2. The system size is the same as in Figs. 1 and 2 except for the probe parts. The position of the gate dip is indicated by two short lines on the bottom edge line. Between these two line the gate dip potential is equal to $-0.6\hbar\omega_c$ and 0 otherwise. The eigenenergy of the state is $0.90301\hbar\omega_c$.

5. Discussion

Electronic states in confined 2D electron systems subject to a strong magnetic field are numerically studied by regarding the probes as part of the system. Those states which are in the bulk region and well away from edges are not affected by probes and edge states which are quite near to edges show bending of their path due to the appearance of probes. In addition to these expected behaviors, we have found the variation of the wave function amplitude for those states located between the bulk and edge states. When the amplitude is enhanced near the probe, a vortex structure is found in the diamagnetic current distribution. The existence of such states will be crucial to understand the effect of probes on the delocalized bulk states realized in the presence of disorder.

The analogy between the effects of the probe and the gate dip has been discussed. From this discussion it is found that the probes yield a similar effect to electrons as an effective attractive potential. It may be worthwhile mentioning that there is a similar analogy between a narrow channel and an effective repulsive potential [16].

We have also studied probes which satisfy the periodic boundary condition at probe ends. This may mimic a very long probe or a current probe. Essential features of the eigenstates as described above were not changed.

Acknowledgements

The author thanks T. Ohtsuki for fruitful discussions, particularly on the effect of a gate dip. He also appreciate B. Kramer for useful discussions. A part of the calculations was performed with HITAC-S820 at the Computor Center of the University of Tokyo. The present work was partially financed by Grant-in-Aid for Scientific Research in Priority Area "Electron Wave Interference Effects in Mesoscopic Structures" from the Ministry of Education, Science and Culture.

References

[1] R. Kubo, J. Phys. Soc. Jpn. **12**, 570 (1957).
[2] R. Kubo, S.J. Miyake and N. Hashitsume, in *Solid State Physics*, edited by F. Seitz and D. Turnbull (Academic Press, New York, 1965), Vol. 17, p.269.
[3] A. Bastin, C. Lewiner, O. Betbeder-Matibet and P. Nozières, J. Phys. Chem. Solids **32**, 1811 (1971).
[4] H. Aoki and T. Ando, Solid State Commun. **38**, 1079 (1981).
[5] P. Středa, J. Phys. C**15**, L717 (1982).
[6] L. Smrcka, J. Phys. C**17**, L63 (1984).
[7] M. Jonson, Phys. Scr. **32**, 435 (1985).
[8] Y. Ono and B. Kramer, Z. Phys. **67**, 341 (1987).
[9] A.M.M. Pruisken, Nucl. Phys. **B235** [FS11], 277 (1984).

[10] J. Hajdu, M. Janßen and O. Viehweger, Z. Phys. B**66**, 433 (1987).

[11] T. Ohtsuki and Y. Ono, Solid State Commun. **68**, 787 (1988), J. Phys. Soc. Jpn. **58**, 956 (1989).

[12] E. Teller, Z. Phys. **67**, 311 (1931).

[13] Y. Ono and T. Ohtsuki, Z. Phys. B**68**, 445 (1987), T. Ohtsuki and Y. Ono, J. Phys. Soc. Jpn. **58**, 2482 (1989).

[15] T. Ohtsuki and Y. Ono, J. Phys. Soc. Jpn. **59**, 637 (1990).

[16] B. Kramer, private communication.

Part VII

Persistent Current and Orbital Magnetism

Spectral Correlations, Symmetry Breaking and Novel Orbital Magnetic Effects in Mesoscopics

Y. Imry

Nuclear Physics Department, The Weizmann Institute of Science, Rehovot 76100, Israel

Abstract. The quasiclassical picture yielding the correlations in the spectrum of an electron diffusing in a mesoscopic "quantum dot" is reviewed. This includes the range where random-matrix theory holds as well as the new universality class found by Altshuler and Shklovskii. Some generalizations are mentioned and the orthogonal-unitary symmetry crossover brought about by a magnetic field is treated. Using a recent thermodynamic relationship between the difference of derivatives of canonical and grand-canonical quantities and grand-canonical fluctuations, a new paramagnetic orbital contribution to the electronic susceptibility is obtained and discussed.

1. Introduction

It is by now firmly established that the energy-level spectra of a finite disordered "mesoscopic" quantum dot obeys random matrix theory (RMT) [1] statistics in certain parameter ranges. In particular, the level density – level density correlation function has been shown by Altshuler and Shklovskii [2] to lead to RMT – type spectral rigidity for energy-differences in the range between the average level spacing and the Thouless energy (see below). At energies above the latter (but below the inverse elastic scattering time) a new universality class was found in Ref. 2. Earlier, Efetov [3] proved that RMT statistics apply to such spectra at low energies using a different method. To obtain physical understanding of how the statistical properties which are relevant to many physical phenomena (see eg. [4]) are generated it is advantageous to use a quasiclassical picture. Indeed, in the related field of "quantum chaology" the observation [5] that the quasiclassical spectrum of a system whose classical motion is chaotic obeys RMT rules, has been substantiated by Berry [6]. Recently, Argaman et al. [7] have used the Berry method to demonstrate that RMT correlations apply to the quasiclassical spectrum of a particle whose classical motion is diffusive, for low energies. All the results of Ref. 2 are reproduced in detail. This, with some generalizations will be reviewd in Section 2. As an application we shall review the crossover between "orthogonal" and "unitary" symmetry brought about by the time-reversal symmetry breaking induced by a magnetic field. This crossover [7] is described best in the energy-magnetic field plane. It will be discussed in Section 3.

Using a recently established relationship between impurity-ensemble canonical averages and grand-canonical fluctuations (Eq. 24), the above correlations result in interesting thermodynamic properties as functions of an external magnetic field. In particular a new type of, possibly large, paramagnetic orbital magnetic response [4,8] will be considered in Section 4. The instructive example of mesoscopic persistent currents which was the impetus for some of the above work will be briefly mentioned as well. Some outstanding remaining questions will be posed and explained in Section 5.

2. Quasiclassical Theory of Spectral Correlations

We consider a metallic particle of volume $V = L^d$ and are interested in its level statistics near the Fermi energy E_F. The characteristic energy scales in the problem are:

The level spacing

$$\Delta \sim \frac{E_F}{k_F^d V}, \tag{1}$$

the energy associated with the elastic scattering rate $1/\tau$, and the Thouless energy:

$$E_c = \frac{\hbar D}{L^2}. \tag{2}$$

Here the D is the electronic diffusion constant in the classical limit, $D \sim \frac{v_F \ell}{d} \sim \frac{v_F^2 \tau}{d}$. We are interested in the metallic limit

$$\Delta \ll E_c \ll \hbar/\tau \ll E_F. \tag{3}$$

The level density is given by (E_j denoting the energy levels and $G(x,x,E)$ the Green's function):

$$n(E) = \sum_j \delta(E - E_j) = -\frac{1}{\pi} Im \int d\vec{x} G(\vec{x}, \vec{x}, E) \qquad (4)$$

By approximating the Feynman sum for the propagator (Fourier transform of G) as a sum over periodic classical paths, j, Gutzwiller [9] obtained the following "trace formula" for a 'chaotic" system, whose classical paths are all unstable,

$$n(E) = \sum_j A_j e^{iS_j/\hbar}, \qquad (5)$$

where S_j is the classical action along the periodic path j (apart from Maslow-index corrections) and the coefficient A_j is due to the gaussian integral around the (unstable) orbit:

$$A_j = \frac{1}{\hbar} T_j |\det(\mathrm{M_j} - \mathrm{I})|^{-\frac{1}{2}}. \qquad (6)$$

T_j is the period of the orbit j and M_j the "monodromy matrix" (giving, in linear approximation, the transformation of the (2d-2)-dimensional vector measuring the deviation from the orbit in phase space after one period). The factor T_j reflects the fact that the path can start and end at all points on the orbit and all such contributions are coherent for a given orbit.

One is interested in the level-density correlation function $K(E,\epsilon)=\langle \delta n(E - \frac{\epsilon}{2}) \delta n(E + \frac{\epsilon}{2}) \rangle$ and in its Fourier transform $\tilde{K}(E,t)$ which can be called "the spectral structure factor" (the averaging in the definition of K can be over a suitable range of E or on the "impurity ensemble" in the case of a disordered system).

There are serious mathematical questions having to do with the Gutzwiller sum not being absolutely convergent. Moreover, it is clear that the quasiclassical approximation fails at long enough times, due to the exponential proliferation of the orbits as t increases. There are indications (but this issue is far from being rigorously settled) that the limiting time is of order $\hbar/\Delta$. We shall thus employ a long time cut-off τ_ϕ (and a small energy cut-off, $\gamma \sim \tau_\phi^{-1}$, due physically to e.g. dephasing processes). Taking

$\Delta \ll \gamma \ll E_c$ appears to be necessary (this is based also on a camparison with Ref. 2). The quasiclassical method is for sure not valid for energy scales below Δ. Since the interesting scales here are $|\epsilon - \epsilon'| \sim \Delta$, E_c which are of order $\hbar^d, \hbar$, respectively, and much smaller than E_F, a quasiclassical approximation relying on $\hbar$ being small is suggested. Berry [6] in fact used the classical expression $T_j = \partial S_j / \partial E$ to find for the spectral structure factor $\tilde{K}(t)$ in the semiclassical approximation (presumably valid for $t \lesssim \hbar/\Delta$) keeping only the "diagonal" terms in the double sum over levels:

$$\tilde{K}(E,t) = \sum_j |A_j|^2 \delta(t - T_j) . \tag{7}$$

Berry used a classical sum-rule due to Hannay and d'Almeida [10] that the classical probability to return to the origin in time t is the sum over all such orbits

$$P_{cl}(E,t) = \sum_j P_j(E,t) = \sum_j \delta(t - T_j) T_j |\mathrm{det M_j - I}|^{-1} , \tag{8}$$

(here the reason for the appearance of the same matrix $(M_j - I)$ as before is the transformation of the volume element in phase space). He then proved that the quasiclassical spectrum of a classically ergodic system satisfies random matrix correlations ($\tilde{K}(\epsilon) \propto \epsilon^{-2}$) over scales smaller than an energy of order $\hbar$ (see below). It has to be noted that for applying the above to a classically ergodic system and for $t \gtrsim \hbar/\Delta$, the (unstable) orbits and gaussian regimes around them should be regarded as separated.

Very recently Argaman, Smilansky and the author [7] rewrote the Berry calculation in the following way. From Eqs. (5), (6), (7) and (8) (where much simplified proofs of Eqs. (6) and (8) will also be presented in Ref. 7) one obtains [7]:

$$\tilde{K}(t) = \frac{t}{h^2} P_{cl}(t) \tag{9}$$

which is an extremely useful expression and remarkably uses the same classical probability as, e.g. the "weak localization" quantum correction to transport [11-13]. For the simple example of a single particle which performs diffusive motion in the classical limit in a volume $V = L^d$, and which is *not* localized in the quantum case (so the dimensionless conductance $g \cong E_c/\Delta \gg 1$) one obtains (noting that the particle diffuses isotropically on an equal energy surface and thus the return probability density in phase space is $1/4\pi$ of that in x-space, we disregard here the question of the normalization of $P(t)$ over the equal energy surface [7])

$$P_{cl}(t) = \begin{cases} 1 & t \gtrsim \hbar/E_c \sim \frac{L^2}{D} \\ \frac{V}{(4\pi Dt)^{d/2}} & t \lesssim \hbar/E_c \,. \end{cases} \tag{10}$$

Fourier transforming (9) and (10), using only times $\ll \hbar/\Delta$ we obtain very simply the Altshuler-Shklovskii results for the spectral correlations:

$$K(\epsilon) \sim \begin{cases} -\epsilon^{-2} & \gamma \lesssim \epsilon \lesssim E_c \\ \frac{V}{(\hbar D)^{d/2}}\epsilon^{d/2-2} & E_c \lesssim \epsilon \lesssim \hbar/\tau_{cl} \,. \end{cases} \tag{11}$$

The semiclassical interpretation of these results is obvious. The smaller energy range is in agreement with Berry and with RMT (for example, to obtain the spectral rigidity one calculates the fluctuations in the number of levels in an interval W. The double integral of $K(\epsilon)$ for $\epsilon \lesssim E_c$ indeed yields the well-known $\log W/\gamma$ result, with the correct coefficient. The cut-off has to be properly introduced, e.g., by multiplying $K(t)$ by $e^{-\gamma|t|}$. This introduces [2] a positive portion of $K(\epsilon)$ for $\epsilon \lesssim \gamma$ which cancels the infrared behavior of $-\frac{1}{\epsilon^2}$. The limiting energy for the RMT behavior, E_c, is of order $\hbar$, as expected and new behavior is obtained for $\epsilon > E_c$. Different power-laws will be obtained for scale-dependent classical diffusion processes [7,14].

It is straightforward to do the quasiclassical argumentation for the ring with an AB flux [15,16]. One has to group [16] all orbits with winding number m and AB phase $e^{im\varphi}$ with all their time-reversed counterparts, having AB phase factors $e^{-im\varphi}$. One can in fact rederive in this way [7] the flux-dependent $K(\epsilon, \phi)$ used for both sample-specific [17,18] and ensemble-averaged [19-22] persistent currents.

For the case of the quantum dot, the magnetic field simply removes the constructive interference between time-reversed orbits and that is how the factor of $\frac{1}{2}$ reduction is obtained in $K(\epsilon)$ and $\langle \delta N^2 \rangle$. It is in full analogy with the removal of the Cooperon contribution in Ref. 2. This is the basis for the paramagnetic susceptibility discussed below. We first treat the symmetry-breaking effect of the magnetic field in the next section.

3. The Orthogonal-Unitary Crossover

In the "diagonal" sum of Eq. 7 it was assumed that there are no degenerate orbits (having exactly the same S_j). For a system with time-reversal ("orthogonal") symmetry the orbits actually come in degenerate pairs – each orbit and its time-reversed counterpart. Since they have exactly the same phase S_j, their contributions to $\tilde{K}(t)$ add coherently and this increases $\tilde{K}(t)$ by a factor of *two* compared to the ("unitary") case with no time-reversal symmetry. (Note that this factor of two and the factor of t in Eq. 9 both follow from constructive quantum interference). Thus, in obvious notation

$$\tilde{K}^u(t) = \frac{1}{2}\tilde{K}^0(t); \quad K^u(\epsilon) = \frac{1}{2}K^0(\epsilon) \tag{12}$$

The question now arises what happens when a time-reversal symmetry breaking agent with a finite strength, such as a magnetic field H, is applied. The magnetic field splits the phase of the above degenerate orbits by

$$\Delta\varphi = 2Ha/\phi_0\,, \tag{13}$$

where a is the oriented area perpendicular to H comprised by the orbit and ϕ_0 is the flux quantum hc/e. Thus the twofold orbit dengeneracy will be effectively broken for orbits large enough so that $\Delta\varphi \sim \pi$. There are now two regimes:

a) For $t \ll \frac{L^2}{D}$, the orbit does not cover the whole system at time t, its area is on the order of Dt (for diffusion perpendicular to H) and unitary symmetry will occur for

$$t \gg \frac{\pi\phi_0}{DH} \sim \frac{H_0}{E_c H} \equiv t_x^>\,, \quad \epsilon \ll \frac{\hbar DH}{\pi\phi_0} \sim E_c\frac{H}{H_0} \equiv \epsilon_x^>\,, \tag{14}$$

where H_0 is the field needed to produce on the order of a flux quantum in the whole system:

$$H_0 L^2 = \pi\phi_0 \tag{15}$$

For these crossover times to be shorter than $\frac{L^2}{D}$ ($\sim \frac{\hbar}{E_c}$) the field for the above to happen must be "strong"

$$H \gg H_0 \tag{16}$$

From a detailed calculation of the area distribution it was found in Ref. 7 that the form of the crossover is:

$$\tilde{K}(t) = \frac{1}{2}\left(1 + \frac{t/t_x^>}{\sin h(t/t_x^>)}\right)\tilde{K}^0(t) \tag{17}$$

where the crossover time is

$$t_x^> = \frac{\phi_0}{4\pi H D} \tag{18}$$

b) For weak fields, $H \ll H_0$, the particle covers the whole area of the system before accumulating a phase of $O(1)$ from the field. For that it has then to diffuse for a much longer time until the typical net number of revolutions is $\frac{\phi_0}{HL^2}$. This time is then the time to cover the system, $\frac{L^2}{D}$, multiplied by $\left(\frac{\phi_0}{HL^2}\right)^2$, since the sense of the revolution is random. Thus, here the crossover time and energy are of the order of:

$$\tau_x^< \sim \frac{L^2}{D}\left(\frac{H_0}{H}^2\right), \qquad \epsilon_x^< \sim \hbar/\tau_x^< \sim E_c\left(\frac{H}{H_0}\right)^2 \tag{19}$$

Since the distribution of oriented areas is gaussian, one finds [7]

$$\tilde{K}(t) = \frac{1}{2}(1 + e^{-t/t_x^<})\tilde{K}^0(t) \qquad (t > 0)\,, \tag{20}$$

Note that for very small fields $H/H_0 \lesssim \gamma/E_c$, $\epsilon_x^< \sim \gamma \gtrsim \Delta$. This is beyond the validity of our treatment and necessitates a more sophisticated one [3,22].

The crossover domains in the $\epsilon - H$ plan are summarized in Fig. 1. These results are consistent with those of Refs. 2 and 23. The

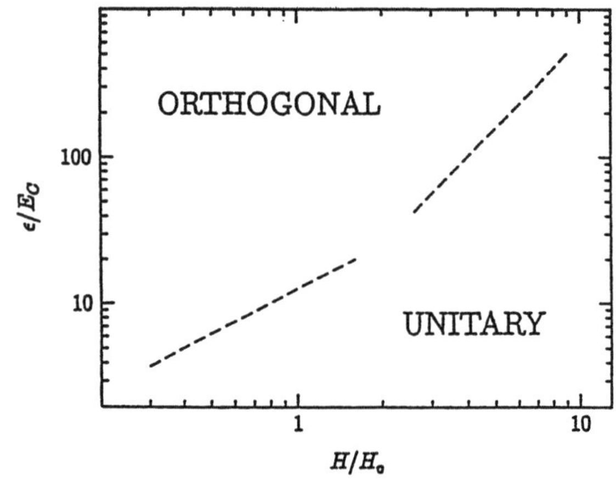

Fig. 1. The regions of approximate orthogonal and unitary symmetry and the crossover regions separating them, in the energy – magnetic field plane for the spectral correlation function $K(\epsilon)$. The dashed lines denote $t_x^<$ and $t_x^>$, and the magnetic field and energy have been normalized by $H_0 = L^2/\phi_0$, and $E_c = \hbar D/L^2$, from Ref. 7.

elimination of the factor of two due to constructive interference is equivalent to the destruction of the double Cooperon contribution to $K(\epsilon)$.

4. The Paramagnetic Orbital Contribution to the Suscepibility

In terms of the density of states correlation function, the fluctuation $\langle \delta N^2 \rangle$ in the number of particles among samples with different impurity realizations in the grand canonical ensemble at $T=0$ can be written as

$$\langle \delta N^2 \rangle = \int_0^{E_F} \int_0^{E_F} \langle \delta n(\epsilon) \delta n(\epsilon') \rangle d\epsilon d\epsilon' \tag{21}$$

This is a huge fluctuation, for noninteracting electrons. It turns out, however, that the dependence of this fluctuation on the magnetic field in the "quantum dot" (the case of the doubly-connected ring was briefly alluded to before) is physically very interesting. The reason for this is a general thermodynamic relationship, first proven in Refs. 3,16,21 and based on arguments in Ref. 20, between the difference of ensemble-averaged derivatives of the free energy at constant N and μ and fluctuations at given μ. Consider a finite system which is constrained to have a constant number of electrons when an external parameter such as the magnetic field is varied ("canonical ensemble"), and the electronic orbital levels are changing (we take the effects of spin to be just additive and will not discuss them here). This constancy of the electron number may be due to the system being electrically isolated, or just to a large "charging energy", $e^2/2C$, which makes changes in N energetically prohibitive [24]. One starts [3,21] by noting that for a fixed N system, when H is varied, the chemical potential changes, $\mu{=}\mu_{eq}(N,H,T){=}\left.\frac{\partial F}{\partial N}\right)_{H,T}$ where $F(N,H,T)$ is the (canonical) free energy. The above relationship is the "equation of state" of the system. The ("grand canonical") thermodynamic potential is $\Omega(\mu,H,T)$ and thermodynamics assures us that

$$\left.\frac{\partial F}{\partial H}\right)_N = \left.\frac{\partial \Omega}{\partial H}\right)_{\mu=\mu_{eq}} = -M(H,T)\,, \tag{22}$$

M being the equilibrium magnetization of the system. Now within the disorder ensemble, μ depends on ϕ and on the ensemble member, we write

$$\mu = \langle\mu\rangle + \delta\mu(\phi) \tag{23}$$

where $\langle\mu\rangle$ is the ensemble average. By expanding the middle member of (21) for small $\delta\mu$, we find

$$\left.\frac{\partial F}{\partial H}\right)_N = \left.\frac{\partial \Omega}{\partial H}\right)_{\mu=\langle\mu\rangle} + \delta\mu\left.\frac{\partial}{\partial\mu}\frac{\partial \Omega}{\partial H}\right)_{\mu=\langle\mu\rangle} \tag{24}$$

By changing the order of derivatives and noting that $N{=}-\left.\frac{\partial\Omega}{\partial\mu}\right)_{H,T}$, we find

$$\left.\frac{\partial F}{\partial H}\right)_N - \left.\frac{\partial \Omega}{\partial H}\right)_{\mu=\langle\mu\rangle} = -\delta\mu\left.\frac{\partial N}{\partial H}\right)_{\mu=\langle\mu\rangle} \tag{25}$$

Since $\left.\frac{\partial\mu}{\partial H}\right)_N=-\left.\frac{\partial\mu}{\partial N}\right)_H\cdot\left.\frac{\partial N}{\partial H}\right)_\mu$, and replacing to lowest order $\frac{\partial\mu}{\partial N}$ by its average value Δ, we obtain for the RHS, $\Delta\cdot\delta N\frac{\partial N}{\partial H}=\Delta\cdot\delta N\frac{\partial\delta N}{\partial H}$, thus [25]:

$$\left\langle\left(\frac{\partial F}{\partial H}\right)_N\right\rangle-\left\langle\left(\frac{\partial\Omega}{\partial H}\right)_{\mu=\langle\mu\rangle}\right\rangle=\frac{\Delta}{2}\frac{\partial}{\partial H}\left\langle\delta N^2\right\rangle_\mu, \tag{26}$$

where we employed ensemble averaging to the last equation which was also valid for each individual system. Eq. 26 is a starting point for our discussion of the new orbital paramagnetic effects. The usual calculations of orbital magnetism are done at constant μ. However, Eq. 26 tells us that if the system is kept at a constant N when H is varied (actually the same is valid for the response to other symmetry-breaking perturbations), there exists a new contribution, the RHS of Eq. 26, which is mainly paramagnetic, as we shall see, and can be sizeable under the appropriate conditions.

The deep reason for the paramagnetic sign of the RHS of Eq. 26 is that the fluctation $\langle\delta N^2\rangle_\mu$ decreases with the magnetic field H, which leads to a paramagnetic contribution χ_p to the susceptibility:

$$\chi_p=-\frac{\Delta}{2}\frac{\partial^2}{\partial H^2}\langle\Delta N^2\rangle. \tag{27}$$

$\langle\Delta N^2\rangle$ decreases with H because the integrand in (21) decreases, as discussed in the last section, by going from the orthogonal to the unitary symmetry. The decrease of $K(\epsilon)$ is by a factor of 2, for ϵ smaller than the crossover energy $\epsilon_x^<$ or $\epsilon_x^>$ of the last section. Thus a rough approximation for the flux-dependent part of ΔN^2 is [26]

$$\langle\Delta N^2(H)\rangle-\langle\Delta N^2(0)\rangle=-\frac{1}{2}\langle\Delta N^2(\epsilon_x(H))\rangle. \tag{28}$$

more complete and accurate calculations covering further cases will be given in Ref. 8.

Thus we can distinguish two main field regimes

a) "Strong" fields, $H\gg H_0$, $\epsilon_c=\epsilon_x^>$ (Eq. 14)

$$\chi_p=\frac{\Delta}{2}\frac{\partial^2}{\partial H^2}\left(\frac{\epsilon_x^>}{E_c}\right)^{d/2}\sim\frac{\Delta d}{4}\frac{H^{d/2-2}}{H_0^{d/2}}. \tag{29}$$

We emphasize that this field range is limited from above by the condition $H \lesssim H_{max}$, or $\ell_H \gg \ell$, where $\ell_H^2 = \frac{\hbar c}{eH}$, thus it is better referred to as the intermediate range [8].
b) Weak (but not too weak – see discussion following Eq. 20) fields, $H \ll H_0$

$$\chi_p = \frac{\Delta}{2}\frac{\partial^2}{\partial H^2}\ell n\frac{\epsilon_c^<}{\gamma} \sim \Delta\frac{\partial^2}{\partial H^2}\ell n\left(\frac{H}{H_0}\right) \sim \frac{\Delta}{H^2} \tag{30}$$

The smallest field for which we can employ this is $H \sim H_0\sqrt{\frac{\gamma}{E_c}}$. Thus the largest value of our paramagnetic orbital term is on the order of [4,8]

$$\chi_{p,m} \sim \frac{E_c}{H_0^2}\frac{\Delta}{\gamma} \lesssim \frac{E_c}{H_0^2} \tag{31}$$

The largest value is achieved for $\gamma \sim \Delta$. At finite temperatures, kT replaces γ for $\gamma \lesssim kT$. Our treatment breaks down for $H \lesssim H_{min} \sim H_0\sqrt{\frac{\gamma}{E_c}}$ and a more sophisticated one is called for. However, there is no reason to expect anything but a rather flat behavior of $\chi_p(H)$ for $H \lesssim H_{min}$. The nonlinear $\chi_p(H)$ is quite interesting: It decreases like $\frac{1}{H^2}$ for $H_{min} < H \lesssim H_0$ and like $\frac{1}{H^{2-d/2}}$ for $H_0 \lesssim H \lesssim H_{max}$. The case d=2 is special: there $\langle \Delta N^2 \rangle$ is proportional to H and thus $M(H)$ is almost a *constant* in the weak-field regime. The *local* χ defined by $\frac{\partial M}{\partial H}$ vanishes. Higher-order terms will be presented in Ref. 8. Note also that the local χ changes sign within the weak-field regime, where the curvature of $\langle \Delta N^2(H) \rangle$ changes sign.

We now estimate the magnitude of χ_p compared with the Landau susceptibility, $-\chi_L$, which is diamagnetic in sign. For this it is useful to note that $\frac{\Delta}{H_0^2} \sim \chi_L(k_F L)^{2(2-d)}$. Thus in a 3D system $\chi_p \ll \chi_L$, but in a purely 2D system $\frac{\chi_{p,m}}{\chi_L} \sim \frac{E_c}{\Delta}$ which is on the order of the dimensionless conductance ($\gg 1$ in the regime of interest). For a slab of finite thickness L_z, the 2D result is divided in general by $\frac{1}{(k_F L_z)^2}$ and at the lowest temperatures, $\gamma \sim \Delta$, by $\frac{1}{k_F L_z}$. Thus, measurable results may be expected for very thin films and small grains. In fact, there may already exist some numerical [27] and possibly experimental [28] support for these results.

5. Conclusions, Questions

We have reviewed the spectral correlations and their modification by a magnetic field in a quantum dot for noninteracting (or Hartree-Fock) fermions, emphasising the quasiclassical picture [6,7]. Similar conclusions are obtained from the Cooperon diffusion picture [2], both approximations are valid for $\frac{1}{\tau} \gtrsim \epsilon \gtrsim \gamma \gtrsim \Delta$. The above considerations are straightforwardly generalizable to include the effects of spin-orbit coupling. Extending these considerations to very small and very large energies/magnetic fields [22] is very desirable. We believe that some of the general results should be valid, with appropriate modifications, also including the very nontrivial effects of electron-electron interactions. Inclusion of these is even more necessary in view of the fact that the independent quasiparticle picture falls short by at least one or two orders of magnitude from explaining ensembe average [19] and sample-specific [18] experimental results for the intimately related persistent current problem. It would be interesting if interactions (whose effect is not understood well enough) would similarly influence the quantum-dot susceptibility. This may be related to older calculations for superconductors above Tc [34-36].

Acknowledgements

The author is grateful to B.L. Altshuler, N. Argaman, Y. Gefen, G. Montambaux and U. Smilansky for collaboration on various parts of this work, and to M. Berry, H. Baranger, E. Doron, O. Entin-Wohlman, D.E. Khmelnitskii, W. Kohn, B. Mühlschlegel, A. Schmid and H. Weidenmüller for discussions. This research was supported by the Minerva Foundation (Münich) and by the German-Israeli Foundation (GIF), Jerusalem. Some of this work was done in the Institute for Theoretical Physics, University of California, Santa Barbara, supported by NSF grant PHY 89-04035.

References

1. See for example M.L. Mehta, Random Matrices (Academic, New York 1967). For the first discussion of the spectra of metallic particles see R. Kubo, J. Phys. Soc. Japan **17** 975 (1962)
2. B.L. Altshuler and B. Shklovskii: Soviet Phys. JETP **64**, 127 (1986).
3. K.B. Efetov: Phys. Rev. Lett. **66**, 2794 (1991)
4. B.L. Altshuler, Y. Gefen and Y. Imry: Phys. Rev. Lett. **66**, 88 (1991)
5. O. Bohigas, M.J. Giannonni and C. Schmidt: Phys. Rev. Lett. **52**, 1 (1984)
6. M.V. Berry: Proc. Roy. Soc. London, Ser. a **400**, 229 (1985)
7. N. Argaman (Freed), U. Smilansky and Y. Imry: to be published
8. B.L. Altshuler, Y. Gefen, Y. Imry and G. Montambaux: In preparation
9. M.C. Gutzwiller: J. Math. Phys. **12**, 343 (1971)
10. J.H. Hannay and A.M. Ozorio de Almeida: J. Phys. **A17** 3429 (1984). In this work the classical sum rule was derived. The physical interpretation in terms of a probability was given by U. Smilansky, S. Tomsovic and O. Bohigas, WIS preprint, 1991 and in Ref. 7.
11. A.I. Larkin and D.E. Khmelnitskii: Usp. Fiz. Nauk **136** 336 (1982) (Sov. Phys. Usp. **25**, 185 (1982)); D.E. Khmelnitskii: Physica **126B**, 235 (1984)
12. G. Bergmann: Phys. Rep. **107**, 1 (1984)
13. S. Chakravarty and A. Schmid: Phys. Rep. **140**, 193 (1986)
14. B. Derrida and Y. Pomeau: Phys. Rev. Lett. **48**, 627 (1982); Ya.G. Sinai: In *Proc. 6th Int. Conf. on Mathematical Physics*, Berlin 1981 (Springer, Berlin 1982)
15. M. Buttiker, Y. Imry and R. Landauer: Phys. Lett. **96A**, 365 (1983)
16. Y. Imry: In Proc. April 1989 NATO ASI on Coherence Effects in Condensed Matter, ed. by B. Kramer (Plenum, New York 1991)

17. H. Cheung, E.K. Riedel and Y. Gefen: Phys. Rev. Lett. **62**, 587 (1989)
18. V. Chandrasekhar, R.A. Webb, M.J. Brady, M.B. Ketchen, W.J. Gallagher and A. Kleinsasser: Phys. Rev. Lett. **67** 3578 (1991)
19. L.P. Levy, G. Dolan, J. Dunsmuir and H. Bouchiat: Phys. Rev. Lett. **64**, 2074 (1990)
20. H. Bouchiat and G. Montambaux: J. Phys. (Paris) **50**, 2695 (1989); G. Montambaux, H. Bouchiat, D. Sigeti and R. Friesner: Phys. Rev. **B42**, 7647 (1990)
21. A. Schmid: Phys. Rev. Lett. **61**, 80 (1991)
22. A. Altland, S. Iida, A. Müller-Groeling and H. Weidenmüller: preprint (1991)
23. A. Pandey and M.L. Mehta: Commun. Math. Phys. **87**, 449 (1983); N. Dupuis and G. Montambaux: Phys. Rev. **B43**, 14390 (1991)
24. R. Kubo in Ref. 1
25. Note a sign error in Eq. (6) of Ref. 3, the similar result in Ref. 16 as well as the physical implications discussed in Ref. 3 are correct. The author thanks V. Ambegaokar for pointing out this error.
26. Note that (21) in an interval W is best evaluated as $\langle \Delta N^2 \rangle = 2\int_0^W X(\omega)d\omega$ where $X(\omega) = \int_0^\omega K(\epsilon)d\epsilon$. Were $K(\epsilon)$ a monotonic finite range function, $X(\omega)$ would be finite for large ω and $\langle \Delta N^2 \rangle$ would be "extensive", i.e. of order W. The logarithmic character of $\langle \Delta N^2 \rangle$ in the RMT regime follows from a delicate cancellation of the "infrared" anomaly following from $K(\epsilon) \sim -\epsilon^{-2}$, due to the change of sign [2] of $K(\epsilon)$ for $\epsilon \lesssim \gamma$. Related cancellations affect $\langle \Delta N^2 \rangle$ in the $W \gtrsim E_c$ regime too. It is the power law (in general) dependence of $X(\omega)$ on ω which causes the validity of (28) for a sharp change of $K(\epsilon)$ at $\epsilon \sim \omega_X$
27. J.M. van Ruitenbeck and D.A. van Leeuwen: Phys. Rev. Lett. **67**, 640 (1991); J.M. van Ruitenbeck, Z. Phys. D, in press.
28. K. Kimura and S. Bandow, Phys. Rev. **37**, 4473 (1988)
29. B.L. Altshuler, D.E. Khmelnitskii and B.Z. Spivak: Solid State Commun. **48**, 841 (1983)

30. V. Ambegaokar and U. Eckern: Phys. Rev. Lett. **65**, 381 (1990); ibid **67**, 3192 (1991) (comments)
31. U. Eckern: Z. Phys. **B82**, 393 (1991)
32. V. Ambegaokar and U. Eckern: Europhys. Lett. **13**, 733 (1990)
33. A. Schmid: Preprint (1991)
34. H. Schmidt: Z. Phys. **216**, 336 (1968)
35. A. Schmid: Phys. Rev. **180**, 627 (1969)
36. L.G. Aslamazov and A.I. Larkin, Zh. Exp. Teor. Fiz. **67**, 647 (1974) [Sov. Phys. JETP **40**, 321 (1975)].

Fluctuation of the Orbital Magnetism

D. Yoshioka[1] *and H. Fukuyama*[2]

[1]Institute of Physics, University of Tokyo,
3-8-1 Komaba, Meguro-ku, Tokyo 153, Japan
[2]Department of Physics, Faculty of Science, University of Tokyo,
7-3-1 Hongo, Bunkyo-ku, Tokyo 113, Japan

Abstract. Orbital magnetization of two-dimensional electron systems in external confining potential is calculated. At low temperature the magnetization show sample specific fluctuation that is much larger than the average. It is shown that at higher temperature the fluctuation decays exponentially, and Landau diamagnetism is realized. It is also shown that average of the fluctuating magnetization at $T = 0$ also gives the Landau diamagnetism.

1. Introduction

It is known that orbital motion of free electrons gives rise to the Landau diamagnetism [1]. In his derivation Landau considered infinite size electron systems, and neglected the boundary effects. Since then there have been a lot of investigations on the orbital magnetism of finite size systems, most of which are devoted to the study of boundary effects on the Landau susceptibility [2,3,4,5,6]. There has been two ways to consider the boundary effects. One is to put electrons in harmonic potentials. The other is to put electrons in containers with hard walls, and imposes the Dirichlet boundary condition. These investigations clarified that the Landau susceptibility, which is obtained by some kind of averaging such as thermal or impurity, is essentially unchanged. It has been clarified, however, that the magnetization at low temperature can fluctuates considerably. In particular, possibility of paramagnetic response has been clarified for samples with rotational symmetry along the axis parallel to the magnetic fields [4,6].

In view of the recent developments of the mesoscopic systems, orbital magnetism of finite size system deserves further investigations. It will become possible to observe the orbital magnetization of quantum dots. However, there has been no systematic investigation on the magnetization of finite size systems. In the previous studies correction to the Landau susceptibility have been the main theme. Therefore, mainly magnetization in the limit of vanishing magnetic field have been investigated. Even when the magnetization was calculated at finite magnetic fields, it was restricted to $T = 0$. Therefore in this paper we investigate the orbital magnetization of a finite size system systematically at finite magnetic field including the effect of finite temperature.

We consider two models for two-dimensional free electrons; harmonic bowl and harmonic groove. In the harmonic bowl electrons move in the xy-plane in an isotropic harmonic potential. In the harmonic groove electrons are confined by a harmonic potential in the y-direction, but are free in the x-direction; periodic boundary condition is imposed in this direction. We will show that in both models the magnetization has large fluctuation at low temperature. It will be shown that the average of such a fluctuation gives the Landau susceptibility. We will show that when temperature becomes higher, the fluctuation decays exponentially, and the Landau susceptibility is recovered.

Recently fluctuation in the susceptibility has been calculated diagrammatically [7,8]. It has been shown that the fluctuation becomes much larger than the susceptibility itself at low temperatures. We discuss possible relationship between our present calculation and the diagrammatically calculated susceptibility. In this paper we consider spinless electrons and neglect the Coulomb interaction among them. We use the unit where $\hbar = k_\mathrm{B} = 1$.

2. Models

We investigate two models: the harmonic bowl and the harmonic groove. The Hamiltonian for the harmonic bowl is

$$\mathcal{H} = \frac{1}{2m_\mathrm{e}}(\mathbf{p}-e\mathbf{A})^2+\frac{1}{2}m_\mathrm{e}\omega_0^2\mathbf{r}^2, \tag{2.1}$$

where $\mathbf{p}$ and $\mathbf{r}$ are 2-d vectors, $\mathbf{A} = (-By/2, Bx/2)$, m_e is the effective mass, and ω_0 is the frequency of the electron in the harmonic potential. The eigenvalue of this Hamiltonian is given by

$$E_{n,m} = \frac{1}{2}m\omega_\mathrm{c}+n\sqrt{\omega_0^2 + \frac{1}{4}\omega_\mathrm{c}^2}, \tag{2.2}$$

where $n = 1, 2, 3, \ldots$ and $m = -(n-1), -(n-3), \ldots, n-3, n-1$ are quantum numbers, and $\omega_\mathrm{c} = eB/m_\mathrm{e}$ is the cyclotron frequency. An electron in these eigenstates has orbital magnetic moment $M_{n,m}$:

$$M_{n,m} = -\frac{\partial E_{n,m}}{\partial B} = -[m+n\frac{\omega_\mathrm{c}}{\sqrt{4\omega_0^2+\omega_\mathrm{c}^2}}]\mu_\mathrm{B}, \tag{2.3}$$

where $\mu_B = e/2m_e$ is the Bohr magneton. Therefore the total magnetic moment of the system with N_e electrons are given by

$$M = \sum_n \sum_m M_{n,m} f(E_{n,m} - \mu). \tag{2.4}$$

Here $f(E) = 1/[e^{E/T}+1]$ is the Fermi distribution function. The chemical potential, μ, in eq.(2.4) is determined as follows as usual:

$$N_e = \sum_n \sum_m f(E_{n,m} - \mu). \tag{2.5}$$

The Hamiltonian for the harmonic groove is given as follows.

$$\mathcal{H} = \frac{1}{2m_e}(p_x+eBy)^2 + \frac{1}{2m_e}p_y^2 + \frac{1}{2}m_e\omega_0^2 y^2. \tag{2.6}$$

The eigenvalue of this Hamiltonian is given by

$$E_{n,m} = \sqrt{\omega_0^2 + \omega_c^2}(n+\frac{1}{2}) + A\frac{\omega_0^3}{\omega_0^2 + \omega_c^2}m^2, \tag{2.7}$$

where $A \equiv (1/2m_e)(2\pi/L_x)^2/\omega_0$ is a dimensionless parameter related to the aspect ratio of the system L_x being the dimension in the x-direction, and n, m are quantum numbers: $n = 0, 1, 2, ...$ and $m = 0, \pm1, \pm2, ...$ The orbital magnetic moment associated with each energy level is

$$M_{n,m} = [-\frac{2\omega_0}{\sqrt{\omega_0^2 + \omega_c^2}}(n+\frac{1}{2}) + A\frac{4\omega_0^4}{(\omega_0^2 + \omega_c^2)^2}m^2]\frac{\omega_c}{\omega_0}\mu_B. \tag{2.8}$$

The total magnetization is calculated by eqs.(2.4) and (2.5) as before.

3. Magnetization

3.1 Harmonic Bowl

When the temperature T is of the order of ω_0 or higher, we can calculate the magnetization analytically to the first order in ω_c. The result is

$$M \simeq -\frac{1}{6}(\frac{\mu}{\omega_0} - \frac{1}{2})\frac{\omega_c}{\omega_0}\mu_B, \tag{3.1}$$

where

$$\mu \simeq \sqrt{2N_e - \frac{1}{3}(\frac{\pi T}{\omega_0})^2}\ \omega_0. \qquad (3.2)$$

If we define the system area S most naturally, namely to be $2\pi\mu/m_e\omega_0^2$, the area where a classical particle with energy μ can exist, M is given as follows.

$$M \simeq -\frac{1}{24\pi}\frac{e^2}{m_e}BS. \qquad (3.3)$$

This is just the value of the Landau diamagnet with area S, since $\chi_L = -(1/24\pi)e^2S/m_e$ per spin. Thus as far as temperature is high enough Landau's result is exact in the harmonic bowl, where the effect of boundary is very important.

At lower temperature the magnetization behaves quite differently. Figures 1 and 2 show the results of numerical calculation of the magnetization as a function of the magnetic field, which is expressed by the ratio ω_c/ω_0. The results for a system with 50 electrons are shown in Fig.1, and those with 55 electrons are shown in Fig.2. In these figures the solid line, the dash-dotted line and the dotted line show the results at $T = 0$, $0.1\omega_0$ and $0.4\omega_0$, respectively. The data at $T = 0.4\omega_0$ almost coincides with the Landau magnetization, eq.(3.3). The Landau magnetization is not shown in the figures, because it overlaps with the data for $T = 0.4\omega_0$.

The magnetization at $T = 0$ differs drastically between the system with 50 electrons and 55 electrons. In the former system, shown in Fig.1, the magnetization has a finite value at $\omega_c = 0$, i.e. the system is ferromagnetic. On the other hand for the latter system the magnetization is zero at $\omega_c = 0$. This is the consequence of the rotational symmetry of the present system and the shell structure of the energy levels. At $\omega_c = 0$ the energy levels are degenerate with respect to the quantum number m. The degeneracy for given quantum number n is n-fold. The degenerate levels has quite different magnetic moments, $M_{n,m} = -m\mu_B$ even at $B = 0$ due to the rotational symmetry. When the number of electrons N_e is given by $n_0(n_0+1)/2$, where n_0 is an integer, the energy levels up to $n = n_0$ are totally occupied forming closed shells. In this case the magnetic moments in each shell are canceled. Thus the system with 55 electrons has zero magnetization at $B = 0$. On the other hand when the highest occupied shell is only partially filled, the magnetization of the shell survives leading to ferromagnetic response. At $T = 0$ and $B = 0$ the magnetization is given as

$$M = (1/4)[2N_e - n_0(n_0-1)][n_0(n_0+1) - 2N_e]\mu_B, \qquad (3.4)$$

when $(n_0 - 1)n_0/2 < N_e < n_0(n_0+1)/2$.

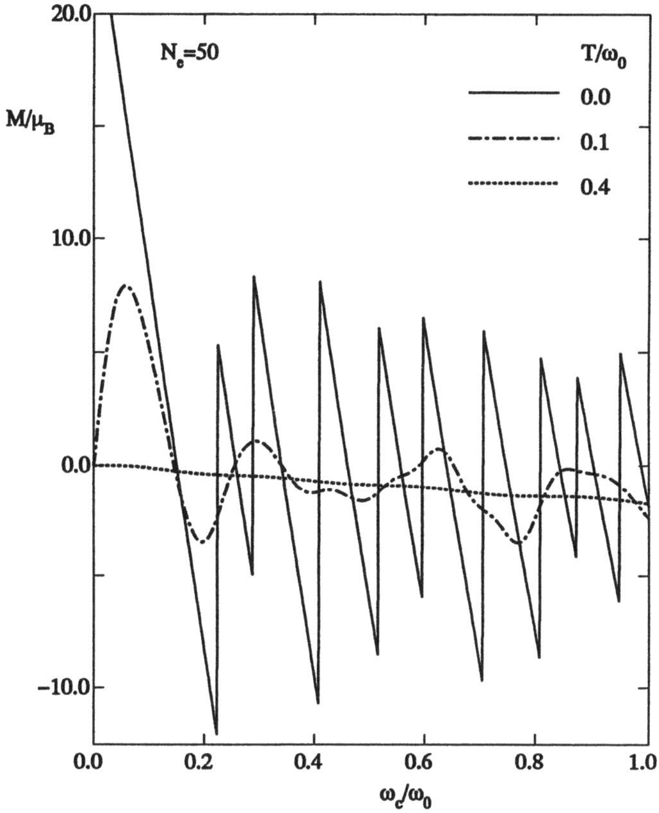

Fig.1 Magnetic field dependence of the magnetization of 50 electrons in isotropic harmonic potential. The solid, dash-dotted, and dotted lines show the magnetization at $T = 0$, $0.1\omega_0$, and $0.4\omega_0$, respectively.

The differential susceptibility at $T = 0$, $\chi_d \equiv \lim_{B\to 0}(\partial M/\partial B)$, is given as follows:

$$\chi_d = -n_0[N_e - \frac{1}{6}(n_0^2-1)]\frac{\mu_B^2}{\omega_0} \simeq -\frac{2}{3}N_e\frac{\mu}{\omega_0^2}\mu_B^2. \tag{3.5}$$

Here $n_0(n_0-1)/2 \leq N_e \leq n_0(n_0+1)/2$ and μ is the chemical potential. This χ_d is about $2N_e$ times larger than the Landau susceptibility. On the other hand at low but finite temperature, $\omega_c << T << \omega_0$, the magnetization is given as follows:

$$M = \left\{ \frac{(n_0^2-1)[2N_e-(n_0-1)n_0][n_0(n_0+1)-2N_e]}{24n_0}\frac{\omega_0}{T} \right.$$
$$\left. -\frac{1}{2}n_0[N_e - \frac{1}{6}(n_0^2-1)] \right\}\frac{\omega_c}{\omega_0}\mu_B. \tag{3.6}$$

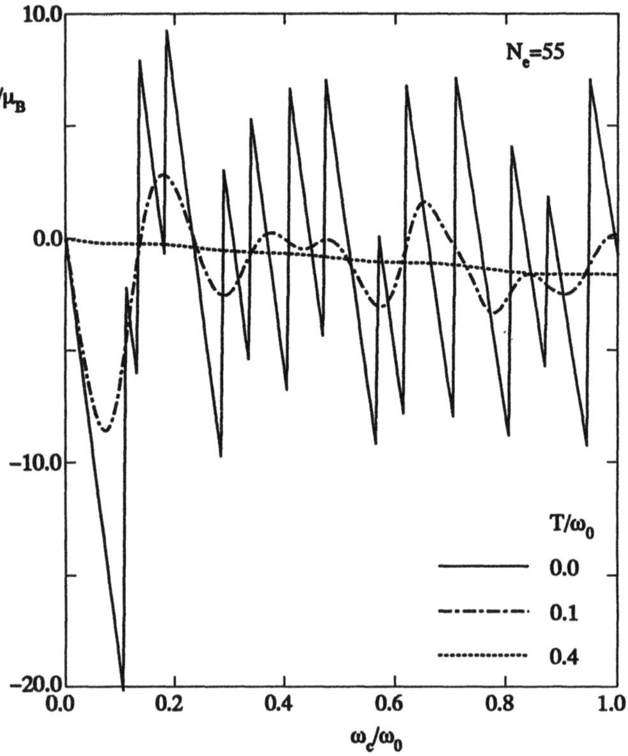

Fig.2 Magnetic field dependence of the magnetization of 55 electrons in isotropic harmonic potential. The solid, dash-dotted, and dotted lines show the magnetization at $T = 0$, $0.1\omega_0$, and $0.4\omega_0$, respectively.

The first term in the curly bracket is paramagnetic Currie term, which take finite value for open shell ($N_e \neq n_0(n_0+1)/2; n_0 = 1, 2, 3, ..$). The second term in the curly bracket is nothing but $\chi_d B$, namely the strong diamagnetic term given in eq.(3.5). At finite temperature the peak of the magnetization at $T = 0$ decays exponentially, $\propto e^{-\pi^2 T/\omega_0}$, and approaches the Landau diamagnetism.

3.2 Harmonic Groove

In this system there is no rotational symmetry. Consequently the magnetic moment of each level, eq.(2.8), is proportional to ω_c at small B, and vanishes as $B \to 0$. In Fig.3 magnetization for a choice of $N_e = 55$ and $A = 0.5$ is shown as a function of B at $T = 0$, $0.25\omega_0$, and ω_0. The magnetization shows large fluctuation at low temperature, which is

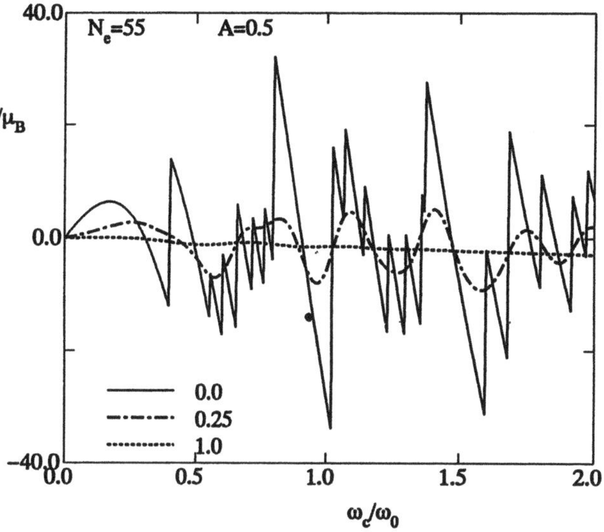

Fig.3 Magnetic field dependence of the magnetization of 55 electrons in a harmonic groove with $A = (1/2m_e)(2\pi/L_x)^2/\omega_0 = 0.5$, where L_x is the size of the system in the x-direction. The solid line, the dash-dotted, and dotted lines show the magnetization at $T = 0$, $0.25\omega_0$, and ω_0, respectively.

similar to that of the harmonic bowls. At $B = 0$ M vanishes. The differential susceptibility at $B = 0$ can become both positive or negative depending on the choice of A and N_e. At higher temperature the magnetization becomes proportional to the magnetic field, $M = \chi B$ as in the cases of harmonic bowls. This χ is almost exactly given by the Landau susceptibility χ_L, where the system area S is given by that of the classical system with the same chemical potential μ: $S = 2L_x\sqrt{2\mu/m_e}/\omega_0$. If $\chi_L B$ were plotted in the figure, the line would almost overlap with the highest temperature results.

4. Fluctuation

We have seen in the previous section that the magnetization at low temperature show fluctuations which is specific to the models and number of electrons. To investigate this fluctuation quantitatively we define average magnetization. Instead of considering average over various samples we here consider average magnetization for each sample. We define average $\overline{M}$ by Gaussian average over the magnetic fields:

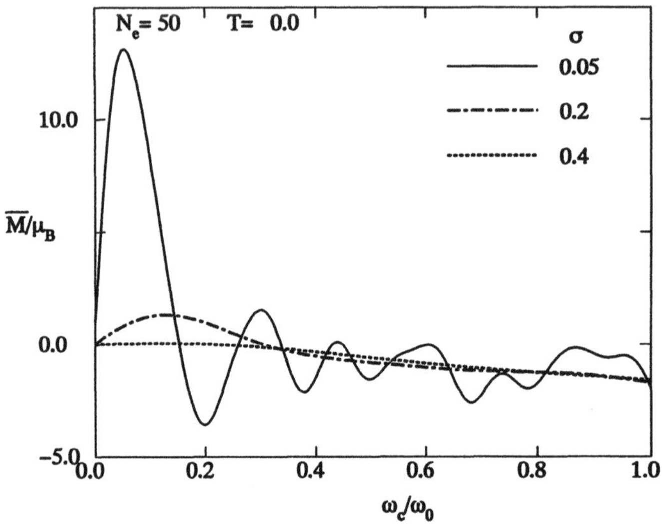

Fig.4 Magnetic field dependence of the Gaussian averaged magnetization of 50 electrons in isotropic harmonic potential. The solid, dash-dotted, and dotted lines show the magnetization at $T = 0$ with $\sigma = 0.05, 0.2$, and 0.4, respectively.

$$\overline{M(B)} \equiv \frac{1}{\sqrt{\pi}\sigma_{\mathrm{B}}} \int_{-\infty}^{\infty} \mathrm{d}B' \mathrm{e}^{-(B'-B)^2/\sigma_{\mathrm{B}}^2} M(B'). \tag{4.1}$$

One example of such average is shown in Fig.4. As shown in the figure, fluctuation in $\overline{M}$ decreases as $\sigma \equiv e\sigma_{\mathrm{B}}/m_e\omega_0$ increases, and for large enough σ ($\sigma > 1$) $\overline{M(B)}$ depends linearly on B. Numerically $\overline{M(B)}$ for large σ coincides with the Landau diamagnetism realized at $T > \omega_0$. This behavior is common both for harmonic bowls and harmonic grooves. However, it should be noted for infinite-size pure 2-d system this averaging does not give the Landau diamagnetism. In this system self-averaging gives twice of the Landau value.

Since the average, eq.(4.1), is given by the Landau diamagnetism, fluctuation in M can be defined as

$$(\delta M)^2 \equiv [M(B) - \chi_{\mathrm{L}} B]^2. \tag{4.2}$$

At low temperature $|M(B)|$ is much larger than $|\chi_{\mathrm{L}} B|$ for almost any value of B, therefore $\delta M \simeq |M(B)|$. At $T = 0$ the peak value of $M(B)$ at finite B is about half of the jump in $M(B)$. In our system this jump in the magnetization happens when two single-electron levels cross each other at the Fermi level. Therefore the jump in M is of the order of the magnetization of a single-electron level that is $O(|m|\mu_B) \simeq O(\sqrt{N_e}\mu_B)$ for the harmonic bowls. The harmonic grooves show similar fluctuations. Thus the fluctuation in M at $T = 0$ is of the order of $\sqrt{N_e}\mu_B$, which is much larger than the Landau diamagnetism, $\sim -(\sqrt{2}/6)(\omega_c/\omega_0)\sqrt{N_e}\mu_B$ for $\omega_c < \omega_0$. This fluctuation decays exponentially, $\simeq e^{-\pi^2 T/\omega_0}$, as the peak in M decays at finite temperature. At higher temperature ($T \simeq \omega_0$) δM vanishes because there $M(B) \simeq \chi_L B$.

The fluctuation considered up to now is sample specific like universal conductance fluctuation. This fluctuation competes with thermal fluctuation at finite temperature. This fluctuation is given as follows,

$$(\delta M_T)^2 \equiv \langle (M - \langle M \rangle)^2 \rangle = \sum_n \sum_m \frac{e^{E_{n,m}/T}}{(e^{E_{n,m}/T}+1)^2} M_{n,m}^2 \tag{4.3}$$

For isotropic quantum bowl this fluctuation is calculated as follows at high temperature and low magnetic field, $T > \omega_0 >> \omega_c$.

$$(\delta M_T)^2 \simeq \frac{T}{3\omega_0}(2N_e)^{3/2}\mu_B^2. \tag{4.4}$$

Hence,

$$\frac{\delta M_T}{\chi_L B} \simeq \sqrt{12}(2N_e)^{1/4}\sqrt{\frac{T\omega_0}{\omega_c^2}}. \tag{4.5}$$

This thermal fluctuation must be smaller than the intrinsic fluctuation for the latter to be observed. For harmonic bowls with 50 electrons thermal fluctuation becomes larger for $T > 0.025\omega_0$, while for harmonic grooves with 55 electrons the condition is $T > 0.05\omega_0$.

5. Discussion

In this paper we have calculated the orbital magnetization of finite electron systems in external potentials. We find that at high temperature ($T > \omega_0$) the system shows Landau diamagnetism, while at low temperature the magnetization fluctuates irregularly.

At low temperature the magnetization fluctuates as B or N_e changes. This is because each level has different magnetic moment and the distribution of electrons depends on B and N_e. This fluctuation should be related to the huge fluctuation of the Landau diamagnetism calculated diagrammatically [7,8]. Since the diagrammatic calculations give differential susceptibility in the limit of $B \to 0$, N_e dependence of the magnetization at $B \to 0$ in the present calculation is more relevant than the

B dependence. Our results show the differential susceptibility can have positive or negative values depending upon the number of electrons or the parameter in the potentials. Of course our system is quite different from infinite system with random short range impurity potentials considered diagrammatically. However, our system can be considered as a model for system where impurity potential is long ranged. In that case electrons are localized around local minima of the impurity potential. The spatial average over such system corresponds to average over the electron number and the potential parameters ω_0. Such average gives Landau diamagnetism with large fluctuations. Another argument is that even for short-ranged impurity potentials electrons are localized within some localization length (which is much larger than the atomic distance). Each localized states have magnetic moments which are different from state to state. Then the situation is similar to that considered here.

The present system is considered here as a solvable model system to investigate the fluctuation in the orbital magnetism. However, experimentally the present system has already been realized as quantum dots [9]. Each quantum dots has size of about 200nm, containing about 50 electrons, and confining potential is known to be parabolic. The value of ω_0 is of the order of 10K. Therefore fluctuation of the magnetization should be realized for $T < 0.1$K. Unfortunately, the magnetization of a quantum dots is quite small: $M < 100\mu_B$. Two dimensional array of the dots may still be not enough at present for the observation of the fluctuation.

Details of the present work will be published elsewhere.

Acknowledgements

The work by DY is partially supported by Grant-in Aid for Scientific Research on Priority Areas " Computational Physics as a New Frontier in Condensed Matter Research" (03247105) from the Ministry of Education, Science and Culture. The work by HF is partially supported by the University-Industry Joint Research Program "Mesoscopic Electrons" and by the Grant-in-Aid for Scientific Research on Priority Area "Electron Wave Interference Effects in Mesoscopic Structures" from the Ministry of Education, Science and Culture.

References

[1] L.D. Landau: Z. Phys. **64**, 629 (1930).
[2] D. Childers and P. Pincus: Phys. Rev. **177**, 1036 (1969).
[3] R.V. Denton: Z. Phys. **265**, 119 (1973).
[4] A.I. Buzdin, O.V. Dolgov and Yu.E. Lozovik: Phys. Lett. **100A**, 261 (1984).
[5] M. Robnik: J. Phys. A: Gen. Phys. **19**, 3619 (1986).
[6] J.M. van Ruitenbeek and D.A. van Leeuwen: Phys. Rev. Lett. **67**, 640 (1991).
[7] H. Fukuyama: J. Phys. Soc. Jpn. **58**, 47 (1989); O.D. Cheishvili: Pis'ma Zh. Eksp. Theor. Fiz. **48**, 206 (1988) [JETP Lett. **48**, 225 (1988).]
[8] R.A. Serota and S. Oh: Phys. Rev. B**41**, 10523 (1990); S. Oh, A.Yu. Zyuzin and R.A. Serota: Phys. Rev. B**44**, 8858 (1991).
[9] D. Heitman, K. Kern, T. Demel, P. Grambow, K. Ploog and Y.H. Zhang: to be published in Surf. Sci. (*Proc. 9th Int. Conf. Electronic Properties of Two-Dimensional Systems, Nara 1991*); A. Lork: *ibid.*

Part VIII

Proximity Effect of Superconductivity

Three "Universal" Mesoscopic Josephson Effects

C.W.J. Beenakker
Instituut-Lorentz, University of Leiden,
P.O. Box 9506, NL-2300 RA Leiden, The Netherlands

Abstract. A recent theory is reviewed for the sample-to-sample fluctuations in the critical current of a Josephson junction consisting of a disordered point contact or microbridge. The theory is based on a relation between the supercurrent and the scattering matrix in the normal state. The root-mean-square amplitude $\mathrm{rms}\, I_c$ of the critical current I_c at zero temperature is given by $\mathrm{rms}\, I_c \simeq e\Delta_0/\hbar$, up to a numerical coefficient of order unity (Δ_0 is the energy gap). This is the superconducting analogue of "Universal Conductance Fluctuations" in the normal state. The theory can also be applied to a ballistic point contact, where it yields the analogue of the quantized conductance, and to a quantum dot, where it describes supercurrent resonances. All three phenomena provide a measurement of the supercurrent unit $e\Delta_0/\hbar$, and are "universal" through the absence of a dependence on junction parameters.

1 Introduction

Nanostructures combining semiconducting and superconducting elements form a new class of systems in which to search for mesoscopic phenomena. The correspondence between transport of normal electrons and transport of Bogoliubov quasiparticles (being the elementary excitations of the superconductor) serves as a useful guide in the search. The appearance of a new length scale — the superconducting coherence length ξ — complicates the correspondence in an interesting way, by introducing two qualitatively different regimes: The *short-junction* regime (junction length $L \ll \xi$) and the *long-junction* regime ($L \gg \xi$). These two regimes appear over and above the transport regimes which depend on the relative magnitude of junction length and mean free path l: The *ballistic* regime ($l \gg L$) and the *diffusive* regime ($l \ll L$). A third transport regime in the normal state, that of *resonant tunneling*, is also subdivided into two new regimes, distinguished by the relative magnitude of ξ and the characteristic length $v_F\tau_{res}$, with τ_{res} the lifetime of the resonant state in the junction and v_F the Fermi velocity.

Our own interest in this field has focused on a set of three phenomena which provide, through the Josephson effect, a measurement of the supercurrent unit $e\Delta_0/\hbar$ (Δ_0 being the superconducting energy gap). This unit of current plays the role of the conductance quantum e^2/h in the normal state. The three phenomena are:

1. Discretization of the critical current of a ballistic point contact [1].
2. Resonant Josephson current through a quantum dot [2, 3].
3. Mesoscopic supercurrent fluctuations in a diffusive point contact [4].

These effects belong, respectively, to the ballistic, resonant-tunneling, and diffusive transport regimes, and within each regime to the short-junction limit.

Effect number 1 is the analogue of the quantized conductance of a quantum point contact [5]–[7]: The critical current I_c of a short and narrow constriction in a superconductor increases stepwise as a function of the constriction width, with step height $e\Delta_0/\hbar$ independent of the properties of the junction [1],

$$I_c = N\frac{e\Delta_0}{\hbar}. \tag{1}$$

The integer N is the number of transverse modes at the Fermi level which can propagate through the constriction. The short-junction limit $L \ll \xi_0$ (where $\xi_0 = \hbar v_F/\pi\Delta_0$ is the ballistic coherence length) is essential: Furusaki et al. [8] studied quantum size effects on the critical current in the long-junction limit and found a geometry-dependent behavior instead of Eq. (1).

Effect number 2 is described by the formula [3]

$$I_c^{res} = \frac{e}{\hbar}\frac{\Delta_0\Gamma_0}{\Delta_0 + \Gamma_0} \tag{2}$$

for the critical current through a bound state at the Fermi level which is coupled with equal tunnel rate $\Gamma_0/\hbar$ to two bulk superconductors. Since $\tau_{res} = \hbar/2\Gamma_0$, the "short-junction" criterion $\xi_0 \gg v_F\tau_{res}$ is equivalent to $\Gamma_0 \gg \Delta_0$. If $\Gamma_0 \gg \Delta_0$, the critical current on resonance becomes [2, 3]

$$I_c^{res} = \frac{e\Delta_0}{\hbar}, \tag{3}$$

which no longer depends on the tunnel rate.

Effect number 3 is the analogue of "Universal Conductance Fluctuations" in disordered normal metals [9]–[11]. The sample-to-sample fluctuations in the critical current of a disordered point contact have root-mean-square value [4]

$$\mathrm{rms}\, I_c \simeq \frac{e\Delta_0}{\hbar}, \tag{4}$$

up to a numerical coefficient of order unity. These mesoscopic fluctuations are *universal* in the sense that they do not depend on the size of the junction or on the degree of disorder, as long as the criteria $l \ll L \ll Nl$ and $L \ll \xi$ for the diffusive, short-junction regime are satisfied. (Here $\xi = (\xi_0 l)^{1/2}$ is the diffusive coherence length.) The short-junction limit $L \ll \xi$ is again essential for universality. The opposite long-junction limit was considered previously by Al'tshuler and Spivak [12], who calculated the supercurrent fluctuations of an SNS junction in the limit $L \gg \xi$ (S=superconductor, N=normal metal, L=separation of the SN interfaces). Their result $\mathrm{rms}\, I_c \simeq ev_F l/L^2$ depends on both l and L and is therefore *not* universal in the sense of UCF.

Table 1: Summary of the mesoscopic Josephson effects described in the text, in the ballistic, resonant-tunneling, and diffusive transport regimes. Universal (junction-independent) behavior is obtained in the short-junction limit, which in terms of the correlation energy E_c is given by $\Delta_0 \ll E_c$.

	Mesoscopic Josephson Effects		
	short-junction	long-junction	E_c
ballistic	$I_c = Ne\Delta_0/\hbar$	geometry dep.	$\hbar v_F/L$
res.tunneling	$I_c^{res} = e\Delta_0/\hbar$	$I_c^{res} = e\Gamma_0/\hbar$	Γ_0
diffusive	$\mathrm{rms}\, I_c \simeq e\Delta_0/\hbar$	$\mathrm{rms}\, I_c \simeq eD/L^2$	$\hbar D/L^2$

We have summarized the three universal short-junction Josephson effects in Table 1, together with their non-universal long-junction counterparts. We have introduced the diffusion constant $D \simeq v_F l$ and the correlation energy (or Thouless energy) $E_c \simeq \hbar/\tau_{\text{dwell}}$, where τ_{dwell} is the dwell time in the junction. The distinction "short-junction" versus "long-junction" may alternatively be given in terms of the relative magnitude of E_c and Δ_0: In the short-junction regime $\Delta_0 \ll E_c$, while in the long-junction regime $\Delta_0 \gg E_c$. The universality is lost in the long-junction regime because E_c (which depends on junction parameters) becomes the characteristic energy for the Josephson effect instead of Δ_0. All these results hold in the zero-temperature limit, or more precisely for $k_BT \ll \min(\Delta_0, E_c)$.

In the present paper we review a scattering theory for the Josephson effect, which provides a unified treatment of the three universal (short-junction) phenomena. In Sec. 2 we show, following Ref. [13], how the supercurrent can be obtained directly from the quasiparticle excitation spectrum of the Josephson junction. This approach is equivalent to the usual method which starts from the finite-temperature Greens function, but is more convenient in the short-junction regime, where — as we shall see — the excitation spectrum has a particularly simple form. In Sec. 3 we follow Ref. [4] in relating the excitation spectrum to the normal-state scattering matrix of the junction. In Sec. 4 we take the short-junction limit and obtain a simple relation between the supercurrent and the eigenvalues of the transmission matrix product $tt^\dagger$. This relation plays the role for the Josephson effect of the Landauer formula for the conductance [7]. We emphasize that the transmission matrix t refers to *normal* electrons. Other scattering approaches to the Josephson effect [14, 15] relate the supercurrent to the scattering matrix for Bogoliubov quasiparticles. The usefulness of the present method is that one can use directly the many properties of t known from normal-state transport theory. Indeed, as we shall see in Sec. 5, the three universal Josephson effects follow as special cases of the general scattering formula of Sec. 4.

2 Supercurrent from Excitation Spectrum

The quasiparticle excitation spectrum of the Josephson junction consists of the positive eigenvalues of the Bogoliubov-de Gennes (BdG) equation [16]. The BdG equation has the form of two Schrödinger equations for electron and hole wavefunctions $\mathrm{u}(\mathbf{r})$ and $\mathrm{v}(\mathbf{r})$, coupled by the pair potential $\Delta(\mathbf{r})$:

$$\begin{pmatrix} \mathcal{H}_0 & \Delta \\ \Delta^* & -\mathcal{H}_0 \end{pmatrix} \begin{pmatrix} \mathrm{u} \\ \mathrm{v} \end{pmatrix} = \varepsilon \begin{pmatrix} \mathrm{u} \\ \mathrm{v} \end{pmatrix}. \tag{5}$$

Here $\mathcal{H}_0 = \mathrm{p}^2/2m + V(\mathbf{r}) - E_\mathrm{F}$ is the single-electron Hamiltonian, containing an electrostatic potential $V(\mathbf{r})$ (zero vector potential is assumed). The excitation energy ε is measured relative to the Fermi energy E_F.

In a uniform system with $\Delta(\mathbf{r}) \equiv \Delta_0 \mathrm{e}^{\mathrm{i}\phi}$, $V(\mathbf{r}) \equiv 0$, the solution of the BdG equation is

$$\begin{aligned} \varepsilon &= \left((\hbar^2 k^2/2m - E_\mathrm{F})^2 + \Delta_0^2\right)^{1/2}, \\ \mathrm{u}(\mathbf{r}) &= \mathcal{V}^{-1/2}(2\varepsilon)^{-1/2}\mathrm{e}^{\mathrm{i}\phi/2}\left(\varepsilon + \hbar^2 k^2/2m - E_\mathrm{F}\right)^{1/2}\mathrm{e}^{\mathrm{i}\mathbf{k}\cdot\mathbf{r}}, \\ \mathrm{v}(\mathbf{r}) &= \mathcal{V}^{-1/2}(2\varepsilon)^{-1/2}\mathrm{e}^{-\mathrm{i}\phi/2}\left(\varepsilon - \hbar^2 k^2/2m + E_\mathrm{F}\right)^{1/2}\mathrm{e}^{\mathrm{i}\mathbf{k}\cdot\mathbf{r}}. \end{aligned} \tag{6}$$

The eigenfunction is normalized to unit probability in a volume $\mathcal{V}$,

$$\int_\mathcal{V} d\mathbf{r}\, \left(|u|^2 + |v|^2\right) = 1. \tag{7}$$

The excitation spectrum is continuous, with excitation gap Δ_0. The eigenfunctions (u, v) are plane waves characterized by a wavevector $\mathbf{k}$. The coefficients of the plane waves are the two coherence factors of the BCS theory. This simple excitation spectrum is modified by the presence of the Josephson junction. The spectrum acquires a discrete part due to the non-uniformities in $\Delta(\mathbf{r})$ near the junction. The discrete spectrum corresponds to bound states in the gap ($0 < \varepsilon < \Delta_0$), localized within a coherence length from the junction. In addition, the continuous spectrum is modified. As we shall show in Sec. 3, in the short-junction limit the supercurrent is entirely determined by the *discrete* part of the excitation spectrum.

Consider an SNS junction with normal region at $|x| < L/2$. Let $\Delta(\mathbf{r}) \rightarrow \Delta_0 \mathrm{e}^{\mp\mathrm{i}\phi/2}$ for $x \rightarrow \pm\infty$. To determine $\Delta(\mathbf{r})$ near the junction one has to solve the self-consistency equation [16]

$$\Delta(\mathbf{r}) = |g(\mathbf{r})| \sum_{\varepsilon>0} \mathrm{v}^*(\mathbf{r})\mathrm{u}(\mathbf{r})[1 - 2f(\varepsilon)], \tag{8}$$

where the sum is over all states with positive eigenvalue, and $f(\varepsilon) = [1 + \exp(\varepsilon/k_\mathrm{B}T)]^{-1}$ is the Fermi function. The coefficient g is the interaction constant of the BCS theory of superconductivity. At an SN interface, g drops abruptly (over atomic distances) to zero. (We assume non-interacting electrons in the normal region.) Therefore, $\Delta(\mathbf{r}) \equiv 0$ for $|x| < L/2$ regardless of Eq. (8).

At the superconducting side of the SN interface, $\Delta(\mathbf{r})$ recovers its bulk value Δ_0 only at some distance from the interface (proximity effect). There exists a class of Josephson junctions where the suppression of $\Delta(\mathbf{r})$ on approaching the SN interface can be neglected, and one can use the step-function model

$$\Delta(\mathbf{r}) = \begin{cases} \Delta_0 e^{i\phi/2} & \text{if } x < -L/2, \\ 0 & \text{if } |x| < L/2, \\ \Delta_0 e^{-i\phi/2} & \text{if } x > L/2. \end{cases} \tag{9}$$

The step-function pair potential is also referred to in the literature as a "rigid boundary-condition". Likharev [17] discusses in detail the conditions under which this model is valid:

1. If the width W of the junction is small compared to the coherence length, the non-uniformities in $\Delta(\mathbf{r})$ extend only over a distance of order W from the junction (because of "geometrical dilution" of the influence of the narrow junction in the wide superconductor). Since non-uniformities on length scales $\ll \xi$ do not affect the dynamics of the quasiparticles, these can be neglected and the step-function model holds. Constrictions in the short-junction limit belong in general to this class of junctions. Note that, if the length and width of the junction are $\ll \xi$, it is irrelevant whether it is made from a normal metal or from a superconductor. In the literature, such a junction is referred to as an ScS junction, without specifying whether the constriction (c) is N or S.
2. Alternatively, the step-function model holds if the resistivity of the junction region is much bigger than the resistivity of the bulk superconductor. This condition is formulated more precisely by Kupriyanov et al. [18], who have also studied deviations from the ideal step-function model due to the proximity effect and due to current-induced suppression of the pair potential. A superconductor — semiconductor — superconductor junction is typically in this second category.

In equilibrium, a phase difference ϕ in the pair potential induces a stationary current I through the junction (DC Josephson effect). The current–phase relationship $I(\phi)$ is 2π-periodic and antisymmetric $[I(\phi + 2\pi) = I(\phi)$, $I(-\phi) = -I(\phi)]$. The maximum $I_c \equiv \max I(\phi)$ is known as the critical current. There is a thermodynamic relation

$$I = \frac{2e}{\hbar} \frac{dF}{d\phi} \tag{10}$$

between the equilibrium current and the derivative of the free energy F with respect to the phase difference. To apply this relation we need to know how to obtain F from the BdG equation. The required formula was derived by Bardeen et al. [19] from the Greens function expression for F. An alternative derivation, directly from the BdG equation, has been given in Ref. [13]. The result is

$$F = -2k_B T \sum_{\varepsilon>0} \ln\left[2\cosh(\varepsilon/2k_B T)\right] + \int d\mathbf{r}\, |\Delta|^2/|g| + \mathrm{Tr}\, \mathcal{H}_0. \tag{11}$$

The first term in Eq. (11) (the sum over ε) can be formally interpreted as the free energy of non-interacting electrons, all of one single spin, in a "semiconductor" with Fermi level halfway between the "conduction band" (positive ε) and the "valence band" (negative ε). This semiconductor model of a superconductor appeals to intuition, but does not give the free energy correctly. The second term in Eq. (11) corrects for a double-counting of the interaction energy in the semiconductor model. The third term $\mathrm{Tr}\,\mathcal{H}_0$ (i.e. the sum of the single-electron eigenenergies) cancels a divergence at large ε of the series in the first term.

From Eqs. (10) and (11) one obtains [13]

$$\begin{aligned} I &= -\frac{2e}{\hbar}\sum_p \tanh(\varepsilon_p/2k_\mathrm{B}T)\frac{d\varepsilon_p}{d\phi} \\ &\quad -\frac{2e}{\hbar}2k_\mathrm{B}T\int_{\Delta_0}^{\infty} d\varepsilon\, \ln\left[2\cosh(\varepsilon/2k_\mathrm{B}T)\right]\frac{d\rho}{d\phi} + \frac{2e}{\hbar}\frac{d}{d\phi}\int d\mathbf{r}\, |\Delta|^2/|g|, \qquad (12) \end{aligned}$$

where we have rewritten $\sum_{\varepsilon>0}$ as a sum over the discrete positive eigenvalues ε_p ($p = 1, 2, \ldots$), and an integration over the continuous spectrum with density of states $\rho(\varepsilon)$. The term $\mathrm{Tr}\,\mathcal{H}_0$ in Eq. (11) does not depend on ϕ, and therefore does not contribute to I. The spatial integral of $|\Delta|^2/|g|$ does contribute in general. If the step-function model for $\Delta(\mathbf{r})$ holds, however, $|\Delta|$ is independent of ϕ so that this contribution can be disregarded. A calculation of the Josephson current from Eq. (12) then requires only knowledge of the eigenvalues, not of the eigenfunctions.

3 Excitation spectrum from scattering matrix

Following Ref. [4], we now show how to relate the excitation spectrum of Bogoliubov quasiparticles to the scattering matrix of normal electrons. The supercurrent then follows from Eq. (12). The model considered is illustrated in Fig. 1. It consists of a disordered normal region between two superconducting regions S_1 and S_2. The disordered region may or may not contain a geometrical constriction. To obtain a well-defined scattering problem we insert ideal

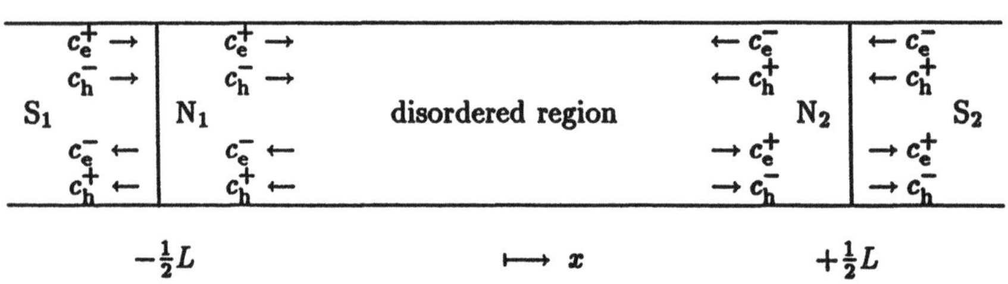

Figure 1: Top: Superconductor — normal-metal — superconductor Josephson junction containing a disordered normal region. Scattering states in the normal and superconducting leads are indicated schematically. Normal scattering (disordered region) and Andreev reflection (SN interface) are spatially separated.

(impurity-free) normal leads N_1 and N_2 to the left and right of the disordered region. The leads should be long compared to the Fermi wavelength λ_F, but short compared to the coherence length ξ_0. The SN interfaces are located at $x = \pm L/2$. We assume that the only scattering in the superconductors consists of Andreev reflection at the SN interfaces, i.e. we consider the case that the disorder is contained entirely within the normal region. The spatial separation of Andreev and normal scattering is the key simplification which allows us to relate the supercurrent directly to the normal-state scattering matrix.

We first construct a basis for the scattering matrix (s-matrix). In the normal lead N_1 the eigenfunctions of the BdG equation (5) can be written in the form

$$\Psi^{\pm}_{n,e}(N_1) = \begin{pmatrix} 1 \\ 0 \end{pmatrix} (k^e_n)^{-1/2}\, \Phi_n(y,z) \exp[\pm i k^e_n (x + \tfrac{1}{2}L)],$$
$$\Psi^{\pm}_{n,h}(N_1) = \begin{pmatrix} 0 \\ 1 \end{pmatrix} (k^h_n)^{-1/2}\, \Phi_n(y,z) \exp[\pm i k^h_n (x + \tfrac{1}{2}L)], \qquad (13)$$

where the wavenumbers k^e_n and k^h_n are given by

$$k^{e,h}_n \equiv (2m/\hbar^2)^{1/2} (E_F - E_n + \sigma^{e,h}\varepsilon)^{1/2}, \qquad (14)$$

and we have defined $\sigma^e \equiv 1$, $\sigma^h \equiv -1$. The labels e and h indicate the electron or hole character of the wavefunction. The index n labels the modes, $\Phi_n(y,z)$ is the transverse wavefunction of the n-th mode, and E_n its threshold energy:

$$[(p_y^2 + p_z^2)/2m + V(y,z)]\Phi_n(y,z) = E_n \Phi_n(y,z). \qquad (15)$$

The eigenfunction Φ_n is normalized to unity, $\int dy \int dz\, |\Phi_n|^2 = 1$. The eigenfunctions in lead N_2 are chosen similarly, but with $x + \frac{1}{2}L$ replaced by $x - \frac{1}{2}L$.

In the superconducting lead S_1, where $\Delta = \Delta_0 \exp(i\phi/2)$, the eigenfunctions are

$$\Psi^{\pm}_{n,e}(S_1) = \begin{pmatrix} e^{i\eta^e/2} \\ e^{-i\eta^e/2} \end{pmatrix} (2q^e_n)^{-1/2} (\varepsilon^2/\Delta_0^2 - 1)^{-1/4} \Phi_n \exp[\pm i q^e_n (x + \tfrac{1}{2}L)],$$
$$\Psi^{\pm}_{n,h}(S_1) = \begin{pmatrix} e^{i\eta^h/2} \\ e^{-i\eta^h/2} \end{pmatrix} (2q^h_n)^{-1/2} (\varepsilon^2/\Delta_0^2 - 1)^{-1/4} \Phi_n \exp[\pm i q^h_n (x + \tfrac{1}{2}L)]. \quad (16)$$

We have defined

$$q^{e,h}_n \equiv (2m/\hbar^2)^{1/2} [E_F - E_n + \sigma^{e,h} (\varepsilon^2 - \Delta_0^2)^{1/2}]^{1/2}, \qquad (17)$$
$$\eta^{e,h} \equiv \tfrac{1}{2}\phi + \sigma^{e,h} \arccos(\varepsilon/\Delta_0). \qquad (18)$$

The square roots are to be taken such that $\mathrm{Re}\, q^{e,h} \geq 0$, $\mathrm{Im}\, q^e \geq 0$, $\mathrm{Im}\, q^h \leq 0$. The function $\arccos t \in (0, \pi/2)$ for $0 < t < 1$, while $\arccos t \equiv -i \ln[t + (t^2 - 1)^{1/2}]$ for $t > 1$. The eigenfunctions in lead S_2, where $\Delta = \Delta_0 \exp(-i\phi/2)$, are obtained by replacing ϕ by $-\phi$ and L by $-L$.

The wavefunctions (13) and (16) have been normalized to carry the same amount of quasiparticle current, because we want to use them as the basis for a unitary s-matrix. Unitarity of the s-matrix is a consequence of the fact that the BdG equation conserves the quasiparticle current density

$$\mathrm{j}_{\mathrm{qp}}(\mathbf{r}) = \frac{1}{m}\mathrm{Re}\begin{pmatrix} \mathrm{u}^* \\ \mathrm{v}^* \end{pmatrix}\begin{pmatrix} \mathbf{p} & \emptyset \\ \emptyset & -\mathbf{p} \end{pmatrix}\begin{pmatrix} \mathrm{u} \\ \mathrm{v} \end{pmatrix}. \tag{19}$$

The quasiparticle current I_{qp} through a lead is given by

$$I_{\mathrm{qp}} = \int dy \int dz\, (\mathrm{j}_{\mathrm{qp}})_x = \frac{\hbar}{m}\mathrm{Re}\int dy \int dz \left(\mathrm{u}^* \frac{\partial}{\mathrm{i}\partial x}\mathrm{u} - \mathrm{v}^* \frac{\partial}{\mathrm{i}\partial x}\mathrm{v}\right). \tag{20}$$

One easily verifies that, provided the wavenumbers $k_n^{\mathrm{e,h}}$ and $q_n^{\mathrm{e,h}}$ are real (i.e. for propagating modes), each of the wavefunctions (13,16) carries the same amount of quasiparticle current $|I_{\mathrm{qp}}|$. (The direction of the current is opposite for $\Psi_{\mathrm{e}}^+, \Psi_{\mathrm{h}}^-$ and $\Psi_{\mathrm{e}}^-, \Psi_{\mathrm{h}}^+$.) We note that the wavefunctions (13) and (16) carry *different* amounts of *charge* current (i.e. electrical current), which is given by

$$I_{\mathrm{charge}} = -e\frac{\hbar}{m}\mathrm{Re}\int dy \int dz \left(\mathrm{u}^* \frac{\partial}{\mathrm{i}\partial x}\mathrm{u} + \mathrm{v}^* \frac{\partial}{\mathrm{i}\partial x}\mathrm{v}\right). \tag{21}$$

This does not contradict the unitarity of the s-matrix constructed from these wavefunctions, because the BdG equation does not conserve the charge of the quasiparticles. (Andreev reflection [20], i.e. the reflection of an electron into a hole at an SN interface, is an example of a non-charge-conserving scattering process.)

A wave incident on the disordered normal region is described in the basis (13) by a vector of coefficients

$$c_{\mathrm{N}}^{\mathrm{in}} \equiv \left(c_{\mathrm{e}}^+(N_1), c_{\mathrm{e}}^-(N_2), c_{\mathrm{h}}^-(N_1), c_{\mathrm{h}}^+(N_2)\right), \tag{22}$$

as shown schematically in Fig. 1. (The mode-index n has been suppressed for simplicity of notation.) The reflected and transmitted wave has vector of coefficients

$$c_{\mathrm{N}}^{\mathrm{out}} \equiv \left(c_{\mathrm{e}}^-(N_1), c_{\mathrm{e}}^+(N_2), c_{\mathrm{h}}^+(N_1), c_{\mathrm{h}}^-(N_2)\right). \tag{23}$$

The s-matrix s_{N} of the normal region relates these two vectors, $c_{\mathrm{N}}^{\mathrm{out}} = s_{\mathrm{N}} c_{\mathrm{N}}^{\mathrm{in}}$. Because the normal region does not couple electrons and holes, this matrix has the block-diagonal form

$$s_{\mathrm{N}}(\varepsilon) = \begin{pmatrix} s_0(\varepsilon) & \emptyset \\ \emptyset & s_0(-\varepsilon)^* \end{pmatrix}, \; s_0 \equiv \begin{pmatrix} r_{11} & t_{12} \\ t_{21} & r_{22} \end{pmatrix}. \tag{24}$$

Here s_0 is the unitary and symmetric s-matrix associated with the single-electron Hamiltonian $\mathcal{H}_0$. The reflection and transmission matrices $r(\varepsilon)$ and $t(\varepsilon)$ are $N \times N$ matrices, $N(\varepsilon)$ being the number of propagating modes at energy ε. (We assume for simplicity that the number of modes in leads N_1 and N_2 is the same.) The dimension of $s_{\mathrm{N}}(\varepsilon)$ is $2N(\varepsilon) + 2N(-\varepsilon)$.

We will make use of two more s-matrices. For energies $0 < \varepsilon < \Delta_0$ there are no propagating modes in the superconducting leads S_1 and S_2. We can then define an s-matrix s_{A} for Andreev reflection at the SN interfaces by $c_{\mathrm{N}}^{\mathrm{in}} = s_{\mathrm{A}} c_{\mathrm{N}}^{\mathrm{out}}$. The elements of s_{A} can be obtained by matching the wavefunctions

(13) at $|x| = L/2$ to the decaying wavefunctions (16). Since $\Delta_0 \ll E_F$ one may ignore normal reflections at the SN interface and neglect the difference between $N(\varepsilon)$ and $N(-\varepsilon)$. This is known as the Andreev approximation [20]. The result is

$$s_A = \alpha \begin{pmatrix} \emptyset & r_A \\ r_A^* & \emptyset \end{pmatrix}, \; r_A \equiv \begin{pmatrix} e^{i\phi/2}\mathbf{1} & \emptyset \\ \emptyset & e^{-i\phi/2}\mathbf{1} \end{pmatrix}, \tag{25}$$

where $\alpha \equiv \exp[-i \arccos(\varepsilon/\Delta_0)]$. The matrices $\mathbf{1}$ and $\emptyset$ are the unit and null matrices, respectively. Andreev reflection transforms an electron mode into a hole mode, without change of mode index. The transformation is accompanied by a phase shift, which consists of two parts: 1. A phase shift $-\arccos(\varepsilon/\Delta_0)$ due to the penetration of the wavefunction into the superconductor; 2. A phase shift equal to plus or minus the phase of the pair potential in the superconductor (*plus* for reflection from hole to electron, *minus* for the reverse process).

For $\varepsilon > \Delta_0$ we can define the s-matrix s_{SNS} of the whole junction, by $c_S^{out} = s_{SNS} c_S^{in}$. The vectors

$$c_S^{in} \equiv \left(c_e^+(S_1), c_e^-(S_2), c_h^-(S_1), c_h^+(S_2)\right), \tag{26}$$
$$c_S^{out} \equiv \left(c_e^-(S_1), c_e^+(S_2), c_h^+(S_1), c_h^-(S_2)\right). \tag{27}$$

are the coefficients in the expansion of the incoming and outgoing wave in leads S_1 and S_2 in terms of the wavefunctions (16) (cf. Fig. 1). By matching the wavefunctions (13) and (16) at $|x| = L/2$, we arrive after some algebra (using again $\Delta_0 \ll E_F$) at the matrix-product expression

$$s_{SNS} = U^{-1}(\mathbf{1} - M)^{-1}(\mathbf{1} - M^\dagger)s_N U, \tag{28}$$
$$U \equiv \begin{pmatrix} r_A & \emptyset \\ \emptyset & r_A^* \end{pmatrix}^{\frac{1}{2}}, \; M \equiv \alpha s_N \begin{pmatrix} \emptyset & r_A \\ r_A^* & \emptyset \end{pmatrix}.$$

One can verify that the three s-matrices defined above (s_N; s_A for $0 < \varepsilon < \Delta_0$; s_{SNS} for $\varepsilon > \Delta_0$) are ***unitary*** ($s^\dagger s = s s^\dagger = \mathbf{1}$) and satisfy the ***symmetry relation*** $s(\varepsilon, \phi)_{ij} = s(\varepsilon, -\phi)_{ji}$, as required by quasiparticle-current conservation and by time-reversal invariance, respectively.

We are now ready to relate the excitation spectrum of the Josephson junction to the s-matrix of the normal region. First the discrete spectrum. The condition $c_{in} = s_A s_N c_{in}$ for a bound state implies $\mathrm{Det}\,(\mathbf{1} - s_A s_N) = 0$. Using Eqs. (24), (25), and the *folding-identity*

$$\mathrm{Det} \begin{pmatrix} a & b \\ c & d \end{pmatrix} = \mathrm{Det}\,(ad - aca^{-1}b) \tag{29}$$

(which holds for arbitrary square matrices a, b, c, d of equal dimension, with $\mathrm{Det}\, a \neq 0$), we find the equation

$$\mathrm{Det}\, \left[\mathbf{1} - \alpha(\varepsilon_p)^2 r_A^* s_0(\varepsilon_p) r_A s_0(-\varepsilon_p)^*\right] = 0, \tag{30}$$

which determines the discrete spectrum. The density of states of the continuous spectrum is related to s_{SNS} by the general relation [21]

$$\rho = \frac{1}{2\pi i}\frac{\partial}{\partial\varepsilon}\ln \mathrm{Det}\, s_{\mathrm{SNS}} + \text{constant}, \tag{31}$$

where "constant" indicates a ϕ-independent term. From Eqs. (28) and (29) we find

$$\frac{\partial\rho}{\partial\phi} = -\frac{1}{\pi}\frac{\partial^2}{\partial\phi\partial\varepsilon}\mathrm{Im}\ln \mathrm{Det}\left[\mathbf{1} - \alpha(\varepsilon)^2 r_{\mathrm{A}}^* s_0(\varepsilon) r_{\mathrm{A}} s_0(-\varepsilon)^*\right], \tag{32}$$

which determines the ϕ-dependence of the continuous spectrum.

4 Short-Junction Limit

In the short-junction limit $L \ll \xi$, the determinantal equations (30) and (32) can be simplified further. As mentioned in the Introduction, the condition $L \ll \xi$ is equivalent to $\Delta_0 \ll E_{\mathrm{c}}$, where the correlation energy $E_{\mathrm{c}} \equiv \hbar/\tau_{\mathrm{dwell}}$ is defined in terms of the dwell time τ_{dwell} in the junction. The elements of $s_0(\varepsilon)$ change significantly if ε is changed by at least E_{c} [22, 23]. We are concerned with ε of order Δ_0 or smaller (since $\rho(\varepsilon,\phi)$ becomes independent of ϕ for $\varepsilon \gg \Delta_0$). For $\Delta_0 \ll E_{\mathrm{c}}$ we may thus approximate $s_0(\varepsilon) \approx s_0(-\varepsilon) \approx s_0(0) \equiv s_0$. Eq. (30) may now be simplified by multiplying both sides by $\mathrm{Det}\, s_0$ and using $s_0^* s_0 = \mathbf{1}$ (unitarity plus symmetry of s_0), as well as the folding identity (29). The result can be written in the form

$$\mathrm{Det}\left[(1-\varepsilon_p^2/\Delta_0^2)\mathbf{1} - t_{12}t_{12}^\dagger \sin^2(\phi/2)\right] = 0. \tag{33}$$

For $\varepsilon > \Delta_0$ one can see that $\mathrm{Det}[\mathbf{1} - \alpha(\varepsilon)^2 r_{\mathrm{A}}^* s_0 r_{\mathrm{A}} s_0^*]$ is a real number. (Use that α is real for $\varepsilon > \Delta_0$ and that the determinant $\mathrm{Det}\,(\mathbf{1} - aa^*)$ is real for arbitrary matrix a.) Eq. (32) then reduces to $\partial\rho/\partial\phi = 0$, from which we conclude that *the continuous spectrum does not contribute to $I(\phi)$ in the short-junction limit.*

Eq. (33) can be solved for ε_p in terms of the eigenvalues T_p ($p = 1, 2, \ldots N$) of the hermitian $N \times N$ matrix $t_{12}t_{12}^\dagger$,

$$\varepsilon_p = \Delta_0\left[1 - T_p \sin^2(\phi/2)\right]^{1/2}. \tag{34}$$

Since $t_{12}t_{12}^\dagger = r_{11}t_{21}^\dagger t_{21} r_{11}^{-1}$ (as follows from unitarity of s_0), the matrices $t_{12}t_{12}^\dagger$ and $t_{21}t_{21}^\dagger$ have the same set of eigenvalues. We can therefore omit the indices of $tt^\dagger$. Unitarity of s_0 also implies that $0 \le T_p \le 1$ for all p.

Substitution of Eq. (34) into Eq. (12) yields the Josephson current

$$I(\phi) = \frac{e\Delta_0}{2\hbar}\sum_{p=1}^{N}\frac{T_p \sin\phi}{[1 - T_p \sin^2(\phi/2)]^{1/2}}\tanh\left(\frac{\Delta_0}{2k_{\mathrm{B}}T}[1 - T_p\sin^2(\phi/2)]^{1/2}\right). \tag{35}$$

Eq. (35) holds for an arbitrary transmission matrix $tt^\dagger$, i.e. for arbitrary disorder potential. It is the *multi-channel* generalization of a formula first obtained by Haberkorn et al. [24] (and subsequently rederived by several authors [25]–[27]) for the *single-channel* case (appropriate for a geometry such as a planar tunnel barrier, where the different scattering channels are uncoupled). A formula of similar generality for the conductance is the multi-channel Landauer formula [23, 28]

$$G = \frac{2e^2}{h}\mathrm{Tr}\, tt^\dagger \equiv \frac{2e^2}{h}\sum_{p=1}^{N} T_p. \tag{36}$$

In contrast to the Landauer formula, Eq. (35) is a *non-linear* function of the transmission eigenvalues T_p. It follows that knowledge of the conductance (i.e. of the sum of the eigenvalues) is *not sufficient* to determine the supercurrent.

5 Universal Josephson Effects

5.1 Quantum Point Contact

Consider the case that the weak link consists of a ballistic constriction ($l \gg L$) with a conductance quantized at $G = 2N_0e^2/h$ (a *quantum point contact* [7]). The integer N_0 is the number of occupied one-dimensional subbands (per spin direction) in the constriction, or alternatively the number of transverse modes at the Fermi level which can propagate through the constriction. Note that $N_0 \ll N$. A quantum point contact is characterized by a special set of transmission eigenvalues, which are equal to either 0 or 1:

$$T_p = \begin{cases} 1 & \text{if } 1 \leq p \leq N_0, \\ 0 & \text{if } N_0 < p \leq N, \end{cases} \tag{37}$$

where the eigenvalues have been ordered from large to small. We emphasize that Eq. (37) is valid whether the transport through the constriction is adiabatic or not. In the special case of adiabatic transport, the transmission eigenvalue T_p is equal to the transmission probability $\mathcal{T}_p$ of the p-th subband. In the absence of adiabaticity there is no simple relation between T_p and $\mathcal{T}_p$.

The discrete spectrum (34) in the short-junction limit contains an N_0-fold degenerate state at energy $\varepsilon = \Delta_0|\cos(\phi/2)|$. Eq. (35) for the supercurrent becomes

$$I(\phi) = N_0\frac{e\Delta_0}{\hbar}\sin(\phi/2)\tanh\left(\frac{\Delta_0}{2k_\mathrm{B}T}\cos(\phi/2)\right), \ |\phi| < \pi. \tag{38}$$

At $T = 0$ the current–phase relationship is given by $I(\phi) = N_0(e\Delta_0/\hbar)\sin(\phi/2)$ for $|\phi| < \pi$, and continued periodically for $|\phi| > \pi$. At $\phi = \pi$ the function $I(\phi)$ has a discontinuity which is smeared at finite temperatures. The ratio $I(\phi)/(\pi\Delta_0G/e)$ at $T = 0$ is plotted in Fig. 2 (solid curve). The critical current $I_c = N_0e\Delta_0/\hbar$ is *discretized* in units of $e\Delta_0/\hbar$. In the classical limit $N_0 \to \infty$ we recover the results of Kulik and Omel'yanchuk [29] for a *classical* ballistic point contact.

Eq. (38) was derived in Ref. [1] under the assumption of adiabatic transport. The present derivation (taken from Ref. [4]) does not assume adiabaticity. As discussed in Ref. [1], Eq. (38) breaks down if the Fermi level lies within Δ_0 of the threshold energy E_n of a subband. In the present con-

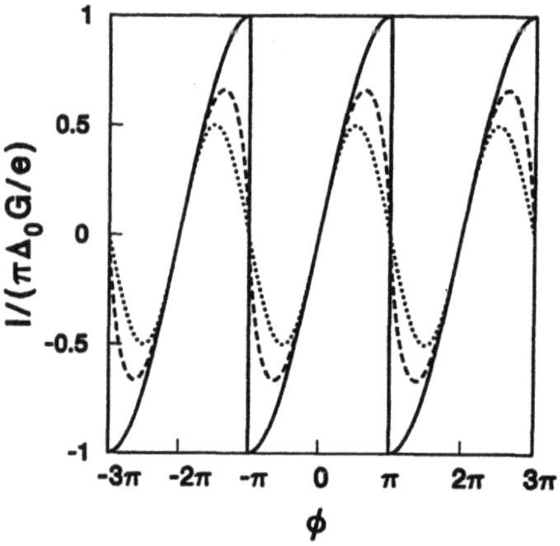

Figure 2: Current–phase relationship at $T = 0$ of a ballistic point contact [solid curve, Eq. (38)], a disordered point contact (ensemble averaged) [dashed curve, Eq. (49)], and a tunnel junction [dotted curve, Eq. (50)]. The current is normalized by the normal-state conductance of the junction.

text this breakdown can be understood from the fact that the dwell time $\tau_{\text{dwell}} = (L/v_{\text{F}})(1 - E_n/E_{\text{F}})^{-1/2}$ becomes infinitely long as E_{F} approaches E_n, so that the "short-junction" condition $\Delta_0 \ll E_c \equiv \hbar/\tau_{\text{dwell}}$ is violated.

5.2 Quantum Dot

Consider a small confined region (of dimensions comparable to the Fermi wavelength), which is weakly coupled by tunnel barriers to two electron reservoirs. We assume that transport through this *quantum dot* occurs via resonant tunneling through a single bound state. Let ε_{res} be the energy of the resonant level, relative to the Fermi energy E_{F} in the reservoirs, and let $\Gamma_1/\hbar$ and $\Gamma_2/\hbar$ be the tunnel rates through the left and right barriers. We denote $\Gamma \equiv \Gamma_1 + \Gamma_2$. If $\Gamma \ll \Delta E$ (with ΔE the level spacing in the quantum dot) and $T \ll \Gamma/k_{\text{B}}$, the conductance G in the case of non-interacting electrons has the form

$$G = \frac{2e^2}{h} \frac{\Gamma_1 \Gamma_2}{\varepsilon_{\text{res}}^2 + \frac{1}{4}\Gamma^2} \equiv \frac{2e^2}{h} T_{\text{BW}}, \tag{39}$$

where T_{BW} is the Breit-Wigner transmission probability at the Fermi level. The normal-state scattering matrix $s_0(\varepsilon)$ which yields this conductance has matrix elements [30]

$$(s_0(\varepsilon))_{nm} = \left(\delta_{nm} - \frac{\mathrm{i}\sqrt{\Gamma_{1n}\Gamma_{2m}}}{\varepsilon - \varepsilon_{\text{res}} + \mathrm{i}\Gamma/2} \right) \mathrm{e}^{\mathrm{i}(\delta_{1n} + \delta_{2m})}, \tag{40}$$

where $\sum_{n=1}^{N} \Gamma_{1n} \equiv \Gamma_1$, $\sum_{n=1}^{N} \Gamma_{2n} \equiv \Gamma_2$. The phases δ_{1n}, δ_{2m}, as well as the basis with respect to which the matrix elements (40) are calculated, need not be further specified. Büttiker [31] has shown how the conductance (39) follows, via the Landauer formula (36), from the Breit-Wigner scattering matrix (40).

In Ref. [3] Van Houten and the author have calculated the supercurrent through the quantum dot from the Breit-Wigner formula (40) in the special case of *single*-channel leads ($N = 1$). Just as for the conductance [31]–[33] one would expect that the results are not changed for multi-channel leads ($N > 1$). We will now demonstrate this explicitly in the limit $\Gamma \gg \Delta_0$ (which for resonant tunneling corresponds to the "short-junction" limit, cf. Sec. 1). The transmission matrix product $t_{12}t_{12}^{\dagger}$ (evaluated at the Fermi level $\varepsilon = 0$) following from the scattering matrix (40) is

$$\left(t_{12}t_{12}^{\dagger}\right)_{nm} = T_{\rm BW}\gamma_n\gamma_m^*, \; \gamma_n \equiv e^{i\delta_n}\sqrt{\Gamma_{1n}/\Gamma_1}. \tag{41}$$

Its eigenvalues are

$$T_p = \begin{cases} T_{\rm BW} & \text{if } p = 1, \\ 0 & \text{if } 2 \leq p \leq N. \end{cases} \tag{42}$$

The single non-zero eigenvalue has eigenvector $u_n = \gamma_n$, the other eigenvectors span the plane orthogonal to the vector γ. Substitution into Eq. (35) yields the current–phase relationship for a wide resonance ($\Gamma \gg \Delta_0$),

$$I(\phi) = \frac{e\Delta_0}{2\hbar}\frac{T_{\rm BW}\sin\phi}{[1 - T_{\rm BW}\sin^2(\phi/2)]^{1/2}}\tanh\left(\frac{\Delta_0}{2k_{\rm B}T}[1 - T_{\rm BW}\sin^2(\phi/2)]^{1/2}\right). \tag{43}$$

The critical current at zero temperature is

$$I_{\rm c} = \frac{e\Delta_0}{\hbar}[1 - (1 - T_{\rm BW})^{1/2}], \text{ if } \Gamma \gg \Delta_0, \tag{44}$$

in agreement with Ref. [3]. Since $T_{\rm BW} = 1$ on resonance ($\varepsilon_{\rm R} = 0$) in the case of equal tunnel rates ($\Gamma_1 = \Gamma_2$), we obtain the result (3) discussed in the Introduction.

For completeness, we also quote the formula for the current–phase relationship in the opposite regime of a narrow resonance (derived in Ref. [3]):

$$I_{\rm c} = \frac{e}{\hbar}(\varepsilon_{\rm res}^2 + \tfrac{1}{4}\Gamma^2)^{1/2}[1 - (1 - T_{\rm BW})^{1/2}], \text{ if } \Gamma, \varepsilon_{\rm res} \ll \Delta_0. \tag{45}$$

As shown in Fig. 3, the lineshapes (44) and (45) of a resonance in the critical current (solid curves) differ substantially from the lorentzian lineshape (39) of a conductance resonance (dotted curve). For $\Gamma_1 = \Gamma_2$, $I_{\rm c}$ has a *cusp* at $\varepsilon_{\rm res} = 0$ (which is rounded at finite temperatures). On resonance, the maximum critical current $I_{\rm c}^{\rm res}$ equals $(2e\Delta_0/\hbar\Gamma)\min(\Gamma_1, \Gamma_2)$ and $(e/\hbar)\min(\Gamma_1, \Gamma_2)$ for a wide and narrow resonance, respectively. An analytical formula, Eq. (2), for the crossover between these two regimes can be obtained for the case of equal tunnel rates [3]. Off-resonance, $I_{\rm c}$ has the lorentzian decay $\propto 1/\varepsilon_{\rm res}^2$ in the case $\Gamma \gg \Delta_0$ of a wide resonance, but a slower decay $\propto 1/\varepsilon_{\rm res}$ in the case $\Gamma, \varepsilon_{\rm res} \ll \Delta_0$. Near $\varepsilon_{\rm res} \simeq \Delta_0$ this linear decay of the narrow resonance crosses over to a quadratic decay (not shown in Fig. 3).

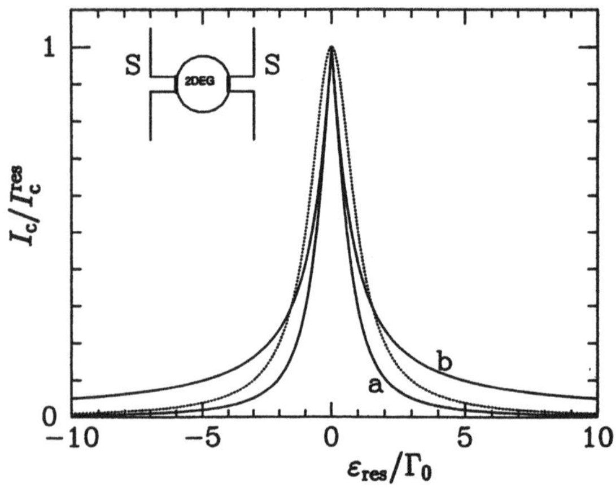

Figure 3: Normalized critical current versus energy of the resonant level, at zero temperature and for equal tunnel barriers ($\Gamma_1 = \Gamma_2 \equiv \Gamma_0$). The two solid curves are the results (44) and (45) for the two regimes $\Gamma_0 \gg \Delta_0$ (curve a) and $\Gamma_0, \varepsilon_{res} \ll \Delta_0$ (curve b). The dotted curve is the Breit-Wigner transmission probability (39). The inset shows schematically the quantum-dot Josephson junction. Taken from Ref. [3].

Since we have assumed non-interacting quasiparticles, the above results apply to a quantum dot with a small charging energy U for double occupancy of the resonant state. Glazman and Matveev have studied the influence of Coulomb repulsion on the resonant supercurrent [2]. The influence is most pronounced in the case of a narrow resonance, when the critical current is suppressed by a factor Γ/Δ_0 (for $U, \Delta_0 \gg \Gamma$). In the case of a wide resonance, the Coulomb repulsion does not suppress the supercurrent, but slightly broadens the resonance by a factor $\ln(\Gamma/\Delta_0)$ (for $U, \Gamma \gg \Delta_0$). The broadening is a consequence of the Kondo effect, and occurs only for $\varepsilon_{res} < 0$, so that the resonance peak becomes somewhat asymmetric [2].

5.3 Disordered Point Contact

We now turn to the regime of diffusive transport through a disordered point contact. We first consider the average supercurrent (averaged over an ensemble of impurity configurations) and then the fluctuations from the average. This section corrects Ref. [4].

5.3.1 Average Supercurrent

The transmission eigenvalue T_p is related to a channel-dependent localization length ζ_p by

$$T_p = \cosh^{-2}(L/\zeta_p). \tag{46}$$

The inverse localization length is uniformly distributed between 0 and $1/\zeta_{min} \simeq 1/l$ for $l \ll L \ll Nl$ [34]. One can therefore write

$$\frac{\left\langle\sum_{p=1}^{N} f(T_p)\right\rangle}{\left\langle\sum_{p=1}^{N} T_p\right\rangle} = \frac{\int_0^{L/\zeta_{\min}} dx\, f(\cosh^{-2} x)}{\int_0^{L/\zeta_{\min}} dx\, \cosh^{-2} x} = \int_0^{\infty} dx\, f(\cosh^{-2} x), \quad (47)$$

where $\langle\ldots\rangle$ indicates the ensemble average and $f(T)$ is an arbitrary function of the transmission eigenvalue such that $f(T) \to 0$ for $T \to 0$. In the second equality in Eq. (47) we have used that $L/\zeta_{\min} \simeq L/l \gg 1$ to replace the upper integration limit by ∞. Note that, in view of the Landauer formula (36), the denominator $\left\langle\sum_{p=1}^{N} T_p\right\rangle$ is just $h/2e^2$ times the average conductance $\langle G\rangle$.

Combining Eqs. (47) and (35) we find for the average supercurrent the expression $(0 < \phi < \pi)$

$$\begin{aligned} \langle I(\phi)\rangle &= \frac{e\Delta_0}{2\hbar}\frac{h}{2e^2}\langle G\rangle \int_0^{\infty} dx\, \frac{(\cosh x)^{-2}\sin\phi}{[1-(\cosh x)^{-2}\sin^2(\phi/2)]^{1/2}} \\ &\quad \times \tanh\left(\frac{\Delta_0}{2k_{\rm B}T}[1-(\cosh x)^{-2}\sin^2(\phi/2)]^{1/2}\right) \\ &= \frac{\pi\Delta_0}{e}\langle G\rangle\cos(\phi/2)\int_{\cos(\phi/2)}^{1} dz\, \frac{\tanh(\Delta_0 z/2k_{\rm B}T)}{[z^2-\cos^2(\phi/2)]^{1/2}}. \end{aligned} \quad (48)$$

At $T=0$ this integral can be evaluated in closed form, with the result

$$\langle I(\phi)\rangle = \frac{\pi\Delta_0}{e}\langle G\rangle\cos(\phi/2)\,\mathrm{arctanh}\,[\sin(\phi/2)], \quad (49)$$

plotted in Fig. 2 (dashed curve). Note that the derivative $dI/d\phi$ diverges at $\phi=\pi$. The critical current is $I_{\rm c}^{(0)} \equiv \max\langle I(\phi)\rangle = 1.32\,(\pi\Delta_0/2e)\langle G\rangle$, reached at $\phi \equiv \phi_{\rm c}^{(0)} = 1.97$.

Eq. (49) for the average supercurrent in a disordered point contact agrees with the result of Kulik and Omel'yanchuk [35]. These authors pointed out that a disordered weak link and a tunnel junction with the *same* conductance exhibit nevertheless a quite *different* Josephson effect. For a tunnel junction with $T_p \ll 1$ for all p one may linearize Eq. (35) in T_p, with the result (at zero temperature)

$$I = \frac{e\Delta_0}{2\hbar}\sin\phi\sum_{p=1}^{N} T_p = \frac{\pi\Delta_0}{2e}G\sin\phi, \quad (50)$$

which is also plotted for comparison in Fig. 2 (dotted curve). The corresponding critical current is $(\pi\Delta_0/2e)G$. The well-known result (50) for a tunnel junction [36] differs from the result (49) for a disordered weak link [35]. Zaïtsev [25] has explained the difference in terms of different boundary conditions for the Greens functions at the interface between the junction and the bulk superconductor. The present work yields an alternative perspective on this issue: Although the tunnel junction and the disordered weak link may have the same conductance (i.e. the same value of $\mathrm{Tr}\, tt^\dagger$), *the supercurrents differ because the distribution of transmission eigenvalues is different.* In a tunnel junction all N eigenvalues are $\ll 1$, while in the disordered system a fraction l/L of the eigenvalues is of order unity, the remainder being exponentially small [34].

5.3.2 Supercurrent Fluctuations

The analysis of Kulik and Omel'yanchuk is based on a diffusion equation for the *ensemble-averaged* Greens function, and can not therefore describe the mesoscopic fluctuations of $I(\phi)$ from the average. In contrast, Eq. (35) holds for a *specific* member of the ensemble of impurity configurations. The statistical properties of the transmission eigenvalues T_n in this ensemble are known from the theory of conductance fluctuations [11]. A general result [34, 37] is that a *linear statistic* $\mathcal{L}(f) \equiv \sum_{n=1}^{N} f(T_n)$ on the eigenvalues (with $f(T) \rightarrow 0$ for $T \rightarrow 0$ and $f(T) = \mathcal{O}(1)$ for $T \lesssim 1$) fluctuates with root-mean-square value of order unity

$$\mathrm{rms}\,\mathcal{L} \equiv \left(\langle \mathcal{L}^2 \rangle - \langle \mathcal{L} \rangle^2\right)^{1/2} = \mathcal{O}(1), \tag{51}$$

independent of N, L, or l as long as $l \ll L \ll Nl$. An additional requirement is that the junction is not much wider than long: $W_y, W_z \lesssim L$ where W_y and W_z are the widths in the y- and z-direction. If the junction is much wider than long in one direction, then $\mathrm{rms}\,\mathcal{L}$ is of order $(W_y W_z/L^2)^{1/2}$. It is not necessary for these results that $f(T)$ is itself a linear function of T. In the special case that $f(T) = T$, the coefficient on the right-hand-side of Eq. (51) is known [10]:

$$\mathrm{rms} \sum_{n=1}^{N} T_n \equiv C_{\mathrm{UCF}} = 0.365, \text{ if } W_y, W_z \ll L. \tag{52}$$

According to Eq. (35) the supercurrent $I(\phi)$ is a linear statistic on T_n, so that we conclude that (at zero temperature) $\mathrm{rms}\, I(\phi) \simeq e\Delta_0/\hbar$ times a function of ϕ which is independent of the length of the junction or the degree of disorder. At small ϕ we may linearize in T and use Eq. (52), with the result

$$\mathrm{rms}\, I(\phi) = \tfrac{1}{2} C_{\mathrm{UCF}} (e\Delta_0/\hbar)\phi \tanh(\Delta_0/2k_{\mathrm{B}}T) + \mathcal{O}(\phi^2). \tag{53}$$

The critical current $I_c \equiv \max I(\phi) \equiv I(\phi_c)$ is not by definition a linear statistic on T_n, since the phase ϕ_c at which the maximum supercurrent is reached depends itself on all the transmission eigenvalues. It is therefore not possible in general to write I_c in the form $\sum_n f(T_n)$, as required for a linear statistic. However, I_c does become a linear statistic in the limit $L/Nl \ll 1$ (or equivalently $\langle G \rangle \gg e^2/h$) of a good conductor (which is the appropriate limit for mesoscopic fluctuations). To see this, we write

$$I(\phi) = \langle I(\phi) \rangle [1 + \epsilon X(\phi)], \quad \epsilon \equiv L/Nl, \tag{54}$$

where the function $X(\phi)$ accounts for the sample-to-sample fluctuations of $I(\phi)$ around the ensemble average $\langle I(\phi) \rangle$. One has $\langle X \rangle = 0$, $\mathrm{rms}\, X = \mathcal{O}(1)$. We now expand I_c to lowest order in ϵ. Defining $\phi_c \equiv \phi_c^{(0)} + \epsilon\phi_c^{(1)}$, where $\max \langle I(\phi) \rangle \equiv \langle I(\phi_c^{(0)}) \rangle$, one can write

$$\begin{aligned} I_c &\equiv \langle I(\phi_c)\rangle [1+\epsilon X(\phi_c)] \\ &= \left\langle I(\phi_c^{(0)}+\epsilon\phi_c^{(1)})\right\rangle \left[1+\epsilon X(\phi_c^{(0)}+\epsilon\phi_c^{(1)})\right] \\ &= \left\langle I(\phi_c^{(0)})\right\rangle \left[1+\epsilon X(\phi_c^{(0)})+\mathcal{O}(\epsilon^2)\right] = I(\phi_c^{(0)}). \end{aligned} \tag{55}$$

In the third equality we have used that, by definition, $d\langle I\rangle/d\phi = 0$ at $\phi = \phi_c^{(0)}$. Since $I(\phi_c^{(0)})$ is a linear statistic on T_n (note that $\phi_c^{(0)}$ is by definition independent of T_n), we conclude that the critical current I_c is a linear statistic on the transmission eigenvalues in the limit $\epsilon \equiv L/Nl \to 0$, and hence that $\mathrm{rms}\, I_c = \mathrm{rms}\, I(\phi_c^{(0)}) \simeq e\Delta_0/\hbar$. The value of the numerical coefficient remains to be calculated.

Experimentally, sample-to-sample fluctuations are not easily studied. Instead, fluctuations in a given sample as a function of Fermi energy E_F are more accessible. Josephson junctions consisting of a two-dimensional electron gas (2DEG) with superconducting contacts allow for variation of E_F in the 2DEG by means of a gate voltage [38]. Point-contact junctions can be defined in the 2DEG either lithographically or electrostatically (using split gates) [7]. For such a system one would expect that if E_F is varied on the scale of E_c, the low-temperature critical current will fluctuate by an amount of order $e\Delta_0/\hbar$, independent of the properties of the junction.

Acknowledgements

Part of the work reviewed in this paper was performed in collaboration with H. van Houten at the Philips Research Laboratories in Eindhoven, The Netherlands. Research at the University of Leiden is supported by the "Stichting Fundamenteel Onderzoek der Materie" (FOM), with financial support from the "Stichting Nederlands Wetenschappelijk Onderzoek" (NWO).

References

[1] C. W. J. Beenakker and H. van Houten, Phys. Rev. Lett. 66, 3056 (1991).

[2] L. I. Glazman and K. A. Matveev, Pis'ma Zh. Eksp. Teor. Fiz. 49, 570 (1989) [JETP Lett. 49, 659 (1989)].

[3] C. W. J. Beenakker and H. van Houten, in: Proc. 4th Int. Conf. on Superconducting and Quantum Effect Devices and their Applications (*SQUID'91*), ed. by H. Koch and H. Lübbig (Springer, Berlin, to be published).

[4] C. W. J. Beenakker, Phys. Rev. Lett. 67, 3836 (1991); Erratum (ibidem, to be published). The assertion made in this Ref. (directly above Eq. 16) that $T_p \ll 1$ for $l \ll L$ is incorrect (see Sec. 5.3 of the present paper).

[5] B. J. van Wees, H. van Houten, C. W. J. Beenakker, J. G. Williamson, L. P. Kouwenhoven, D. van der Marel, and C. T. Foxon, Phys. Rev. Lett. 60, 848 (1988).

[6] D. A. Wharam, T. J. Thornton, R. Newbury, M. Pepper, H. Ahmed, J. E. F. Frost, D. G. Hasko, D. C. Peacock, D. A. Ritchie, and G. A. C. Jones, J. Phys. C **21**, L209 (1988).

[7] A review of quantum transport in semiconductor nanostructures is: C. W. J. Beenakker and H. van Houten, Solid State Phys. **44**, 1 (1991).

[8] A. Furusaki, H. Takayanagi, and M. Tsukada, Phys. Rev. Lett. **67**, 132 (1991); Phys. Rev. B (to be published).

[9] B. L. Al'tshuler, Pis'ma Zh. Eksp. Teor. Fiz. **41**, 530 (1985) [JETP Lett. **41**, 648 (1985)].

[10] P. A. Lee and A. D. Stone, Phys. Rev. Lett. **55**, 1622 (1985).

[11] A review of mesoscopic fluctuation phenomena is: B. L. Al'tshuler, P. A. Lee, and R. A. Webb, eds., *Mesoscopic Phenomena in Solids* (North-Holland, Amsterdam, 1991).

[12] B. L. Al'tshuler and B. Z. Spivak, Zh. Eksp. Teor. Fiz. **92**, 609 (1987) [Sov. Phys. JETP **65**, 343 (1987)].

[13] C. W. J. Beenakker and H. van Houten, in: *Nanostructures and Mesoscopic Systems*, ed. by W. P. Kirk and M. A. Reed (Academic, New York, to be published).

[14] A. Furusaki and M. Tsukada, Solid State Comm. **78**, 299 (1991).

[15] B. J. van Wees, K.-M. H. Lenssen, and C. J. P. M. Harmans, Phys. Rev. B **44**, 470 (1991).

[16] P. G. de Gennes, *Superconductivity of Metals and Alloys* (Benjamin, New York, 1966).

[17] K. K. Likharev, Rev. Mod. Phys. **51**, 101 (1979).

[18] M. Yu. Kupriyanov and V. F. Lukichev, Fiz. Nizk. Temp. **8**, 1045 (1982) [Sov. J. Low Temp. Phys. **8**, 526 (1982)]; A. A. Zubkov and M. Yu. Kupriyanov, Fiz. Nizk. Temp. **9**, 548 (1983) [Sov. J. Low Temp. Phys. **9**, 279 (1983)].

[19] J. Bardeen, R. Kümmel, A. E. Jacobs, and L. Tewordt, Phys. Rev. **187**, 556 (1969).

[20] A. F. Andreev, Zh. Eksp. Teor. Fiz. **46**, 1823 (1964); **49**, 655 (1965) [Sov. Phys. JETP **19**, 1228 (1964); **22**, 455 (1966)].

[21] E. Akkermans, A. Auerbach, J. E. Avron, and B. Shapiro, Phys. Rev. Lett. **66**, 76 (1991).

[22] A. D. Stone and Y. Imry, Phys. Rev. Lett. **56**, 189 (1986).

[23] Y. Imry, in: *Directions in Condensed Matter Physics*, Vol. 1, ed. by G. Grinstein and G. Mazenko (World Scientific, Singapore, 1986).

[24] W. Haberkorn, H. Knauer, and J. Richter, Phys. Status Solidi A **47**, K161 (1978).

[25] A. V. Zaĭtsev, Zh. Eksp. Teor. Fiz. **86**, 1742 (1984) [Sov. Phys. JETP **59**, 1015 (1984)].

[26] G. B. Arnold, J. Low Temp. Phys. **59**, 143 (1985).

[27] A. Furusaki and M. Tsukada, Physica B **165&166**, 967 (1990).

[28] D. S. Fisher and P. A. Lee, Phys. Rev. B **23**, 6851 (1981).

[29] I. O. Kulik and A. N. Omel'yanchuk, Fiz. Nizk. Temp. **3**, 945 (1977); **4**, 296 (1978) [Sov. J. Low Temp. Phys. **3**, 459 (1977); **4**, 142 (1978)].

[30] L. D. Landau and E. M. Lifshitz, *Quantum Mechanics* (Pergamon, Oxford, 1977).

[31] M. Büttiker, IBM J. Res. Dev. **32**, 63 (1988).

[32] V. Kalmeyer and R. B. Laughlin, Phys. Rev. B **35**, 9805 (1987).

[33] W. Xue and P. A. Lee, Phys. Rev. B **38**, 3913 (1988).

[34] A. D. Stone, P. A. Mello, K. A. Muttalib, and J.-L. Pichard, in Ref. [11].

[35] I. O. Kulik and A. N. Omel'yanchuk, Pis'ma Zh. Eksp. Teor. Fiz. **21**, 216 (1975) [JETP Lett. **21**, 96 (1975)].

[36] V. Ambegaokar and A. Baratoff, Phys. Rev. Lett. **10**, 486 (1963); **11**, 104(E) (1963).

[37] Y. Imry, Europhys. Lett. **1**, 249 (1986).

[38] H. Takayanagi and T. Kawakami, Phys. Rev. Lett. **54**, 2449 (1985).

DC Josephson Effect of Superconducting Quantum Point Contacts

A. Furusaki

Department of Applied Physics, University of Tokyo,
Bunkyo-ku, Tokyo 113, Japan

Abstract. The dc Josephson effect of a superconducting quantum point contact is investigated theoretically. It is shown that the critical current increases stepwise as a function of the width of the point contact. The step height generally depends on both the superconducting gap and junction parameters. The resonant Josephson effect is also discussed within the adiabatic approximation.

1. Introduction

Transport phenomena in mesoscopic systems have attracted great attention, and our understanding has been very much improved for the last ten years [1]. Most studies have been concerned with the normal-state properties of mesoscopic conductors, such as universal conductance fluctuations [2], the Aharanov-Bohm (AB) effect [3], and the conductance quantization [4,5]. Recent theoretical and experimental findings on mesoscopic physics are related mainly to quantum-size effects in quantum point contacts and quantum dots, in which electrons travel ballistically [6].

Compared with normal-state properties, mesoscopic superconducting systems have attracted less attention, but very recently several theoretical investigations have appeared by analogy with the mesoscopic effects in the normal state: conductance fluctuations in a normal metal–superconductor–normal metal (N-S-N) junction [7], the AB effect in an S-N-S ring [8,9], discretization of the critical Josephson current in a superconducting quantum point contact (SQPC) [10,11], and quantum fluctuations in the critical current of an SQPC [12,13].

In this paper we investigate theoretically the quantum-size effect in SQPCs: the discretization of the critical current and resonant Josephson current through two barriers. A superconductor–two-dimensional electron gas–superconductor (S-2DEG-S) junction is one of the most promising candidates for the observation of these effects. In fact the Josephson current through a 2DEG has been observed in Nb/p-InAs/Nb junctions [14], and in near future it will be possible to fabricate high-mobility S-2DEG-S Josephson junctions, in which one can control the Josephson current by the voltage on a gate electrode. If the size of the junction is made small enough, quantum-size effects would manifest itself. In particular, if the normal-state conductance of the link (constriction) in a 2DEG, which connects two superconductors and through which supercurrent flows, is quantized in units of $e^2/\pi\hbar$, then it is very likely that the critical Josephson current of the S-2DEG-S junction is also quantized (discretized); this is the subject of the present paper.

2. Formula for the Josephson Current

We calculate the dc Josephson current by solving the Bogoliubov-de Gennes equation (BdGE), treating an SQPC as a two-dimensional system. It is assumed that the transport of quasiparticles with an effective mass m is ballistic and that no magnetic field is applied. The motion of quasiparticles is described by the BdGE,

$$\begin{pmatrix} H(x,y) & \Delta(x,y) \\ \Delta^*(x,y) & -H(x,y) \end{pmatrix} \begin{pmatrix} u(x,y) \\ v(x,y) \end{pmatrix} = E \begin{pmatrix} u(x,y) \\ v(x,y) \end{pmatrix}, \tag{1}$$

where $H(x,y)$ is the single-electron Hamiltonian,

$$H(x,y) = -\frac{\hbar^2}{2m}\left(\frac{\partial^2}{\partial x^2} + \frac{\partial^2}{\partial y^2}\right) + U(x,y) - E_F, \tag{2}$$

and $\Delta(x,y)$ is the pair potential. The net supercurrent flows in the x-direction, and the width of the system in the y-direction varies only in the normal region, 2DEG ($0 < x < L$), and is constant, W_S, in the superconductor ($x < 0$ and $x > L$). We adopt hard-wall boundary conditions.

In real S-2DEG-S junctions the effective mass, m, and the Fermi energy, E_F, of the 2DEG are different from those of the superconductor, which results in the scattering at the S-2DEG interfaces. In this paper, however, we assume that m and E_F are common in the 2DEG (semiconductor) and the superconductor. Instead the scattering effect is included in our theory by introducing the diagonal potential,

$$U(x,y) = V\{\delta(x) + \delta(x-L)\}. \tag{3}$$

The off-diagonal potential, $\Delta(x,y)$, is assumed as

$$\Delta(x,y) = \Delta_0\{\Theta(-x) + \exp(i\varphi)\Theta(x-L)\}. \tag{4}$$

This functional form is a good approximation for the case where wide superconductors are linked by a narrow constriction.

Solutions of the BdGE are linear combinations of plane waves in the superconducting region. The wave function, representing the process whereas an electron-like quasiparticle is injected from the left superconductor, is written as

$$\begin{aligned}\psi(x,y) = {} & e^{ip_j^+ x} \sin\left\{q_j\left(y+\frac{W_S}{2}\right)\right\}\begin{pmatrix} u_0 \\ v_0 \end{pmatrix} \\ & + \sum_{k=1}^{\infty}\left\{a_{1jk}e^{ip_k^- x}\begin{pmatrix} v_0 \\ u_0 \end{pmatrix} + b_{1jk}e^{-ip_k^+ x}\begin{pmatrix} u_0 \\ v_0 \end{pmatrix}\right\}\sin\left\{q_k\left(y+\frac{W_S}{2}\right)\right\} \end{aligned} \tag{5}$$

if $x < 0$, and

$$\begin{aligned}\psi(x,y) = {} & \sum_{k=1}^{\infty}\left\{c_{1jk}e^{ip_k^+ x}\begin{pmatrix} u_0\exp(i\varphi/2) \\ v_0\exp(-i\varphi/2) \end{pmatrix} + d_{1jk}e^{-ip_k^- x}\begin{pmatrix} v_0\exp(i\varphi/2) \\ u_0\exp(-i\varphi/2) \end{pmatrix}\right\} \\ & \times \sin\left\{q_k\left(y+\frac{W_S}{2}\right)\right\} \end{aligned} \tag{6}$$

if $x > L$, where j is a positive integer,

$$p_k^{\pm} = \left\{\frac{2m}{\hbar^2}(E_F \pm \Omega) - q_k^2\right\}^{1/2}, \quad q_k = \frac{k\pi}{W_S}, \quad u_0 = \left\{\frac{1}{2}\left(1 + \frac{\Omega}{E}\right)\right\}^{1/2},$$
$$v_0 = \left\{\frac{1}{2}\left(1 - \frac{\Omega}{E}\right)\right\}^{1/2}, \quad \text{and} \quad \Omega = \sqrt{E^2 - \Delta_0^2}. \tag{7}$$

The coefficients a_{1jk}, b_{1jk}, c_{1jk}, and d_{1jk} are functions of the excitation energy E and the phase difference φ across two superconductors, and they are determined by the matching conditions at S-2DEG interfaces.

The dc Josephson current I at temperature T can be calculated from the temperature Green's function $G_{\omega_n}(x, y; x', y')$ which is constructed by appropriately combining four independent eigenfunctions of the BdGE [15]. Taking into account the charge-conservation law, we calculate the current at $x = 0$:

$$I = \frac{e\hbar}{2im}\int_{-W_S/2}^{W_S/2} dy \left(\frac{\partial}{\partial x'} - \frac{\partial}{\partial x}\right)\frac{1}{\beta}\sum_{\omega_n}\mathrm{Tr}\{G_{\omega_n}(x, y; x', y')\}\Big|_{x=x'=0,\, y=y'}$$
$$= \frac{e\Delta_0}{2\hbar\beta}\sum_{\omega_n}\frac{1}{\Omega_n}\sum_j (p_j^+ + p_j^-)\left(\frac{a_{1jj}(\varphi, i\omega_n)}{p_j^+} - \frac{a_{2jj}(\varphi, i\omega_n)}{p_j^-}\right), \tag{8}$$

where $\omega_n = \pi k_B T(2n+1)$ $(n = 0, \pm 1, \pm 2, \ldots)$, $\beta = 1/(k_B T)$, and $\Omega_n = (\omega_n^2 + \Delta_0^2)^{1/2}$. The coefficient $a_{1jj}(\varphi, i\omega_n)$ is obtained from $a_{1jj}(\varphi, E)$ in Eq. (5) and is the scattering amplitude for the process that an electron-like quasiparticle of the j-th mode injected from the left-hand side is reflected as a hole-like quasiparticle of the j-th mode, whereas the coefficient a_{2jj} is the scattering amplitude for the reverse process that a hole-like quasiparticle is reflected as an electron-like quasiparticle.

Equation (8) shows that the dc Josephson current is related to the scattering amplitude of the Andreev reflection [16]. Noting that the Andreev reflection of an electron-like quasiparticle into a hole-like quasiparticle involves creation of a Cooper pair and that the reverse process gives rise to destruction of a Cooper pair (Fig. 1), it is naturally understood why the dc Josephson current is given

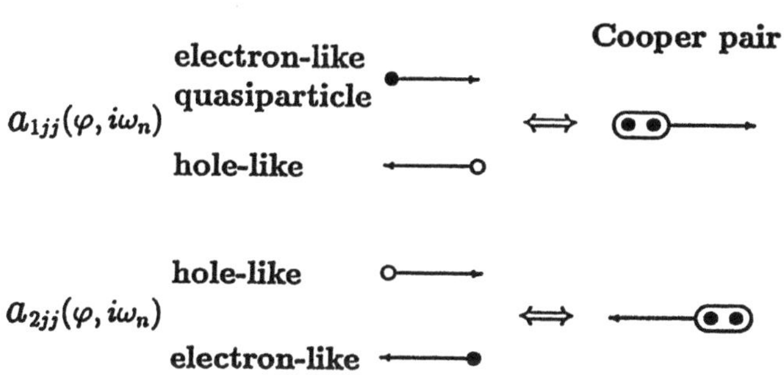

Figure 1: Relation between the Andreev reflection and motion of a Cooper-pair.

by the difference between $a_{1jj}(\varphi, i\omega_n)$ and $a_{2jj}(\varphi, i\omega_n)$. We also note that these scattering amplitudes satisfy a kind of detailed-balance relation,

$$p_j^- a_{1jj}(\varphi) = p_j^+ a_{2jj}(-\varphi), \tag{9}$$

so that the current is an odd function of φ as it should be.

3. Results

Using the formula for the dc Josephson current, we will investigate the dc Josephson effect of SQPCs for three cases. We first consider an ideal S-2DEG-S junction, in which no reflection except the Andreev reflection occurs at S-2DEG interfaces, in adiabatic-transport regime and in non-adiabatic-transport regime. We then consider the effect of normal reflections at S-2DEG interfaces, i.e., the resonant tunneling of Cooper pairs.

Let us consider an ideal SQPC ($U(x,y) = 0$) in which the width of a normal link, $W(x)$, varies slowly compared with the Fermi wave length. In this case the scattering between the modes can be neglected, and thus the Andreev-scattering amplitude a_{ijk} is nonzero only for $j = k$. Solving the BdGE within the WKB approximation, we obtain the scattering amplitude as

$$a_{1jj}(\varphi, i\omega_n) = \frac{i\Delta_0\{\exp(i\varphi) - \exp(-\Phi_j)\}}{(\omega_n + \Omega_n)\exp(i\varphi) - (\omega_n - \Omega_n)\exp(-\Phi_j)}, \tag{10}$$

where

$$\Phi_j = 2\,\mathrm{Im}\int_0^L dx \left[\frac{2m}{\hbar^2}(E_F + i\omega_n) - \left(\frac{j\pi}{W(x)}\right)^2\right]^{1/2}. \tag{11}$$

Substituting Eq. (10) into Eq. (8), we obtain the dc Josephson current as

$$I = \frac{2e\Delta_0^2}{\hbar\beta}\sum_{j=1}^{N}\sum_{\omega_n} \frac{\sin\varphi}{(\omega_n^2 + \Omega_n^2)\cosh\Phi_j + 2\omega_n\Omega_n\sinh\Phi_j + \Delta_0^2\cos\varphi}, \tag{12}$$

where N is the number of open modes through which the Josephson current can flow.

Equation (12) can be simplified in two limits of interest. First, in the short-junction limit where the length of the normal region L is much shorter than the superconducting coherence length ξ, Eq. (12) reduces to

$$\begin{aligned} I &= N\frac{2e\Delta_0^2}{\hbar\beta}\sum_{\omega_n}\frac{\sin\varphi}{\omega_n^2 + \Omega_n^2 + \Delta_0^2\cos\varphi} \\ &= N\frac{e\Delta_0}{\hbar}\sin(\varphi/2)\tanh\left[\frac{\Delta_0}{2k_BT}\cos(\varphi/2)\right]. \end{aligned} \tag{13}$$

The Josephson current is a multiple of the supercurrent carried by a single mode [11]. Thus the critical current increases stepwise as the constriction becomes wider, and the step height is $e\Delta_0/\hbar$ at zero temperature.

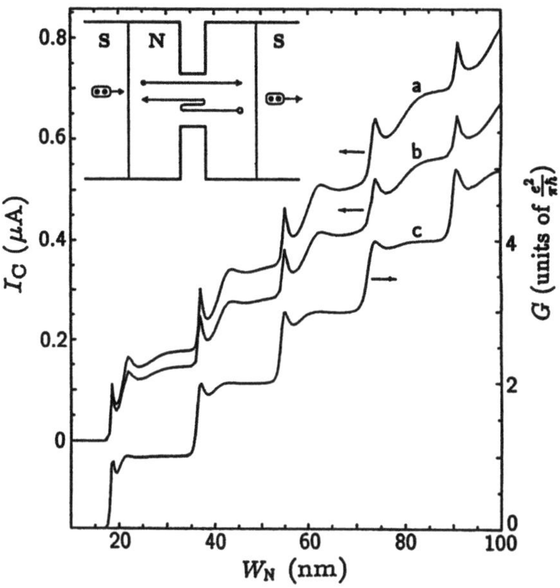

Figure 2: The critical current, (a) and (b), and the normal-state conductance, (c), as a function of the width of the constriction. Inset: Illustration of supercurrent through the junction.

On the other hand, if $L \gg \xi$, Eq. (12) can be written at zero temperature as

$$I \approx \frac{e}{\pi L}\sum_{j=1}^{N} v_j \varphi, \tag{14}$$

where v_j is a velocity of the j-th mode and is defined by

$$(v_j)^{-1} = \frac{1}{L}\int_0^L dx \frac{m}{\hbar}\left[\frac{2mE_F}{\hbar^2} - \left(\frac{j\pi}{W(x)}\right)^2\right]^{-1/2}. \tag{15}$$

In this limit the critical current depends on junction parameters such as L and v_j, and is not exactly proportional to the number of open modes N [10]. Thus, the critical current does not always increase stepwise.

Next we shall consider a different situation in which the width of the constriction changes abruptly. Again we assume that S-2DEG interfaces are ideal. The geometry of the sytem is shown in the inset of Fig. 2; the width is W_S for $|x - L/2| > D/2$ and is W_N for $|x - L/2| < D/2$. In this case scatterings between the modes are important, and we calculate the dc Josephson current by solving the BdGE fully numerically.

The critical current I_C at $T = 0.5\,\mathrm{K}$ is shown in Fig. 2 (curve (a) and (b)). We set $L = 100\,\mathrm{nm}$ for the curve (a) and $200\,\mathrm{nm}$ for the curve (b), $D = 50\,\mathrm{nm}$, $W_S = 120\,\mathrm{nm}$, $E_F = 0.05\,\mathrm{eV}$, $\Delta_0 = 1\,\mathrm{meV}$, and $m = 0.024 m_e$ with m_e as the electronic mass. With these paprameters, an 18nm-increase in W_N

amounts to the opening of another channel, and the coherence length ξ is 180 nm. The curve (c) in Fig. 2 is the normal-state conductance at zero temperature which is calculated from the Landauer formula. It is clear that these three curves show common resonant structures, which are due to the resonances in the transmission amplitude of electrons and holes going through the narrow constriction ($|x - L/2| < D/2$); see the inset of Fig. 2.

So far we have assumed that the S-2DEG interfaces are ideal ones. In real samples, however, there often exists a Schottky barrier at the interface, and the Fermi momentum is different between a superconductor and a 2DEG. Both cause normal reflections, in addition to the Andreev reflection, at the S-2DEG interfaces: electron(-like)→electron(-like) and hole(-like)→hole(-like). The effect of the normal reflection can be incorporated in our theory by keeping V in Eq. (3) nonzero.

We calculate the dc Josephson current in the adiabatic approximation used for the derivation of Eq. (12). Solving the BdGE in the WKB approximation, and then using Eq. (8), we obtain the dc Josephson current as

$$I = \frac{2e\Delta_0^2}{\hbar\beta} \sum_{j=1}^{N} \sum_{\omega_n} \frac{\sin\varphi}{\Gamma}, \tag{16}$$

where

$$\begin{aligned} \Gamma &= \{\omega_n^2 + \Omega_n^2 + 4Z_j^2(1+Z_j^2)\Omega_n^2\}\cosh\Phi_j + 2\omega_n\Omega_n(1+2Z_j^2)\sinh\Phi_j \\ &\quad +4Z_j^2(1-Z_j^2)\Omega_n^2\cos\Psi_j + 8Z_j^3\Omega_n^2\sin\Psi_j + \Delta_0^2\cos\varphi, \\ \Psi_j &= 2\,\mathrm{Re}\int_0^L dx \left[\frac{2m}{\hbar^2}(E_F + i\omega_n) - \left(\frac{j\pi}{W(x)}\right)^2\right]^{1/2}, \\ Z_j &= \frac{mV}{\hbar^2}\left(\frac{2mE_F}{\hbar^2} - q_j^2\right)^{-1/2}. \end{aligned} \tag{17}$$

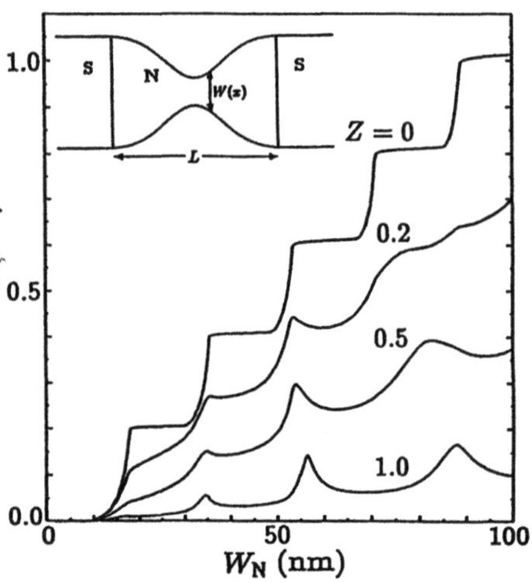

Figure 3: The critical current as a function of the constriction width for various barrier strengths Z. Inset: Schematic of the junction.

The two terms in Γ, which are proportional to $\cos\Psi_j$ and $\sin\Psi_j$, come from the interference of electrons or holes which go and return between the S-2DEG interfaces. Figure 3 shows the critical current at zero temperature as a function of the minimum width W_N for various values of $Z = (mV^2/2\hbar^2 E_F)^{1/2}$. We assumed that the constriction width varies as

$$W(x) = \frac{1}{1-e^{-4}}\left\{W_S - W_N e^{-4} - (W_S - W_N)\exp\left[-16\left(\frac{x}{L}-\frac{1}{2}\right)\right]\right\} \tag{18}$$

as shown in the inset of Fig. 3. We chose $W_S = 200\,\text{nm}$, and the other parameters are the same as those used in Fig. 2. Figure 3 clearly shows that $I_C - W_N$ curve begins to have resonances as the parameter Z increases, i.e., as the normal reflection dominates the Andreev reflection.

4. Conclusion

The dc Josephson effect of SQPCs has been investigated both in the adiabatic-transport regime and in the non-adiabatic-transport regime. The Josephson current is calculated from the scattering amplitude of the Andreev reflection; this method can be regarded as a variation of Landauer's approach to the electron transport in normal conductors in that the electric current is expressed in terms of scattering data. It is shown that for an ideal SQPC the critical current increases stepwise as a function of the constriction width, and that the step height generally depends on Δ_0, L, and v_j. When S-2DEG interfaces are not ideal ones, the normal reflection begins to play an important role. As the normal reflection dominates the Andreev reflection, the nature of the Josephson effect changes from the transport of Cooper-pairs through all the open channels to the resonant tunneling through one channel on resonance. These two effects, quasiquantization of I_C and the resonant tunneling, will be observed experimentally in the near future when high-mobility S-2DEG-S junctions are fabricated.

Acknowledgement. The author would like to thank Prof. Masaru Tsukada and Dr. Hideaki Takayanagi for fruitful discussions and their encouragement during the course of this work. This work is supported in part by a Grant-in-Aid from Ministry of Education, Science, and Culture of Japan, and numerical calculations were performed on HITAC S-820 at the computer center of the University of Tokyo.

References

[1] Y. Imry, in *Directions in Condensed Matter Physics*, eds. G. Grinstein and E. Mazenko (World Scientific, Singapore, 1986).
[2] P. A. Lee and A. D. Stone, Phys. Rev. Lett. **55**, 1622 (1985).
[3] R. A. Webb, S. Washburn, C. P. Umbach, and R. B. Laibowitz, Phys. Rev. Lett. **54**, 2696 (1985).

[4] B. J. van Wees, H. van Houten, C. W. J. Beenakker, J. G. Williamson, L. P. Kouwenhoven, D. van der Marel, and C. T. Foxon, Phys. Rev. Lett. **60**, 848 (1988).
[5] D. A. Wharam, T. J. Thornton, R. Newbury, M. Pepper, H. Ajmed, J. E. F. Frost, D. G. Hasko, D. C. Peacock, D. A. Ritchie, and G. A. C. Jones, J. Phys. C **21**, L209 (1988).
[6] C. W. J. Beenakker and H. van Houten, Solid State Physics **44**, 1 (1991).
[7] Y. Takane and H. Ebisawa, J. Phys. Soc. Jpn. **60**, 3130 (1991).
[8] M. Büttiker and T. M. Klapwijk, Phys. Rev. B **33**, 5114 (1986).
[9] H. Yoshioka and H. Fukuyama, J. Phys. Soc. Jpn. **59**, 3065 (1990).
[10] A. Furusaki, H. Takayanagi, and M. Tsukada, Phys. Rev. Lett. **67**, 132 (1991).
[11] C. W. J. Beenakker and H. van Houten, Phys. Rev. Lett. **66**, 3056 (1991).
[12] C. W. J. Beenakker, this symposium.
[13] On the issue of quantum fluctuations in the critical current, there exists an *old* work by Al'tshuler and Spivak, Sov. Phys. JETP **65**, 343 (1987). However, no further studies have followed this paper until very recently.
[14] H. Takayanagi and T. Kawakami, Phys. Rev. Lett. **54**, 2449 (1985).
[15] A. Furusaki and M. Tsukada, Solid State Commun. **78**, 299 (1991).
[16] The term, Andreev reflection, usually refers to the scattering of an electron into a hole. Here it is generalized to refer also to the scattering of an electron-like quasiparticle into a hole-like quasiparticle.

Fluctuations of Josephson Currents in Mesoscopic Systems

H. Yoshioka and H. Fukuyama

Department of Physics, Faculty of Science, University of Tokyo,
7-3-1 Hongo, Bunkyo-ku, Tokyo 113, Japan

Abstract. Various aspects of fluctuations in Josephson currents are studied for superconductor - normal metal - superconductor (SNS) systems with mesoscopic normal regions. In the system with diffusive normal region, relative magnitude of the fluctuation to the average is calculated. Effects of pair breaking on fluctuations are investigated especially in the systems with spin glass ordering, where it is found that the fluctuation remains finite even in the strong depairing limit although the average vanishes. In the system with Aharonov - Bohm ring geometry, we find hc/e oscillation in the critical current besides the well known $hc/2e$ oscillation.

1. Introduction

One of the most remarkable features in mesoscopic systems is the clear manifestation of interference effects of electron waves due to the persistence of the phase coherence of electrons. For example, the conductance undergoes random time - independent fluctuations as a function of the magnetic field that varies from sample to sample but is reproducible (magnetofingerprint)[1] and the amplitude of the fluctuation is of the order of e^2/h (Universal Conductance Fluctuation)[2,3]. Furthermore, the conductance in Aharonov - Bohm (AB) ring geometry shows the periodic oscillation with AB flux besides the aperiodic one mentioned above, which is known as the AB effect and the most remarkable example showing persistence of the phase coherence of electron waves in mesoscopic systems[1].

Although such fluctuations are usually of a fraction of the average, these can be comparable to or even larger than the average in some cases. The typical example is the fluctuation of orbital magnetism[4]. Even in good metallic samples, its fluctuation can be far bigger than the average (Landau diamagnetism) in isolated systems[5] and in systems attached to the perfect leads[6,7].

The other systems where the phase coherence is maintained are superconductors. Their phase coherence, however, is due to the macroscopic condensation of Cooper pairs and persists over a macroscopic scale. If a superconductor (S) has a boundary with a normal metal (N), the amplitude of Cooper pair condensate in the superconductor can penetrate into the normal region, which is known as proximity effects leading to finite Josephson currents in SNS systems[8,9]. Hence, if the N region in an SNS system is of mesoscopic scale, interesting interplay of two different kinds of phase coherence will be expected, which we will explore in this paper. This problem seems to be important as a first step to understand the relationship of phases between mesoscopic systems and superconductors.

The mesoscopic fluctuation in Josephson currents has been discussed by Al'tshuler *et al.*[10] and Beenakker[11] for wire geometry. In this paper, we will study a similar problem with particular emphasis on the effects of spin glass ordering. Moreover, a similar problem in the AB ring structure is also studied.

We take a unit of $k_B = \hbar = 1$.

2. Disordered N Region with Wire Geometry

First we will study the effect of mesoscopic fluctuations on the Josephson current in simple wire geometries shown in Fig.1.

2-1) The Case of Normal Impurity Scattering

We begin with the formula of Josephson currents given in the lowest order in $\Delta^l(\epsilon_n)$ and $\Delta^r(\epsilon_n)$ at S_l and S_r (S_l and S_r are superconductors attached to the left and the right side of the N region, respectively) :

$$I = 4ieT \sum_{\epsilon_n>0} \int d\vec{r}_{1\perp} d\vec{r}_{2\perp} \Big[X(\vec{r}_1, \vec{r}_2; \epsilon_n) \Delta^l(\epsilon_n) \Delta^{r*}(\epsilon_n) - X(\vec{r}_2, \vec{r}_1; \epsilon_n) \Delta^r(\epsilon_n) \Delta^{l*}(\epsilon_n) \Big], \quad (1)$$

where $\epsilon_n = (2n+1)\pi T$ and $\vec{r}_1 = (0, \vec{r}_{1\perp})$, $\vec{r}_2 = (L, \vec{r}_{2\perp})$ with 0 and L being the x coordinates of S_lN and NS_r interfaces, respectively. In eq.(1), $X(\vec{r}_1, \vec{r}_2; \epsilon_n) = G(\vec{r}_1, \vec{r}_2; \epsilon_n) G(\vec{r}_1, \vec{r}_2; -\epsilon_n)$ represents the penetration of Cooper pairs into the N region and $\Delta^{(l,r)}(\epsilon_n) = C\Delta \exp\{i\theta_{(l,r)}\} / \sqrt{\epsilon_n^2 + \Delta^2}$[12] is essentially the amplitude of the pair condensate at the S_lN and NS_r interface, $F^{(l,r)}(\vec{r}_{(1,2)}, \vec{r}_{(1,2)}; \epsilon_n)$, and C is a constant depending on the details of the SN interface. The Δ and $\theta_{(l,r)}$ are the gap and the phase of superconductors, respectively. Noting time reversal symmetries, we obtain the average, $< I(\varphi) >$, and the fluctuation, $< \delta I(\varphi)^2 >$, of Josephson currents as

$$< I(\varphi) >= 8eT \sum_{\epsilon_n>0} |\Delta^l(\epsilon_n)||\Delta^r(\epsilon_n)| \int d\vec{r}_{1\perp} d\vec{r}_{2\perp} < X(\vec{r}_1, \vec{r}_2; \epsilon_n) > \sin\varphi, \quad (2)$$

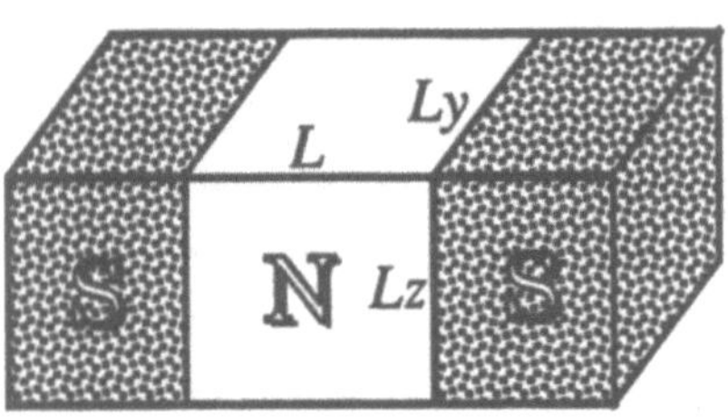

Fig.1
SNS junction
with simple wire geometry.

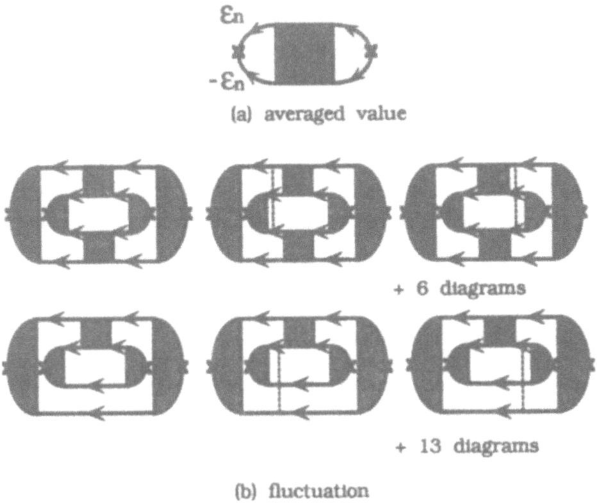

Fig.2 Diagrams for averaged value and fluctuation.
A dotted line means a single impurity scattering.

$$< \delta I(\varphi)^2 >= 64e^2T^2 \sum_{\epsilon_n, \epsilon_n' > 0} |\Delta^l(\epsilon_n)||\Delta^r(\epsilon_n)||\Delta^l(\epsilon_n')||\Delta^r(\epsilon_n')|$$

$$\int \mathrm{d}\vec{r}_{1\perp}\mathrm{d}\vec{r}_{2\perp}\mathrm{d}\vec{r}_{1\perp}'\mathrm{d}\vec{r}_{2\perp}' < X(\vec{r}_1, \vec{r}_2; \epsilon_n)X(\vec{r}_1', \vec{r}_2'; \epsilon_n') > \sin^2\varphi, \quad (3)$$

where φ is the phase difference between S_l and S_r, $\varphi \equiv \theta_l - \theta_r$, and $< \cdots >$ implies impurity averaging. Here the phase difference, φ, is taken to be fixed. To the lowest order in $1/\epsilon_F\tau$, ϵ_F and τ being the Fermi energy and the life time of elastic scattering in the N region, the average and the fluctuation are given by the diagrams shown in Figs.2 (a) and (b), respectively.

In this calculation, two kinds of cooperons are needed; one connecting the same system and the other connecting different systems. In the high temperature regime, where $L > \xi_D$ with $\xi_D = \sqrt{D/2\pi T}$ being the coherence length in the N region in the diffusive regime with the diffusion constant, D, the average current and the relative intensity of the fluctuation to the average are given as follows:

$$< I >= 32eT\sin\varphi|\Delta^l(\pi T)||\Delta^r(\pi T)|\pi N_0 S\frac{\xi_D}{D}\exp(-L/\xi_D), \quad (4)$$

$$\frac{< \sqrt{\delta I^2} >}{< I >} \simeq \frac{1}{N_0 SD/L}\sqrt{c_d}, \quad (5)$$

where N_0 is the density of states per unit volume in the N region and $S = L_y L_z$ represents the cross sectional area. The c_d's are constants depending on the dimensionality of the cooperons and are given as follows: $c_1 \simeq 1$ for $L > \xi_D > L_y, L_z$; $c_2 \simeq L_y/\sqrt{\xi_D L}$ for $L, L_y > \xi_D > L_z$ and $c_3 \simeq S/(\xi_D L)$ for $L, L_y, L_z > \xi_D$. The factor $N_0 SLD/L^2$ is the effective channel number $N_{c,eff}$, so L/N_0SD is of the order of $1/<g>$ with g being the conductance of the N region in units of e^2/h.

Finally we will discuss the observability of this fluctuation. The fluctuations have been calculated for the fixed phase difference, which will be realized, *e.g.*, by changing the gate voltage which controls the carrier density. On the other hand, the fluctuation will be introduced by the applied magnetic field as well. However, it will be difficult to observe the fluctuation since the correlation range in the fluctuation is comparable to the period of the damped oscillation in the Fraunhofer pattern.

2-2) Effects of Pair Breaking due to Localized Impurity Spins

We discuss effects of magnetic scattering on mesoscopic fluctuation of the Josephson current. This scattering is known as a pair breaking mechanism of superconductivity and generally reduces supercurrent.

In the case where impurity spins are in a paramagnetic state, there is no magnetic correlation between two measurements because the spin configuration will vary in a time scale of an experiment[13]. Formally speaking, the two measurements cannot be connected by the spin scattering. Then the spin scattering rate, $1/\tau_s$, can be regarded as an effective dephasing rate also for the fluctuation. Namely, this rate becomes the mass of cooperon and diffuson: $1/(D\vec{q}^2+\omega_n) \rightarrow 1/(D\vec{q}^2+\omega_n+1/\tau_s)$. Therefore, magnetic scatterings play the role of pair breaking on fluctuation and then the fluctuations as well as the averages will be reduced in strong magnetic scattering.

In spin glass systems where the spin configurations are frozen, however, the situation is different. This is because there now exist magnetic correlations between the different measurements by the fixed spin orientation[14] and then the two measurements are connected by the spin scattering. Namely, if the spin configuration remains fixed on the time scale of one measurement, the random spins act as random field. In this case the cooperons and the diffusons shown in Figs.3 are given by[15]:

$$C^{++++} = \frac{1}{2\pi N_0\tau^2}\frac{1}{D\vec{q}^2+\omega_n+2/3\tau_s},$$
$$C^{++--} = \frac{1}{4\pi N_0\tau^2}[\frac{1}{D\vec{q}^2+\omega_n+2/3\tau_s}+\frac{1}{D\vec{q}^2+\omega_n+2/\tau_s}],$$
$$C^{+--+} = \frac{1}{4\pi N_0\tau^2}[\frac{1}{D\vec{q}^2+\omega_n+2/3\tau_s}-\frac{1}{D\vec{q}^2+\omega_n+2/\tau_s}],$$
$$D^{++++} = \frac{1}{4\pi N_0\tau^2}[\frac{1}{D\vec{q}^2+\omega_n}+\frac{1}{D\vec{q}^2+\omega_n+4/3\tau_s}],$$

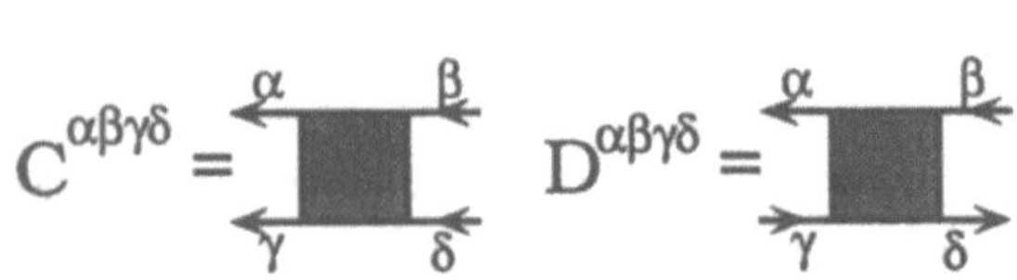

Fig.3 (a) cooperon (b) diffuson

$$D^{+-+-}=\frac{1}{4\pi N_0\tau^2}\left[\frac{1}{D\vec{q}^2+\omega_n}-\frac{1}{D\vec{q}^2+\omega_n+4/3\tau_s}\right],$$
$$D^{++--}=\frac{1}{2\pi N_0\tau^2}\frac{1}{D\vec{q}^2+\omega_n+4/3\tau_s}. \tag{6}$$

There exists a reversal symmetry with respect to spin index; for example $C^{++++}=C^{----}$.

It should be noticed that only the singlet diffuson is without mass and the others are massive. Therefore, the process including this singlet diffuson will be most effective in the presence of strong spin scattering limit, $\tau_s\to 0$. This conclusion is not affected even if the Zeemann energy is taken into account because the singlet diffuson is independent of it.

Using these results, we can evaluate the fluctuation of the Josephson current as follows:

$$\begin{aligned}<\delta I^2(\varphi)> = 8e^2T^2 \sum_{\epsilon_n,\epsilon_n'>0} |\Delta^l(\epsilon_n)||\Delta^r(\epsilon_n)||\Delta^l(\epsilon_n')||\Delta^r(\epsilon_n')| \int d\vec{r}_{1\perp}d\vec{r}_{2\perp}d\vec{r}_{1\perp}'d\vec{r}_{2\perp}' \\ \times\, \sigma^x_{\alpha\gamma}\sigma^x_{\beta\delta}\sigma^x_{\alpha'\gamma'}\sigma^x_{\beta'\delta'}\big[<X^{\alpha\beta\gamma\delta}(\vec{r}_1,\vec{r}_2;\epsilon_n)X^{\beta'\alpha'\delta'\gamma'}(\vec{r}_2',\vec{r}_1',\epsilon_n')> \\ -<X^{\alpha\beta\gamma\delta}(\vec{r}_1,\vec{r}_2;\epsilon_n)X^{\alpha'\beta'\gamma'\delta'}(\vec{r}_1',\vec{r}_2';\epsilon_n')>\cos 2\varphi\big],\end{aligned} \tag{7}$$

where $X^{\alpha\beta\gamma\delta}$ means $G_{\alpha\beta}G_{\gamma\delta}$ and the Pauli matrix σ^x projects out singlet components. In eq.(7), the first and the second terms, except for the factor $\cos 2\varphi$, are generally different, since the first term includes diffuson processes besides cooperon processes, whereas the second term has only cooperon processes. Hence eq.(7) does not vanish even if $\varphi=0$, *i.e.* the fluctuation persists even if the phase difference between the two superconductors, φ, is zero in contrast to the normal impurity case. Especially, in the strong depairing limit, *i.e.* $\tau_s\to 0$, the average is obviously vanishing but the fluctuation survives and is independent of the phase difference because only the first term of the r.h.s of eq.(7) survives. In this limit, processes given in Fig.4 determine the magnitude of fluctuations.

If $L>\xi_D$, these processes gives the result that the relative intensity of the fluctuation to the critical current in the case of $\tau_s^{-1}=0$ is given by $\sqrt{c_d}/(N_0Sv_F)$, where N_0Sv_F corresponds to the channel number, N_c, and the c_d is defined above. Hence

$$\frac{\sqrt{<\delta I^2(\varphi)>_{\tau_s\to 0}}}{\sqrt{<\delta I^2(\varphi)>_{\tau_s^{-1}=0}}}\simeq\frac{N_{c,eff}}{N_c}\simeq\frac{\ell}{L}, \tag{8}$$

where ℓ is the mean free path and $\ell\ll L$ in the present diffusive regime. The reason why the fluctuation is suppressed in this limit compared to the case without the spin scattering is that the cooperon among the same system does not contribute to the fluctuation.

Similar phenomena can be observed in persistent currents in disordered metallic rings with spin glass ordering. Namely, the persistent current fluctuates even if there

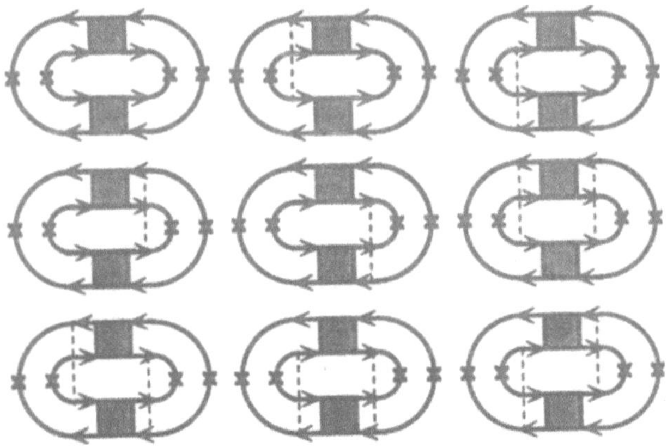

Fig.4 Diagrams in the lowest order of $1/\epsilon_F\tau$ in strong depairing limit.

is no AB flux and the intensity is independent of the flux in the strong spin scattering limit[15].

Such a remarkable phenomenon of the finite fluctuation with the vanishing average has been noticed in RKKY interaction in disordered media[16].

In calculation of cooperons and diffusons, we did not take into account the effect of the magnetic field on spin glass systems. In general, spin glass systems have a finite susceptibility, so we need to consider the effect of the spin polarization due to the applied magnetic field[15], and we find that the result in eq.(8) is not modified except for a numerical factor.

3. AB Ring Geometry

The Josephson current of an SNS system with the N region of the AB ring structure as shown in Fig.5 is of particular interest, since the transport current through the AB ring is known to exhibit AB oscillation.

We have investigated this problem[17] and the Josephson current was shown to be given as follows in the case, $L_1 + L_2 + R\theta, L_1 + L_2 + R(2\pi - \theta) > \xi$ with $\xi = \xi_D$ in the dirty case and $\xi = \xi_C \equiv v_F/2\pi T$ in the clean case[18]:

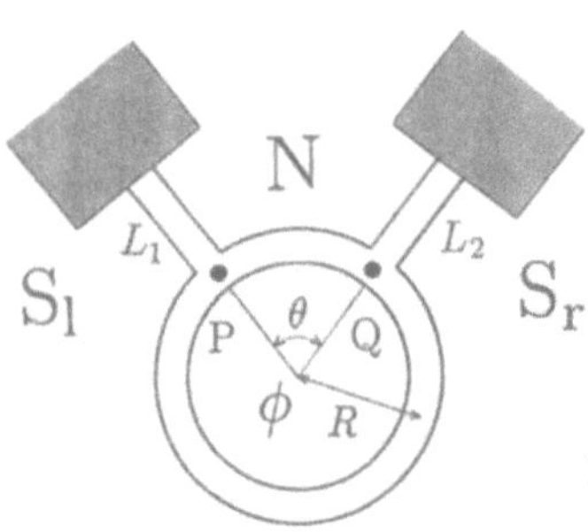

Fig.5 Superconductor - Normal Metal - Superconductor junction with an AB ring in the N region.

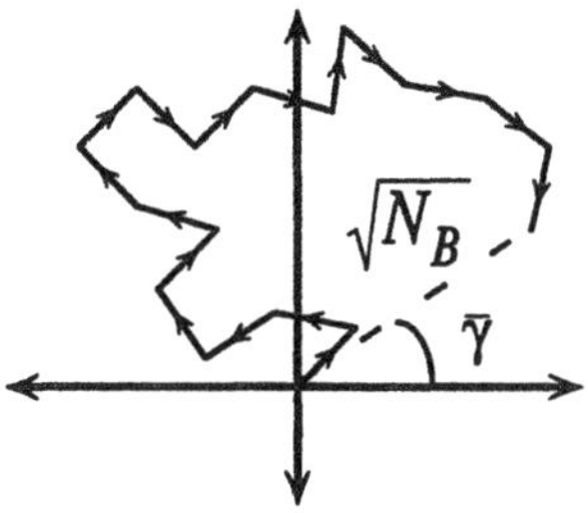

Fig.6
Definition of $\bar{\gamma}$.

$$I = I_d \exp[-\frac{L_1 + L_2}{\xi}]|\Delta^l(\pi T)||\Delta^r(\pi T)|$$
$$[A_1 \sin(\Psi - 4\pi\phi/\phi_0) + A_2 \sin\Psi + 2\sqrt{A_1A_2/N_B}\sin(\Psi - 2\pi\phi/\phi_0)\cos\bar{\gamma}], \quad (9)$$

where $A_1 = \exp[-R(2\pi - \theta)/\xi]$, $A_2 = \exp[-R\theta/\xi]$, Ψ and ϕ_0 are the gauge invariant phase difference between S_l and S_r[19] and the unit flux, hc/e, respectively, whereas I_d is a constant depending on the spatial dimension and the temperature[20]. In eq.(9) N_B is the effective number of Cooper pairs contributing to the proximity effect. The first and the second terms in eq.(9) are due to processes that two electrons forming a single Cooper pair go through the same path (process 1), while the third term results from the process that the two electrons go through the different path separately (process 2). The impurity scattering or boundary roughness results in the random phase shifts of electron waves. In process 1, these phase shifts can be canceled because two electrons forming a single Cooper pair can be scattered by the same impurity. However, such a cancellation cannot occur in process 2. The factor $\cos\bar{\gamma}$ in the third term results from the summation of these random phase shifts and is defined in Fig.6. Therefore, contributions from processes 1 and 2 are of the order of N_B and $\sqrt{N_B}$, respectively. This is the reason why the factor N_B appears.

From this formula, we find the critical current[18];

$$I_M = I_d \exp[-\frac{L_1 + L_2}{\xi}]F(2\pi\phi/\phi_0), \quad (10)$$
$$F(x) = [A_1^2 + A_2^2 + 4A_1A_2\cos^2\bar{\gamma}/N_B + 2A_1A_2\cos 2x + 4(A_1 + A_2)\sqrt{A_1A_2/N_B}\cos\bar{\gamma}\cos x]^{1/2}. \quad (11)$$

It is to be noted that there exists ϕ_0 oscillation in addition to the well-known $\phi_0/2$ oscillation. This ϕ_0 oscillation results from the interference between two electron waves forming a single Cooper pair in process 2 and then is the effect of mesoscopic fluctuation. This oscillation is one example indicating the fact that phases of mesoscopic systems and those in superconductors are entangled[17]. The relative intensity of the former to the latter is given by $2(\sqrt{A_1/A_2} + \sqrt{A_2/A_1})/\sqrt{N_B}$ besides a random factor $\cos\bar{\gamma}$ representing the fact that this is the mesoscopic fluctuation effect and depends on the geometry and the effective number N_B. It is important to estimate the actual value of N_B, which is a new concept, in order to assess the observability of the ϕ_0 oscillation.

However, the relative intensity of the ϕ_0 and $\phi_0/2$ oscillation discussed above is nothing but that of the fluctuation to the average. Therefore, neglecting the factor from ring geometry, the value, $1/\sqrt{N_B}$, will be of the same order as that in the wire system studied in sec.2. In contrast to the simple wire geometry studied in sec.2.1 in which mesoscopic fluctuation will not easily be observed on the top of Fraunhofer pattern, this ϕ_0 oscillation can be clearly observed as the Fourier power spectrum in the critical current.

4. Summary and Discussion

In this paper we have studied the effects of mesoscopic fluctuations on Josephson current in various SNS systems. In wire systems with normal impurities, the relative intensity of the fluctuation to the average, $\delta I/ < I >$, is of the same order as $1/ < g >$ in the high temperature regime, $L > \xi_D$, with $< g >$ being the averaged conductance in units of e^2/h. We also examined effects of pair breaking on the fluctuation. In the presence of paramagnetic impurities, magnetic scattering affects the fluctuation in a similar way as the average as a pair breaking mechanism. However, in the spin glass ordering, the fluctuation does not vanish even if the average is greatly reduced. Especially in the strong depairing limit, the fluctuation is suppressed but remains finite and it is independent of the difference of the phases of the superconducting order parameter. All the results are due to the fact that the singlet diffuson among the different systems is massless. In Aharonov - Bohm ring geometry, we have found hc/e oscillation in critical currents in addition to the well-known $hc/2e$ oscillation. This oscillation results from the interference between the electrons forming a single Cooper pair and is characteristic in mesoscopic systems.

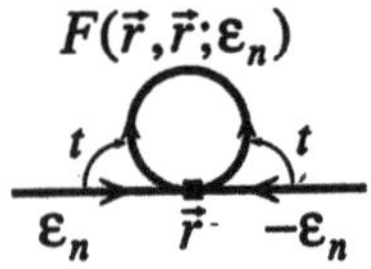

Fig.7
Schematic description of the model we treat.

Our method, which is introduced by Kresin, corresponds to the model where an incident electron is reflected into a hole by an effect of transfer t between N and S region as in the study by Aslamazov *et al.*[21], which is schematically shown in Fig.7, where C, defined in sec.2, is given by t^2/v_{Fs} with v_{Fs} being the Fermi velocity in the S region. Our treatment is different from that by Al'tshuler *et al.* and Beenakker in which they considered a model where an incident electron is reflected into a hole by Andreev reflection.

Acknowledgement

The authors would like to thank H. Kohno and O. Narikiyo for valuable discussions. This work is financially supported by Ministry - Industry Joint Research program "Mesoscopic Electronics" and by a Grant - in - Aid for Scientific Research on Priority Area, "Electron Wave Interference Effects in Mesoscopic Structure" from the Ministry of Education, Science and Culture.

References

[1] See reviews by S. Washburn and R. A. Webb: Adv. Phys. **35** (1986) 375; Y. Imry: in *Directions in Condensed Matter Physics*, eds. G. Grinstein and E. Mazenko (World Scientific, Singapore, 1986).

[2] B. L. Al'tshuler: JETP Lett. **41** (1985) 648.

[3] P. A. Lee and A. D. Stone: Phys. Rev. Lett. **55** (1985) 622.

[4] D. Yoshioka and H. Fukuyama: this issue.

[5] R. A. Serota and S. Oh: Phys. Rev. B **41** (1990) 10523.

[6] O. D. Cheishvili: JETP Lett. **48** (1988) 225.

[7] H. Fukuyama: J. Phys. Soc. Jpn. **58** (1989) 47.

[8] H. Takayanagi and T. Kawakami: Phys. Rev. Lett. **54** (1985) 2449.

[9] T. Nishino, M. Miyake Y. Harada and U. Kawabe: IEEE Electron Devices Lett. **6** (1985) 297.

[10] B. L. Al'tshuler and B. Z. Spivak: Sov. Phys. JETP **65** (1987) 343.

[11] C. W. J. Beenakker: Phys. Rev. Lett. **67** (1991) 3836 and this issue.

[12] V. Z. Kresin: Phys. Rev. B **34** (1986) 7587.

[13] P. A. Lee, A. D. Stone and H. Fukuyama: Phys. Rev. B **35** (1987) 1039.

[14] S. Hershfield: Phys. Rev. B **44** (1991) 3320.

[15] H. Yoshioka and H. Fukuyama: to be published.

[16] A. Yu. Zyuzin and B. Z. Spivak: JETP Lett. **43** (1986) 234.

[17] H. Yoshioka and H. Fukuyama: J. Phys. Soc. Jpn. **59** (1990) 3065.

[18] H. Fukuyama and H. Yoshioka: to appear in *New horizons in low dimensional electron systems* (Kluwer Academic Publishers).

[19] M. Tinkham: *Introduction to Superconductivity* (McGraw - Hill, 1975) Chap. 6.1.

[20] Detailed formula for I_d are given in Physica C **185-189** (1991) 1625 by H. Yoshioka and H. Fukuyama.

[21] L. G. Aslamazov, A. I. Larkin and Yu. N. Ovchinnikov: Sov. Phys. JETP **28** (1969) 171 .

Phase Coherence of Normal Electrons in Mesoscopic Composite Systems with Normal and Superconducting Segments

H. Ebisawa

Department of Engineering Science, Tohoku University, Aoba, Aramaki, Aoba-ku, Sendai 980, Japan

Abstract The phase memory of a normal electron is shown to vanish in a superconductor when its size is larger than ξ. Effects of the Andreev reflection on the conductance fluctuation and the A-B type oscillation in NSN composite wire systems are discussed. A simple conductance formula of the Landauer type is presented for a small system with a single superconducting segment.

1. Introduction

The coherence of quantum mechanical phases of electrons plays a central role in mesoscopic phenomena. In normal metals, the distance of coherence is limited by the inelastic relaxation time of single electron states. In superconductors, however, the phase of the condensate wave function is coherent over macroscopic lengthes. It is interesting to clarify the relationship and study the interplay of those phases. There have been several studies on the mesoscopic nature of the quantum transport in the normal metals (N) which are placed between two large superconductors (S), i.e. in the SNS systems.[1,2,3]

Here we are concerned about the mesoscopic properties of the whole system that contains normal and superconducting segments, i.e. the NSN system of mesoscopic size. Some examples are shown in Fig. 1. For the mesoscopic characteristics in such systems, the coherence of the phases of normal electron states over the whole system is essential. For example, the conductance fluctuation of the sample of Fig.1(c) depends on whether the fluctuations of its two N segments correlated, i.e. on whether the phases of normal electron states in those two N segments are mutually coherent. Similarly, the period of A-B oscillation of conductance in the magnetic flux of the sample of Fig. 1(b) depends on the penetration of the memories of phases of normal electron states through the S segment.

We will see that the phase memories of single electrons in a normal (N) segment practically cannot penetrate at all into an adjacenct superconducting (S) segment,[4,5] in section 2. The mesoscopic properties of the NSN composite systems are then determined[4,5] by the Andreev re-

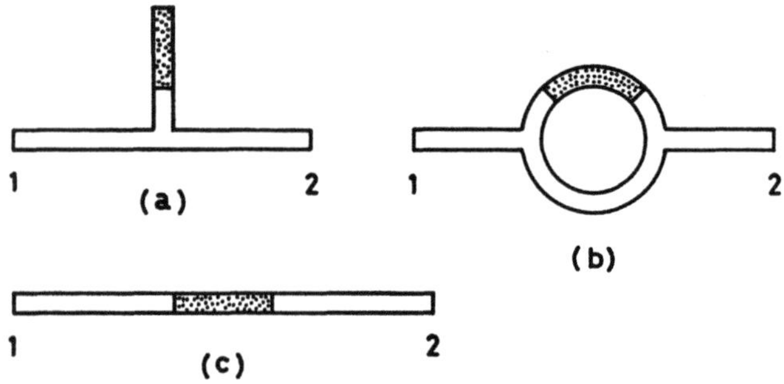

Fig. 1. Examples of **NSN** composite wire systems. Dotted segments are superconductors.

flection,[6] as will be shown in section 3. Although we can find those characteristic features in the microscopic theory, the numerical studies[7] of such properties should give us valuable informations. For such purposes, we will present a conductance formula,[8] which is an extension of the Landauer formula,[9,10] in section 4.

2. Decay of the Phase Coherence of Single Electron States in Superconductors

The phase coherence length of normal electron states is a key scale for the mesoscopic effects. In normal metal systems, the coherence length is the decay length of the diffusion propagators which determines the conductance variance of such systems.[11,12,13] In composite systems like those in Fig. 1, one has to deal with the diffusion propagators in inhomogeneous systems to calculate the variance. However, limiting oneself to the estimation of the coherence legth, one needs only the diffusion propagators of an infinite superconductors, i.e. of the constant order parameter Δ, as far as the proximity effect is negligible, i.e. the superconducting coherence length is short enough.

The diagrammatic expression of the diffusion propagator is a sum of ladder type digrams in the expansion with respect to the impurity scattering potential.[11,12,13] In the superconducting state, we take anomalous Green's function into account. Thus there appear four types of ladder sums that provide a set of components of the diffusion propagators. The denominator of the diffusion propagators is found to be commonly given by[4]

$$2\tau\Delta + \tau D q^2 \left(1 - 6\tau\Delta\right), \tag{1}$$

where D is the diffusion constant of electrons and τ is the relaxation time due to the impurity scattering. Thus the decay length of the duffusion propagator, which we denote as ξ_c, is found to be the coherence length of dirty superconductors, i.e.

$$\xi_c \sim \sqrt{\frac{\hbar D}{k_B T_c}}. \tag{2}$$

As the characteristic length ξ_c is of the same order as that of the spatial variation of the order parameter near the SN boundary, the proximity effect is actually not completely negligible contrarily to our expect. Thus the estimation of ξ_c is correct up to the order of magnitude.

This coherence lengh ξ_c is much shorter than L_ϕ, the one in the normal state. This means that the pairing interaction breaks the phase coherence strongly in contrast to the usual effect of the electron-electron interactions that do not destroy the coherence so much at low temperatures. Intuitively this can be understood as the condensate provides a strong constraint to the single particle phase variation in superconductors, because every single particle state is coupled with its time reversed state in such a way that the quantum phase of all pair wave functions are the same. As the effect of the pairing is so strong, reduction of L_ϕ should be seen even in the fluctuation regime above the superconducting transition temperature.[5] In order to see that we calculate the diffusion popagator just above T_c, taking account of the fluctuation propagator of the Cooper pairs in the diagrammatic expressions. Only the first order terms are retained in the unit structure of the impurity ladders, as shown in Fig. 2. The calculations are essentially the same as those by Keller and Korenman[14] for the derivation of the intrinsic cutoff of the divergence of so called Maki-Thompson term[15,16] in the fluctuation conductivity above T_c.

The phase coherence length L_ϕ is then shown to decrease when the temperature decreases towards T_c. The correction to the normal value of $1/{L_\phi}^2$ in the d-dimensional system at $t = (T - T_c)/T_c$ is found to be[5]

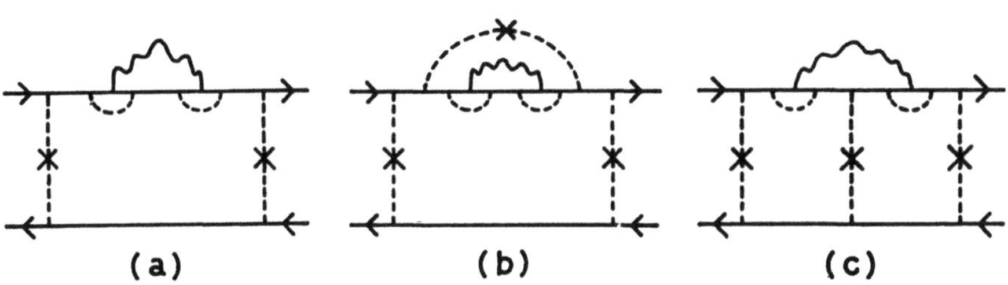

Fig. 2. Diagrams for the correction to the diffusion propagators due to the superconducting fluctuation. Wavy lines are the fluctuation propagators.

$$\delta\left(\frac{1}{L_\phi{}^2}\right) = A_d \frac{k_B T}{2\pi\hbar^2 N(0) D^2} w^{d-3} \left(\frac{k_B T}{\hbar D}\right)^{d/2-1} t^{d/2-2}, \tag{3}$$

where w is the thickness of the system for $d = 1$ and 2, A_d is numerical factors given as

$$A_3 = \sqrt{\frac{8}{\pi}}, \quad A_2 = 1 \quad \text{and} \quad A_1 = \sqrt{\frac{\pi^3}{2}}, \tag{4}$$

and $N(0)$ is the density of states at the fermi level.

3. Roles of the Andreev Reflection

As the phase memory does not penetrate into the S segment beyond ξ_c, the conductance fluctuations of the two N segments in the geometry of Fig. 1(c) are independent so one could expect that the total conductance fluctuation has less variance than in purely normal metal systems. But the Andreev reflection[6] at the NS bounary plays essential roles. In the Andreev reflection, incident particles from N to S results in reflected holes whose quantum phase depends on that of the incident particles as well as that of the condensate. This brings out two effects.

Firstly, the diffusion propagator has two components which are coupled with each other because of the correlation of particles and holes. This amounts to that the NS boundary serves a sort of open-end boundary condition for the corresponding diffusion equation.[1,2] Thus the fluctuation has a larger contribution to its variance, and the contribution from the simplest diagrammatic expression is found to be about 6 times as large as that in systems of a single N segment.[4] Secondly, in the system shown in Fig. 1(b), the reflected hole which carries the phase memory of the condensate in the S segment at one boundary can come to the other side of the S segment, where the Andreev reflection again takes place and the coming out particle has then the phase difference between both ends of the S segment, which can thus give an effect to the phase coherent properties of the N segment.

There have been made several predictions on effects of the Andreev reflection on properties of normal systems as functions of the condensate phase difference between two adjacent superconductors.[1,2,3] Here we make a prediction on its effect when the applied magnetic flux pierces the NS ring. The phase difference of the diffusion propagator in the N segment due to the vector potential of the magnetic field is determined by the charge e for electrons and $-e$ for reflected holes moving to the

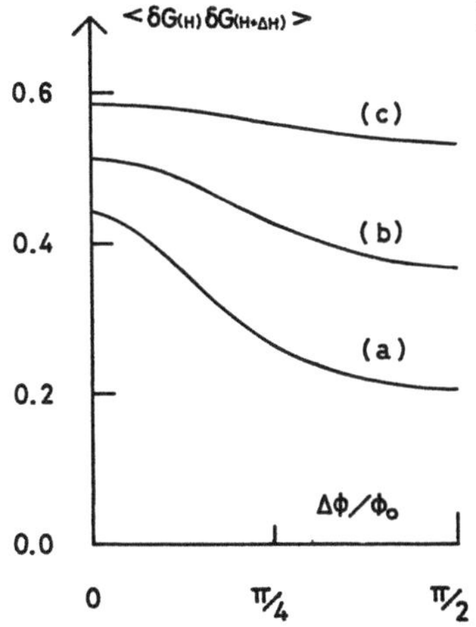

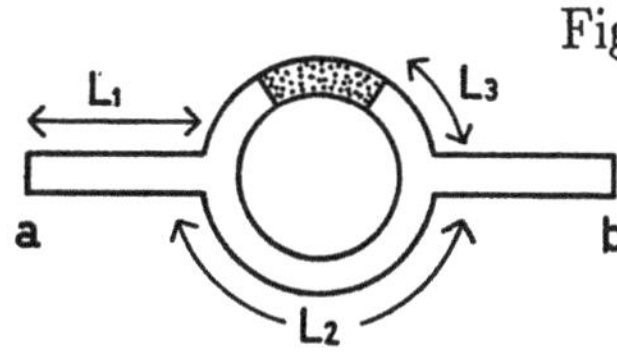

Fig. 3. The conductance correlation function of the magnetic flux in unit of $\phi_0/2\pi = \hbar c/e$ is plotted for a half period, being normalized by the conductance variation for the normal wire of length $2L_1 + L_2$. Parameters are $L_1 = 0.4L_\phi, L_2 = 0.1L_1$ and (a) $L_3 = 0.1L_1$, (b) $L_3 = 0.4L_1$ and (c) $L_3 = L_1$.

Fig. 4. The size parameters L_1, L_2 and L_3 are defined for an NSN ring structure.

opposite direction. They are coupled with each other due to the Andreev reflection. Thus the sum of these phases and the phase difference in the S segment mentioned above determines the conductance. This consideration suggests the Aharonov-Bohm type oscillation with the period $hc/2e$. The explicit evaluation of the condutance correlation as a function of the magnetic flux is given in refs.[4] and [5] to yield, for example, some results as shown in Fig. 3 for several size parameters, which are defined in Fig. 4. The calculations have been done for only the contributions of the simplest diagrammatic expression in order to observe the qualitative feature.

We note here that we need the system of the mesoscopic size for N segments but the S segment can be macroscopically long in order to observe this half period, because the S segment needs to carry no phase memory of normal electrons in the present situation. We note also here that the idea of half period was mentioned in the context of the localization for a purely one dimensional ring system, a part of which is superconducting, the period being $hc/4e$.[17]

4. A Conductance Formula for Composite Systems

For normal metal systems, the Landauer formula[9] has been frequently used for an expression of a conductance. The formula is essentially derived in the context of the linear response theory in terms of the Green's functions by use of the formal scattering theory.[18,19,20] An expression of the formula we will study is of the form like

$$I_m = \sum_n g_{mn} V_n \quad (m, n = 1, 2), \tag{5}$$

where m and n are the indices of the leads attached to the sample, of arbitrary numbers in general, I_m is the current flowing through the m-th lead, V_n is the electric potential of the n-th lead, and g_{mn} is the conductance coefficient among leads which is expressed in terms of the transmission and the reflection coefficients, t_{mn} and r_{mn}, respectively, of electrons at the fermi level.

The present NSN systems shown in Fig. 1 are examples of the two lead system shown in Fig. 5, which contains N and S segments. Then the Adreev reflection takes place so that the coefficents t_{mn}'s and r_{mn}'s within only the particle channel are not enough and one needs those which contains also the hole channel elements. In the present study, we present the expressions of conductance coefficients g_{mn} in terms of such t_{mn}'s and r_{mn}'s. In the derivation, we follow the formulation of ref. [20] but the Nambu space represention is used to take account of the Andreev reflection explicitly. For the sake of simplicity, the electric potential at the S segment, which is assumed only one in the system, is set to be zero. Under this convention of the potential, the two-terminal conductance is expressed in terms of g_{mn}, m and n being 1 or 2, as followings,[8]

$$\Gamma = \frac{g_{11}g_{22} - g_{12}g_{21}}{g_{11} + g_{22} - g_{12} - g_{21}}. \tag{6}$$

The first key relation in the derivation of the formula is the expression of the conductance coefficient in terms of the Green's function which is now 2×2 matrix. Using the expression of the current density at m-th lead in terms of the local electric field and the nonlocal conductivity $\sigma(\mathbf{r}, \mathbf{r}')$, which is expressed in terms of the Green's function in the context

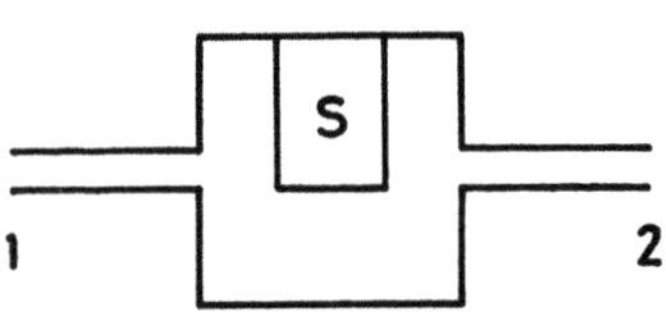

Fig. 5. A two-probe structure. Two leads are assumed of perfect conductors. There is a single superconducting part in the structure.

of the Kubo formula, integrating over the cross section to obtain I_m with the help of the partial integration and comparing the result with Eq. (5), one obtains the desired expression as[8]

$$g_{mm} = -\frac{e^2\hbar^2}{32\pi m^2}\int_{c_m} d\rho \int_{c_m} d\rho'$$
$$\times \operatorname{tr}\left\{\left[\mathbf{G}^+(\mathbf{r},\mathbf{r}') - \mathbf{G}^-(\mathbf{r},\mathbf{r}')\right]\overleftrightarrow{\mathbf{D}}_z\overleftrightarrow{\mathbf{D}}_z' \left[\mathbf{G}^+(\mathbf{r}',\mathbf{r}) - \mathbf{G}^-(\mathbf{r}',\mathbf{r})\right]\right\}, \quad (7)$$

where

$$\mathbf{D}_z = \begin{pmatrix} \overleftrightarrow{D}_z & 0 \\ 0 & -\overleftrightarrow{D}_z \end{pmatrix}, \qquad \mathbf{G}^\pm = \begin{pmatrix} G^\pm & F^\pm \\ F^{\pm *} & -G^{\pm *} \end{pmatrix},$$

and

$$g_{mn} = \frac{e^2\hbar^2}{32\pi m^2}\int_{c_m} d\rho \int_{c_n} d\rho' \operatorname{tr}\left[\mathbf{G}^+(\mathbf{r},\mathbf{r}')\overleftrightarrow{\mathbf{D}}_z\overleftrightarrow{\mathbf{D}}_z'\mathbf{G}^-(\mathbf{r}',\mathbf{r})\right] \quad (8)$$

for $m \neq n$, where c_n's denote the cross sectional areas of the n-th lead, $\overleftrightarrow{D}_z$'s are spatial derivative operator operating both on the right and the left Green's functions $\mathbf{G}^\pm$, $\pm$ indicating the two kinds of analyticities, and the trace is over the Nambu space. The local coordinate is chosen so that the lead is running in z-direcion. Here $\overleftrightarrow{D}_z$'s can be free of the vector potential at leads by an appropriate choice of the gauge.[8]

Here we introduce four kinds of transmission and reflection coefficients for an incident electron from the left, by which we call the lead 1. We denote the reflection coefficient as an electron to the left as $r^L_{e,e}$ and the transmission coefficient as an electron to the right as $t^L_{e,e}$, as usual. Furthermore, due to the Andreev reflection, we need the reflection coefficient to the left as a hole denoted as $r^L_{h,e}$ and the transmission coefficient to the right as a hole denoted as $t^L_{h,e}$. We notice here that no other waves like an incoming hole or electron from the right, by definition.

The second key relation is the asymptotic expression of the Green's functions when the arguments are in the lead region. By use of the formal scattering theory, one can derive them, some of which are as follows.[8] When $\mathbf{r}$ is in the lead 1 and $\mathbf{r}'$ is in the lead 2, for example, $G^+(\mathbf{r},\mathbf{r}')$ is proportional to $t^L_{e,e}$ with a normalization factor dependig on the states related with the degree of freedom in the cross section, and $F^+(\mathbf{r},\mathbf{r}')$, that is the off-diagonal element in the Nambu space, is proportional to $t^L_{h,e}$. When both of $\mathbf{r}$ and $\mathbf{r}'$ are in the lead 1, $G^+(\mathbf{r},\mathbf{r}')$ consists of the reflected wave with the coefficient $r^L_{e,e}$ and the incident wave.

Combining those two key relations, one can finally obtain the result[8]

$$g_{12} = \frac{e^2}{h} \sum_{a,b} \left\{ |t^L_{be,ae}|^2 - |t^L_{bh,ae}|^2 \right\}, \tag{9}$$

$$g_{11} = \frac{e^2}{h} \sum_{a,b} \left\{ \delta_{ab} - |r^L_{be,ae}|^2 + |r^L_{bh,ae}|^2 \right\}, \tag{10}$$

while g_{21} and g_{22} are given by the similar expressions with other coefficients for the incident electrons from the right, i.e. $t^R_{be,ae}$ etc. In these expressions, we denoted explicitly the index for the channel, a and b.

We note that the reduction of the conductance due to the hole transmission and the enhancement due to the hole reflection in this form are understood reasonable as pointed out by Blonder et al.[21] in a simple model system of the NS boundary with a potential barrier.

Acknowledgements

The author thanks to his collaborator Mr. Yositake Takane who made major contributions to obtain the results reported here and read critically this manuscript. The work is partially supported by a Grant-in-Aid from the Ministry of Educations, Science and Culture.

References

[1] B. Z. Spivak and D. E. Khmel'nitskii, JETP Lett. **35**, 412 (1982).
[2] B. L. Al'tshuler and B. Z. Spivak, JETP **65**, 343 (1987).
[3] H. Yoshioka and H. Fukuyama, J. Phys. Soc. Jpn. **59**, 3065 (1990); Physica C **185-189**, 1625 (1991).
[4] Y. Takane and H. Ebisawa, J. Phys. Soc. Jpn. **60**, 3130 (1991).
[5] H. Ebisawa and Y. Takane, Physica C **185-189**, 2703 (1991).
[6] A. F. Andreev, JETP **19**, 1228 (1964).
[7] Y. Takane, unpublished.
[8] Y. Takane and H. Ebisawa, J. Phys. Soc. Jpn. (submitted).
[9] R. Landauer, Philos. Mag. **21**, 863 (1970).
[10] M. Büttiker, Y. Imry, R. Landauer and S. Pinhas, Phys. Rev. **B31**, 6207 (1985).
[11] P. A. Lee and A. D. Stone, Phys. Rev. Lett. **55**, 1622 (1985); P. A. Lee, A. D. Stone and H. Fukuyama, Phys. Rev. **B35**, 1039 (1987).
[12] B. L. Al'tshuler, JETP Lett. **41**, 648 (1985); B. L. Al'tshuler and D. E. Khmel'nitskii JETP Lett. **42**, 359 (1985).

[13] Y. Isawa, H. Ebisawa and S. Maekawa, J. Phys. Soc. Jpn. **55**, 2523 (1986); S. Maekawa, Y. Isawa and H. Ebisawa, J. Phys. Soc. Jpn. **56**, 25 (1987).
[14] J. Keller and V. Korenmann, Phys. Rev. Lett. **27**, 1270 (1971); Phys. Rev. **B5**, 4367 (1972).
[15] K. Maki, Prog. Theor. Phys. **39**, 897 (1968); *ibid.* **40**, 193 (1968).
[16] R. S. Thompson, Phys. Rev. **B1**, 327 (1970); Physica **55**, 296 (1971).
[17] M. Büttiker and T. M. Klapwijk, Phys. Rev. **B33**, 5114 (1986).
[18] D. S. Fisher and P. A. Lee, Phys. Rev. **B23**, 6851 (1981).
[19] A. D. Stone and A. Szafer, IBM J. Res. Dev. **32**, 384 (1988).
[20] H. U. Baranger and A. D. Stone, Phys. Rev. **B40**, 8169 (1989).
[21] G. E. Blonder, M. Tinkham and T. M. Klapwijk, Phys. Rev. **B25**, 4515 (1982).

Index of Contributors

Springer Series in Solid-State Sciences

Editors: M. Cardona P. Fulde K. von Klitzing H.-J. Queisser

1 **Principles of Magnetic Resonance** 3rd Edition By C. P. Slichter
2 **Introduction to Solid-State Theory** 2nd Printing By O. Madelung
3 **Dynamical Scattering of X-Rays in Crystals** By Z. G. Pinsker
4 **Inelastic Electron Tunneling Spectroscopy** Editor: T. Wolfram
5 **Fundamentals of Crystal Growth I** Macroscopic Equilibrium and Transport Concepts. 2nd Printing By F. Rosenberger
6 **Magnetic Flux Structures in Superconductors** By R. P. Huebener
7 **Green's Functions in Quantum Physics** 2nd Edition By E. N. Economou
8 **Solitons and Condensed Matter Physics** 2nd Printing Editors: A. R. Bishop and T. Schneider
9 **Photoferroelectrics** By V. M. Fridkin
10 **Phonon Dispersion Relations in Insulators** By H. Bilz and W. Kress
11 **Electron Transport in Compound Semiconductors** By B. R. Nag
12 **The Physics of Elementary Excitations** By S. Nakajima, Y. Toyozawa, and R. Abe
13 **The Physics of Selenium and Tellurium** Editors: E. Gerlach and P. Grosse
14 **Magnetic Bubble Technology** 2nd Edition By A. H. Eschenfelder
15 **Modern Crystallography I** Symmetry of Crystals. Methods of Structural Crystallography By B. K. Vainshtein
16 **Organic Molecular Crystals** Their Electronic States. By E. A. Silinsh
17 **The Theory of Magnetism I** Statics and Dynamics. 2nd Printing By D. C. Mattis
18 **Relaxation of Elementary Excitations** Editors: R. Kubo and E. Hanamura
19 **Solitons.** Mathematical Methods for Physicists. 2nd Printing By G. Eilenberger
20 **Theory of Nonlinear Lattices** 2nd Edition By M. Toda
21 **Modern Crystallography II** Structure of Crystals By B. K. Vainshtein, V. M. Fridkin, and V. L. Indenbom
22 **Point Defects in Semiconductors I** Theoretical Aspects By M. Lannoo and J. Bourgoin
23 **Physics in One Dimension** Editors: J. Bernasconi, T. Schneider
24 **Physics in High Magnetic Fields** Editors: S. Chikazumi and N. Miura
25 **Fundamental Physics of Amorphous Semiconductors** Editor: F. Yonezawa
26 **Elastic Media with Microstructure I** One-Dimensional Models. By I. A. Kunin
27 **Superconductivity of Transition Metals** Their Alloys and Compounds By S. V. Vonsovsky, Yu. A. Izyumov, and E. Z. Kurmaev
28 **The Structure and Properties of Matter** Editor: T. Matsubara
29 **Electron Correlation and Magnetism in Narrow-Band Systems** Editor: T. Moriya
30 **Statistical Physics I** Equilibrium Statistical Mechanics 2nd Edition By M. Toda, R. Kubo, N. Saito
31 **Statistical Physics II** Nonequilibrium Statistical Mechanics 2nd Edition By R. Kubo, M. Toda, N. Hashitsume
32 **Quantum Theory of Magnetism** 2nd Edition By R. M. White
33 **Mixed Crystals** By A. I. Kitaigorodsky
34 **Phonons: Theory and Experiments I** Lattice Dynamics and Models of Interatomic Forces. By P. Brüesch
35 **Point Defects in Semiconductors II** Experimental Aspects By J. Bourgoin and M. Lannoo
36 **Modern Crystallography III** Crystal Growth 2nd Edition By A. A. Chernov
37 **Modern Crystallography IV** Physical Properties of Crystals Editor: L. A. Shuvalov
38 **Physics of Intercalation Compounds** Editors: L. Pietronero and E. Tosatti
39 **Anderson Localization** Editors: Y. Nagaoka and H. Fukuyama
40 **Semiconductor Physics** An Introduction 5th Edition By K. Seeger
41 **The LMTO Method** Muffin-Tin Orbitals and Electronic Structure By H. L. Skriver
42 **Crystal Optics with Spatial Dispersion, and Excitons** 2nd Edition By V. M. Agranovich and V. L. Ginzburg
43 **Structural Analysis of Point Defects in Solids** An Introduction to Multiple Magnetic Resonance Spectroscopy By J.-M. Spaeth, J. R. Niklas, and R. H. Bartram
44 **Elastic Media with Microstructure II** Three-Dimensional Models By I. A. Kunin
45 **Electronic Properties of Doped Semiconductors** By B. I. Shklovskii and A. L. Efros
46 **Topological Disorder in Condensed Matter** Editors: F. Yonezawa and T. Ninomiya
47 **Statics and Dynamics of Nonlinear Systems** Editors: G. Benedek, H. Bilz, and R. Zeyher
48 **Magnetic Phase Transitions** Editors: M. Ausloos and R. J. Elliott
49 **Organic Molecular Aggregates** Electronic Excitation and Interaction Processes Editors: P. Reineker, H. Haken, and H. C. Wolf

Springer Series in Solid-State Sciences

Editors: M. Cardona P. Fulde K. von Klitzing H.-J. Queisser

50 **Multiple Diffraction of X-Rays in Crystals**
By Shih-Lin Chang

51 **Phonon Scattering in Condensed Matter**
Editors: W. Eisenmenger, K. Laßmann, and S. Döttinger

52 **Superconductivity in Magnetic and Exotic Materials** Editors: T. Matsubara and A. Kotani

53 **Two-Dimensional Systems, Heterostructures, and Superlattices**
Editors: G. Bauer, F. Kuchar, and H. Heinrich

54 **Magnetic Excitations and Fluctuations**
Editors: S. Lovesey, U. Balucani, F. Borsa, and V. Tognetti

55 **The Theory of Magnetism II** Thermodynamics and Statistical Mechanics By D. C. Mattis

56 **Spin Fluctuations in Itinerant Electron Magnetism** By T. Moriya

57 **Polycrystalline Semiconductors,**
Physical Properties and Applications
Editor: G. Harbeke

58 **The Recursion Method and Its Applications**
Editors: D. Pettifor and D. Weaire

59 **Dynamical Processes and Ordering on Solid Surfaces** Editors: A. Yoshimori and M. Tsukada

60 **Excitonic Processes in Solids**
By M. Ueta, H. Kanzaki, K. Kobayashi, Y. Toyozawa, and E. Hanamura

61 **Localization, Interaction, and Transport Phenomena** Editors: B. Kramer, G. Bergmann, and Y. Bruynseraede

62 **Theory of Heavy Fermions and Valence Fluctuations** Editors: T. Kasuya and T. Saso

63 **Electronic Properties of Polymers and Related Compounds**
Editors: H. Kuzmany, M. Mehring, and S. Roth

64 **Symmetries in Physics** Group Theory Applied to Physical Problems
By W. Ludwig and C. Falter

65 **Phonons: Theory and Experiments II**
Experiments and Interpretation of Experimental Results By P. Brüesch

66 **Phonons: Theory and Experiments III**
Phenomena Related to Phonons
By P. Brüesch

67 **Two-Dimensional Systems: Physics and New Devices**
Editors: G. Bauer, F. Kuchar, and H. Heinrich

68 **Phonon Scattering in Condensed Matter V**
Editors: A. C. Anderson and J. P. Wolfe

69 **Nonlinearity in Condensed Matter**
Editors: A. R. Bishop, D. K. Campbell, P. Kumar, and S. E. Trullinger

70 **From Hamiltonians to Phase Diagrams** The Electronic and Statistical-Mechanical Theory of sp-Bonded Metals and Alloys By J. Hafner

71 **High Magnetic Fields in Semiconductor Physics**
Editor: G. Landwehr

72 **One-Dimensional Conductors**
By S. Kagoshima, H. Nagasawa, and T. Sambongi

73 **Quantum Solid-State Physics**
Editors: S. V. Vonsovsky and M. I. Katsnelson

74 **Quantum Monte Carlo Methods** in Equilibrium and Nonequilibrium Systems Editor: M. Suzuki

75 **Electronic Structure and Optical Properties of Semiconductors** 2nd Edition
By M. L. Cohen and J. R. Chelikowsky

76 **Electronic Properties of Conjugated Polymers**
Editors: H. Kuzmany, M. Mehring, and S. Roth

77 **Fermi Surface Effects**
Editors: J. Kondo and A. Yoshimori

78 **Group Theory and Its Applications in Physics**
By T. Inui, Y. Tanabe, and Y. Onodera

79 **Elementary Excitations in Quantum Fluids**
Editors: K. Ohbayashi and M. Watabe

80 **Monte Carlo Simulation in Statistical Physics**
An Introduction 2nd Edition
By K. Binder and D. W. Heermann

81 **Core-Level Spectroscopy in Condensed Systems**
Editors: J. Kanamori and A. Kotani

82 **Introduction to Photoemission Spectroscopy**
By S. Hüfner

83 **Physics and Technology of Submicron Structures**
Editors: H. Heinrich, G. Bauer, and F. Kuchar

84 **Beyond the Crystalline State** An Emerging Perspective By G. Venkataraman, D. Sahoo, and V. Balakrishnan

85 **The Fractional Quantum Hall Effect**
Properties of an Incompressible Quantum Fluid
By T. Chakraborty and P. Pietiläinen

86 **The Quantum Statistics of Dynamic Processes**
By E. Fick and G. Sauermann

87 **High Magnetic Fields in Semiconductor Physics II**
Transport and Optics Editor: G. Landwehr

88 **Organic Superconductors**
By T. Ishiguro and K. Yamaji

89 **Strong Correlation and Superconductivity**
Editors: H. Fukuyama, S. Maekawa, and A. P. Malozemoff

MIX
Papier aus verantwortungsvollen Quellen
Paper from responsible sources
FSC® C105338

If you have any concerns about our products,
you can contact us on
ProductSafety@springernature.com

In case Publisher is established outside the EU,
the EU authorized representative is:
Springer Nature Customer Service Center GmbH
Europaplatz 3, 69115 Heidelberg, Germany

Printed by Libri Plureos GmbH
in Hamburg, Germany